VOLUME EIGHTY SEVEN

ADVANCES IN
BOTANICAL RESEARCH
Membrane Transport in Plants

ADVANCES IN BOTANICAL RESEARCH

Series Editor

Jean-Pierre Jacquot
Professor, Membre de l'Institut Universitaire de France, Unité Mixte de Recherche INRA, UHP 1136 "Interaction Arbres Microorganismes", Université de Lorraine, Faculté des Sciences, Vandoeuvre, France

VOLUME EIGHTY SEVEN

ADVANCES IN BOTANICAL RESEARCH

Membrane Transport in Plants

Volume Editor

CHRISTOPHE MAUREL

Biochimie et Physiologie Moléculaire des Plantes (BPMP), Université de Montpellier, CNRS, INRA, SupAgro, Montpellier, France

ELSEVIER

ACADEMIC PRESS

An imprint of Elsevier

Academic Press is an imprint of Elsevier
125 London Wall, London, EC2Y 5AS, United Kingdom
The Boulevard, Langford Lane, Kidlington, Oxford OX5 1GB, United Kingdom
50 Hampshire Street, 5th Floor, Cambridge, MA 02139, United States
525 B Street, Suite 1650, San Diego, CA 92101, United States

First edition 2018

Notices

Knowledge and best practice in this field are constantly changing. As new research and experience broaden our understanding, changes in research methods, professional practices, or medical treatment may become necessary.

Practitioners and researchers must always rely on their own experience and knowledge in evaluating and using any information, methods, compounds, or experiments described herein. In using such information or methods they should be mindful of their own safety and the safety of others, including parties for whom they have a professional responsibility.

To the fullest extent of the law, neither the Publisher nor the authors, contributors, or editors, assume any liability for any injury and/or damage to persons or property as a matter of products liability, negligence or otherwise, or from any use or operation of any methods, products, instructions, or ideas contained in the material herein.

ISBN: 978-0-12-809390-0
ISSN: 0065-2296

Publisher: Zoe Kruze
Acquisition Editor: Sam Mahfoudh
Editorial Project Manager: Teresa Pons-Ferrer
Production Project Manager: James Selvam
Cover Designer: Miles Hitchen

Typeset by SPi Global, India

CONTENTS

CONTRIBUTORS

Rashed Abualia
IST Austria, Klosterneuburg, Austria

Nadia Bazihizina
Tasmanian Institute for Agriculture, College of Science and Engineering, University of Tasmania, Hobart, TAS, Australia; Department of Agrifood Production and Environmental Sciences, University of Florence, Florence, Italy

Eva Benkova
IST Austria, Klosterneuburg, Austria

Katia E. Bonaldi
Division of Biological Sciences, University of California, San Diego, CA, United States

Marc Boutry
Louvain Institute of Biomolecular Science and Technology, Université catholique de Louvain, Louvain-la-Neuve, Belgium

Monica Calvo-Polanco
Biochimie et Physiologie Moléculaire des Plantes (BPMP), Université de Montpellier, CNRS, INRA, SupAgro, Montpellier, France

Tracey Ann Cuin
Tasmanian Institute for Agriculture, College of Science and Engineering, University of Tasmania, Hobart, TAS, Australia

John W.C. Dunlop
Morphophysics, Department of Chemistry and Physics of Materials, University of Salzburg, Salzburg, Austria

Seckin Eroglu
Department of Genetics and Bioengineering, Izmir University of Economics, Izmir, Turkey

Kevin Garcia
Department of Crop and Soil Sciences, North Carolina State University, Raleigh, NC, United States

Alain Gojon
Biochimie et Physiologie Moléculaire des Plantes (BPMP), Université de Montpellier, CNRS, INRA, SupAgro, Montpellier, France

Carmen Guerrero-Galán
Biochimie et Physiologie Moléculaire des Plantes (BPMP), Université de Montpellier, CNRS, INRA, SupAgro, Montpellier, France

Gabriella Houdinet
Biochimie et Physiologie Moléculaire des Plantes (BPMP), Université de Montpellier, CNRS, INRA, SupAgro, Montpellier, France

Tetsuya Ishikawa
Tasmanian Institute for Agriculture, College of Science and Engineering, University of Tasmania, Hobart, TAS, Australia

Peter Lackner
Immunoinformatics, Department of Biosciences, University of Salzburg, Salzburg, Austria

Benoît Lacombe
Biochimie et Physiologie Moléculaire des Plantes (BPMP), Université de Montpellier, CNRS, INRA, SupAgro, Montpellier, France

François Lefèvre
Louvain Institute of Biomolecular Science and Technology, Université catholique de Louvain, Louvain-la-Neuve, Belgium

Laurence Lejay
Biochimie et Physiologie Moléculaire des Plantes (BPMP), Université de Montpellier, CNRS, INRA, SupAgro, Montpellier, France

Nathalie Leonhardt
Laboratoire de Biologie du Développement des Plantes, Commissariat à l'Energie Atomique et aux Energies Alternatives (CEA), CNRS, Aix-Marseille Université, UMR 7265, Institut de Biosciences et Biotechnologies d'Aix-Marseille, CEA Cadarache, Saint-Paul-lez-Durance, France

Christophe Maurel
Biochimie et Physiologie Moléculaire des Plantes (BPMP), Université de Montpellier, CNRS, INRA, SupAgro, Montpellier, France

Gerhard Obermeyer
Molecular Plant Biophysics and Biochemistry, Department of Biosciences, University of Salzburg, Salzburg, Austria

Jeppe Thulin Østerberg
Department of Plant and Environmental Sciences, University of Copenhagen, Frederiksberg, Denmark

Michael Palmgren
Department of Plant and Environmental Sciences, University of Copenhagen, Frederiksberg, Denmark

Heidi Pertl-Obermeyer
Molecular Plant Biophysics and Biochemistry, Department of Biosciences, University of Salzburg, Salzburg, Austria

Baptiste Pierman
Louvain Institute of Biomolecular Science and Technology, Université catholique de Louvain, Louvain-la-Neuve, Belgium

Brice Roux
Laboratoire de Biologie du Développement des Plantes, Commissariat à l'Energie Atomique et aux Energies Alternatives (CEA), CNRS, Aix-Marseille Université, UMR 7265, Institut de Biosciences et Biotechnologies d'Aix-Marseille, CEA Cadarache, Saint-Paul-lez-Durance, France

Sergey Shabala
Tasmanian Institute for Agriculture, College of Science and Engineering, University of Tasmania, Hobart, TAS, Australia; Department of Horticulture, Foshan University, Foshan, China

Lionel Verdoucq
Biochimie et Physiologie Moléculaire des Plantes (BPMP), Université de Montpellier, CNRS, INRA, SupAgro, Montpellier, France

Sabine Dagmar Zimmermann
Biochimie et Physiologie Moléculaire des Plantes (BPMP), Université de Montpellier, CNRS, INRA, SupAgro, Montpellier, France

PREFACE

Membrane transport in plants contributes to key aspects of their growth and development, and to their adaptation to multiple and ever changing environments. From the 1950s, membrane transport has emerged as a main topic in plant physiology based on robust biophysical and theoretical grounds. From the 1990s, this field of research has experienced an amazing explosion of molecular data through the identification and functional characterization of numerous families of membrane transport proteins. Current research aims at combining this molecular and physiological knowledge to draw an integrative view of transport functions in plants.

Because of the multiplicity of transport systems and physiological contexts, assembling of volume on this topic is a challenge, for sure! As we clearly cannot be comprehensive, we decided to look at this topic through three different angles. The resulting views obviously overlap, and the reader will hopefully take these overlaps as signs that we are approaching a multidimensional understanding of plant membrane transport complexity.

The first chapters take the classical approach of focusing on a given family of membrane transporters. The main message is that a same class of proteins can fulfil multiple transport functions; much more than initially foreseen. For instance, the study of ABC transporters (Chapter 1) has revealed an unsuspected range of metabolites that can be transported in plants, from cuticle precursors to glutathione conjugates and chlorophyll metabolites. Many more are to be discovered. The aquaporins (Chapter 2) which were initially seen as water channels can actually transport important gases (carbon dioxide) or signalling molecules (hydrogen peroxide). P-type ATPase pumps are essential for energization of membrane transport in plants. In Chapter 3, Østerberg and Palmgren focus on the P1B ATPases which are present in different subcellular membranes of plants and control heavy metal homeostasis in plants. Phylogenetic analyses provide interesting insights into how plant membrane functions can be used to track the paths of plant evolution.

The converse approach was to consider a class of transported molecules, not transporters. Eroglu (Chapter 4) follows up on metal transport and discusses the specific context of seeds, showing how an impressive series arrangement of transport systems permits translocation of metals from phloem to embryo. While auxin transport was long identified as crucial for tropic responses of plants, Abualia et al. show in Chapter 5 that all major groups of plant hormones do have specific transport systems for controlling their distribution. The role of membrane transport in plant mineral nutrition is addressed by Lejay and Gojon (Chapter 6) through the emblematic example of nitrate uptake by roots. Here, the identification of transceptors, membrane proteins that serve both as transporters and receptors, identifies another interesting connection between transport and signalling.

Ultimately, we need to embrace the contribution of various types of membrane transport in a whole process. This is the third approach taken by several contributors of this volume. How to talk about membrane transport in plants without touching on guard cells? As discussed by Roux and Leonhardt in Chapter 7, this cell type continues to bring new knowledge on ion transport regulation in the plasma and vacuolar membranes, to integrate multiple endogenous signals and environmental stimuli. Pollen vegetative cells and pollen tubes represent another spectacular cell type of plants. Pertl-Obermeyer et al. present in Chapter 8 an overview of the transportome of pollen tubes and propose a speculative model to explain how heterogeneous distribution of transporters in the plasma membrane could explain the electrical current pattern observed in these cells. Understanding transport processes occurring in less accessible organs or cell type remains a great challenge. In these respects, Ishikawa et al. present in Chapter 9 a prospective view of mechanisms at work during xylem loading in plants under stress. Finally, it should be realized that plants in natural environments interact with a myriad of microbes and fungi, which balance between pathogenic and beneficial effects. Guerrero-Galán et al. (Chapter 10) explain how plant mineral nutrition a traditional topic in membrane transport most often occurs in symbiotic interaction with fungi.

This first overview spans over many fascinating facets of plant biology. I invite the reader to discover many more in this volume.

Christophe Maurel

CHAPTER ONE

The ABC of ABC Transporters

Baptiste Pierman[2], Marc Boutry[1], François Lefèvre[3]
Louvain Institute of Biomolecular Science and Technology, Université catholique de Louvain, Louvain-la-Neuve, Belgium
[1]Corresponding author: e-mail address: marc.boutry@uclouvain.be

Contents

Abstract

ATP-binding cassette (ABC) proteins are mainly active primary transporters and, in plants, form a large family of more than 120 members, which are organized into eight subfamilies. In this chapter, we first briefly describe the general structural and enzymatic properties of ABC transporters. We review the major functions of plant ABC transporters such as hormone transport, response to biotic stress, building surface structures, or cell detoxification. In each case, we discuss how the substrates were identified. We then raise several issues that should be addressed in future research: the requirement of an active transporter for hydrophobic substrates, the problem of identifying the substrates in the absence of transport assays, and the existence of pleiotropic plant ABC transporters with multiple structurally unrelated substrates.

1. STRUCTURAL AND ENZYMATIC PROPERTIES

The family of ATP-binding cassette (ABC) proteins is very large and is present in all kingdoms of life. Members of this family are mainly active primary transporters, i.e., using ATP as an energy source to transport substrates

[2] Present address: GlaxoSmithKline Vaccines, 1330 Rixensart, Belgium.
[3] Present address: Boyce Thompson Institute, Ithaca, NY 14850, United States.

Advances in Botanical Research, Volume 87
ISSN 0065-2296
https://doi.org/10.1016/bs.abr.2018.09.005

against their concentration gradient. The family is divided into nine subfamilies (from ABCA to ABCI) according to their structural organization and their phylogenetic relationships (Fig. 1). Every subfamily but ABCH is represented in plants (Rea, 2007; Verrier et al., 2008). More trivial names are also used to

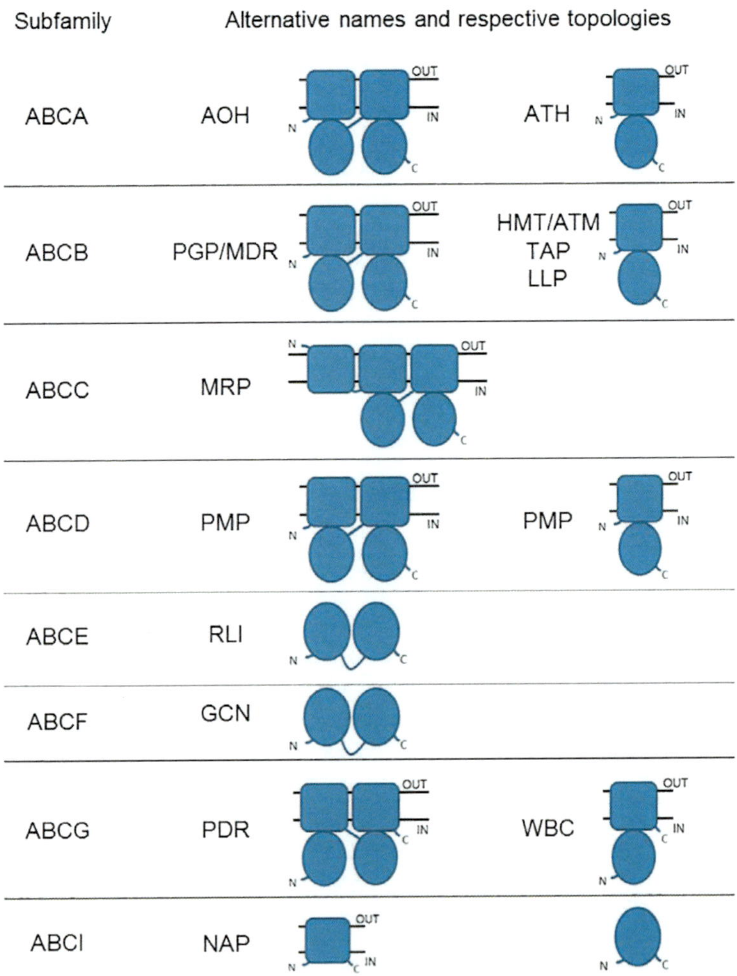

Fig. 1 Schematic representation of the predicted topologies of plant ABC proteins. Squares represent TMDs while ovals represent NBDs. 'C', 'N', 'IN', and 'OUT' refer to the C-terminal end, N-terminal end, cytosolic side, and extracellular side, respectively. AOH: ABC one homologues; ATH: ABC 2 homologues; PGP: P-glycoprotein; MDR: multidrug resistance; HMT: heavy metal tolerance; ATM: ABC transporters of the mitochondrion; TAP: transporters associated with antigen processing; LLP: prokaryotic lipid A-like transporters; MRP: multidrug resistance associated proteins; PMP: peroxisomal membrane protein; RLI: RNAse L inhibitor homologues; GCN: general control nonrepressive homologues; PDR: pleiotropic drug resistance; WBC: white–brown complex; NAP: nonintrinsic ABC proteins.

refer to them and usually relate to a function associated with the first transporter identified in a subfamily, e.g., MDR for multidrug resistance (ABCB) or WBC for white–brown complex (ABCG). However, caution is required because a function identified for a particular transporter in a specific subfamily does not mean that all members of the same subfamily have the same function.

ABC transporters are composed of two different core units. On the one hand, the transmembrane domain (TMD) plays the role of hydrophobic anchor of the protein into the membrane and forms the region where the transported substrates bind and cross the lipid bilayer. On the other hand, the nucleotide binding domain (NBD) possesses the capacity of both binding and hydrolysing ATP. These two domains must be present twice for the transporter to be active. They can be encoded by a single gene (full-size *ABCs*), two different genes (half-size) or even four different genes (for the prokaryotic-like ABCI subfamily) (Higgins, 2001; Rea, 2007; Verrier et al., 2008). In addition, ABC transporters can be forward oriented, meaning that one finds a TMD at the N-terminal part of the protein (which is thus TMD–NBD–TMD–NBD) or reverse oriented, in which case the transporter is as follows: NBD–TMD–NBD–TMD. The different topologies of ABC transporters according to their subfamily are displayed in Fig. 1.

The transmembrane domains are usually composed of six alpha helices (yet it can vary between 5 and 10 helices) and no relevant sequence identities were found for these domains between distant organisms (Licht & Schneider, 2011). This observation fits the fact that the TMDs are responsible for the substrate specificity, which might considerably vary according to the organism and the transporter.

On the other hand, the NBDs, with about 200 amino acid residues, are much more conserved (between 30% and 40% of sequence identity between distant members). They consist of two domains: a catalytic one (RecA-like) and a helical one. Different motifs are well conserved within the NBDs (Fig. 2). These include the Walker A box (GX4GK[ST]), which is also referred to as P-loop, the Walker B box ((hydrophobic)4[DE]) and the ABC signature motif (also named C motif, C-loop or LSGGQ box with the consensus sequence [LIVMFY]S[SG]GX3[RKA][LIVMYA]X[LIVMF][AG]) which are implicated in ATP binding and hydrolysis (Licht & Schneider, 2011; Rea, 2007).

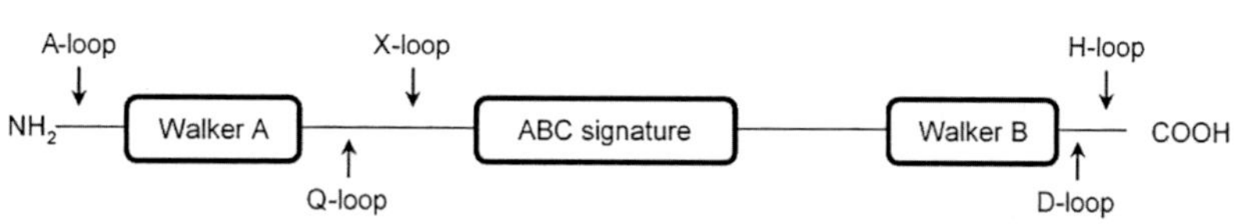

Fig. 2 Linear representation of conserved motifs of an NBD (not drawn to scale).

The tridimensional structures have been obtained for several ABC transporters, most of them from prokaryotes and a few from animal organisms, but none from plants (Aller et al., 2009; Lee et al., 2016; Liu, Zhang, Csanady, Gadsby, & Chen, 2017; Taylor et al., 2017). These transporters share a similar general structure although clear differences can be seen according to the subfamily (Fig. 3, compare A and C). In addition, the structures obtained in the presence or absence of a bound substrate and ATP showed important differences (Fig. 3, compare A and B). Several regions of the NBDs are implicated in the binding and hydrolysis of ATP. Actually, both NBDs are needed to perform this task since a single molecule of ATP is surrounded by the two NBDs and interacts with particular regions of both (Fig. 2D). To do so, the two NBDs are in a head

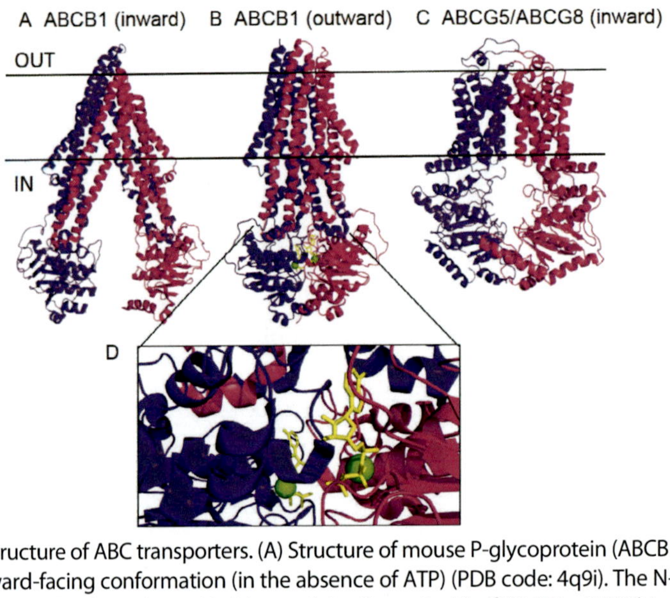

Fig. 3 Structure of ABC transporters. (A) Structure of mouse P-glycoprotein (ABCB1, MDR1) in an inward-facing conformation (in the absence of ATP) (PDB code: 4q9i). The N-terminal half (TMD1+NBD1) is coloured in blue and the C-terminal half (TMD2+NBD2) is coloured in pink. (B) Structure of human P-glycoprotein (ABCB1, MDR1) in the ATP-bound, outward-facing conformation (PDB code: 6c0v). The N-terminal half (TMD1+NBD1) is coloured in blue and the C-terminal half (TMD2+NBD2) is coloured in pink. Two ATP molecules (yellow) and two Mg^{2+} ions (green) are located at the NBD interface. (C) Structure of the human ABCG5/ABCG8 dimer in an inward-facing conformation (PDB code: 5DO7). The two half-size transporters are coloured in blue and pink. (D) Enlargement of the structure displayed in B showing the two ATP molecules (yellow) and Mg^{2+} ions (green) at the interface between the two NBD. The two black lines represent the membrane limits. The structures were drawn with PyMOL.

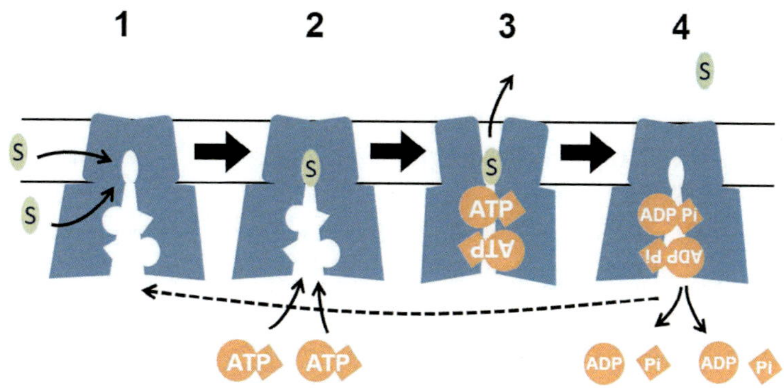

Fig. 4 Model of transport catalytic cycle of an ABC transporter. (1) The substrate (green) coming from the cytosol or already in the membrane binds the protein in an inward-facing conformation. (2) Substrate binding triggers ATP binding. (3) ATP binding generates an important structural modification (outward-facing conformation), which results in substrate release. (4) ATP hydrolysis brings the protein back to its initial inward-facing conformation while ADP and phosphate (Pi) are released. The two horizontal black lines delimit the membrane.

to tail conformation that creates two nucleotide binding sites, in which two ATP molecules can be sandwiched.

Structural as well as enzymatic data led to define a general catalytic cycle (Fig. 4). However, it still has to be considered as a working model, which is likely not valid in every situation. For instance, whether one or two ATPs are hydrolysed per catalytic cycle is still unclear and might depend on the transporter.

2. SUBSTRATES AND FUNCTIONS

Plant genomes contain a large amount of *ABC* genes, with more than 120 members identified in both *Arabidopsis thaliana* and *Oryza sativa* genomes. In contrast, only 49 *ABC* genes are encoded in the human chromosomes (Sanchez-Fernandez, Davies, Coleman, & Rea, 2001; Vasiliou, Vasiliou, & Nebert, 2009; Verrier et al., 2008). This observation has been related to the very productive secondary metabolism of plants as well as their sessile nature, which implies the need for efficient detoxification and stress resistance mechanisms (Hwang et al., 2016).

The subcellular localization of several ABC transporters has now been determined. As a general rule, ABCB and ABCG transporters are targeted to the plasma membrane and ABCC transporters to the tonoplast. Members of the ABCD subfamily have been localized to the peroxisomal membrane while some ABCI transporters are found in mitochondria or the chloroplast. However, there are exceptions. For instance, AtABCB25 localizes to the inner mitochondrial membrane (Schaedler et al., 2014) and AtABCC4 localizes to the plasma membrane (Klein et al., 2004). There is still the possibility that some transporters display a dual localization, especially in the secretory pathway. It is therefore mandatory to carefully check the subcellular localization of each individual transporter in order to assess its function.

The physiological roles of plant ABC transporters have been mainly determined using forward and reverse genetic tools (e.g. T-DNA insertion or gene silencing), which led to various phenotypic modifications in the development or the response of plants to abiotic and biotic stresses (e.g. heavy metals, nutrient deficiency, microbial pathogens, insects). This topic has been largely reviewed (see, for example, Borghi, Kang, Ko, Lee, & Martinoia, 2015; Do, Martinoia, & Lee, 2017; Hwang et al., 2016; Kretzschmar, Burla, Lee, Martinoia, & Nagy, 2011; Lane et al., 2016; Park, Lee, Martinoia, & Geisler, 2017). In particular, a comprehensive list of all plant ABC transporters that have been functionally characterized, along with their putative substrate(s), has appeared recently (table S1 from Lefèvre & Boutry, 2018). In this section, without being exhaustive, we will highlight significant examples to illustrate the large diversity of functions and substrates of ABC transporters.

2.1 Hormone Transport

Long-distance transport of hormones is usually complex and might involve transporters in cells where hormones are synthesized, in those where they fulfil their biological function, and possibly in cells in-between. Although we are still far from having a clear and detailed view of all the transport steps involved in hormone distribution, several protein families involved in hormone transport have been identified. Among these, ABC transporters from different subfamilies have been shown to play a role in the translocation of various hormones such as auxins, cytokinins, abscisic acid, strigolactones, or jasmonic acid.

Auxins are one of the most studied classes of phytohormones as they largely contribute to plant growth and development. Auxin transport is complex as it requires cell to cell movement achieved by the cooperation of multiple transporters belonging to different families, i.e., PIN, AUX, and ABC. In this context, members of the ABCB subfamily were found to play a predominant role in Arabidopsis. Indeed, AtABCB1, 4, 14, 15, and 19 have been shown to be related to transport of indole-3-acetic acid (IAA) and/or 1-naphthalene acetic acid (NAA) (for review, see Cho & Cho, 2013; Geisler, Aryal, Di Donato, & Hao, 2017; Geisler & Murphy, 2006). An intriguing observation is that AtABCB4 and AtABCB21 were shown to act as both importers and exporters, possibly depending on the auxin gradient across the plasma membrane (Cho, Lee, & Cho, 2007; Kamimoto et al., 2012; Kim et al., 2010; Kubes et al., 2012; Yang & Murphy, 2009). This is unusual for ATP-driven transporters and certainly merits further characterization at the biochemical level. Nevertheless, it should be mentioned that auxin transport is not restricted to the ABCB subfamily as ABCG36 (Strader & Bartel, 2009) and ABCG37 (Ruzicka et al., 2010) were shown to be involved in the transport of indole 3-butyric acid (IBA).

Cytokinins are another class of phytohormones which require ABC transporters to fulfil their growth-promoting function *in planta*. Indeed, an *AtABCG14* loss-of-function mutant displayed severe growth retardation, which was found to be related to an impaired translocation of cytokinin from the root to the aerial organs (Ko et al., 2014; Zhang et al., 2014). However, comprehensive assays are yet to be conducted to demonstrate direct transport of cytokinin by AtABCG14.

Abscisic acid (ABA) is critical for the plant response to various abiotic stresses, including heat, salinity, cold, and drought. In the latter case, vascular parenchyma cells synthesize ABA which, in Arabidopsis, is exported by AtABCG25 and then imported by AtABCG40 in the guard cells where it triggers stomatal closure (Kang et al., 2010; Kuromori et al., 2016, 2010; Kuromori, Sugimoto, & Shinozaki, 2014). These two transporters, together with AtABCG30 and AtABCG31, also intervene in maintaining seed dormancy. AtABCG25 and AtABCG31 export ABA from the endosperm while AtABCG30 and AtABCG40 import the hormone in the embryos in order to suppress seed germination until conditions are favourable (Kang et al., 2015).

Strigolactones control the development and hyphal branching of mycorrhizal fungi and ABC transporters were found to be essential for their release in the rhizosphere. Indeed, the *Petunia axillaris* ABCG transporter PaPDR1 was found to be expressed in hypodermal passage cells, the entry point of mycorrhizal fungi, whereas a loss-of-function mutant secreted less strigolactones and displayed altered mycorrhization (Kretzschmar et al., 2012). PaPDR1 is also expressed in the root tip and cortex where the transporter might initiate strigolactone transfer to the aerial tissues, in agreement with the bushy phenotype of the PaPDR1 mutant (Sasse et al., 2015).

Finally, jasmonate also regulates many plant processes, among which the response to biotic stress and in particular herbivory attack. Two ABC transporters so far have been shown to be related to jasmonate. First, AtABCD1, which is localized to the peroxisome where the final steps of jasmonate synthesis takes place, has been suggested to transport 12-*oxo*-phytodienoic acid, a precursor of jasmonic acid, in addition to fatty acyl-CoA (Theodoulou et al., 2005). Then, AtABCG16 (also called jasmonate transporter A) appears to have a dual function at two different subcellular localizations: nuclear influx of the bioactive form jasmonoyl isoleucine and cellular efflux of jasmonic acid (Li et al., 2017). As AtABCG16 is a half-size transporter, it is expected to homo- or heterodimerize. Whether the same form is involved in both transport functions is unknown.

So far, no ABC transporter has been shown to be involved in the transport of other hormones such as gibberellins, ethylene, salicylic acid, and brassinosteroids.

2.2 Response to Biotic Stresses

Historically, the first example of an ABC transporter involved in the response to a biotic stress is that of NpPDR1/NpABCG1 (originally named NpABC1) from *Nicotiana plumbaginifolia*. Silencing of this gene, which is constitutively expressed in roots and in leaf trichomes, resulted in higher susceptibility to pathogens such as *Botrytis cinerea* and *Pseudomonas syringae*. The encoded protein was shown to transport the antifungal diterpene sclareol (Jasinski et al., 2001; Stukkens et al., 2005). Since these initial reports, several transporters from multiple plant species were associated with resistance to pathogens.

NtPDR1, although not the closest homologue of NpPDR1 in *Nicotiana tabacum*, was also shown to transport antimicrobial diterpenes (Crouzet et al., 2013). Two analogues from *Nicotiana benthamiana*, NbABCG1 and NbABCG2, were identified as essential for resistance to the potato late blight pathogen *Phytophthora infestans*. Silencing of NbABCG1/2 reduced secretion of the sesquiterpene capsidiol. Besides this post-invasion response, NbABCG1/2 is also involved in pre-invasion defence, probably by exporting diterpenes (Shibata et al., 2016). In agreement with these data both diterpenes and sesquiterpenes were shown to increase the activity of NtPDR1, upon its reconstitution in asolectin liposomes, suggesting its involvement in both constitutive and induced defences (Pierman et al., 2017).

The wheat transporter TaABCG1 (Lr34/Yr18/Sr57/Pm38/Ltn1) has been extensively studied for its biotechnological potential in plant pathogen defence. Although the substrate of this transporter has not yet been identified, *lr34* mutants have been isolated more than 50 years ago for their ability to show resistance against fungal pathogens in wheat breeding programs. In fact, the role of this protein is peculiar as different genetic alleles confer either resistance or susceptibility to pathogens (Krattinger et al., 2009). Yet, resistance was associated with heterologous expression of the *lr34* mutant gene in rice (Risk et al., 2013), emmer wheat (Rinaldo et al., 2017), barley (Boni et al., 2018), and maize (Sucher et al., 2017).

In *Coptis japonica*, the alkaloid berberine is synthesized in roots and transported to the rhizome, where it accumulates. The ABCG transporter CjMDR1, expressed mainly in xylem tissues of the rhizome, was shown to import berberine when expressed in *Xenopus* oocytes, suggesting that this transporter is responsible for berberine translocation from the root to the rhizome (Shitan et al., 2003).

LjABCG1 from *Lotus japonicus* was first reported as being highly expressed during the early stages of nodulation. However, knockdown plants did not show any clear phenotypic differences in nodule formation as compared to a wild type. Still, heterologous expression of *LjABCG1* in the Arabidopsis *pdr8* mutant significantly reduced the penetration frequency of *P. infestans*, suggesting a role in plant defence (Sugiyama et al., 2015).

In *Medicago truncatula*, the silencing of MtABCG10 expression resulted in a reduced amount of the phenylpropanoid-derived phytoalexin, medicarpin in roots as well as in root exudates (Banasiak et al., 2013). It was later shown by metabolic analysis as well as in vitro transport assays that MtABCG10 transports 4-coumarate and liquiritigenin, two precursors of medicarpin in the phenylpropanoid pathway (Biala, Banasiak, Jarzyniak, Pawela, & Jasinski, 2017).

In Arabidopsis, at least four full-size ABCG transporters were associated with the plant's responses to biotic stresses. First of all, AtBACG30 (AtPDR2) was studied for its impact on soil microbiota. Mutant *atabcg30* plants showed altered root exudates, thereby impacting fungal and bacterial communities underground. Although phenolics and sugars were found to be the most impacted phytochemicals by the mutation, no clear substrate(s) could be formally identified (Badri et al., 2009). AtABCG34 (AtPDR6) was first suggested to be involved in the root exudation of organic acids (Badri et al., 2008). However, a recent report showed its implication in the defence against necrotrophic fungi by secreting camalexin, a major Arabidopsis phytoalexin, at the leaf surface (Khare et al., 2017). Mutant plants displayed increased sensitivity to *Alternaria brassicicola* and *Botrytis cinerea* and secreted less camalexin than the wild type. The expression of AtABCG34 in *N. tabacum* BY-2 cells conferred resistance to camalexin, thereby further confirming *in planta* data.

AtABCG36 (AtPDR8/AtPEN3) is one of the best described examples of PDR transporter for which multiple substrates have been proposed. Among these are compounds involved in plant defence against pathogens. AtABCG36 was found to block the penetration of nonhost fungal pathogens (Campe et al., 2016; Kobae et al., 2006; Stein et al., 2006). It is localized at the plasma membrane, recruited to the infection site (Underwood & Somerville, 2013), and involved in glucosinolate-dependent pathogen defence (Clay, Adio, Denoux, Jander, & Ausubel, 2009). The identity of the chemicals secreted by AtABCG36 is still unknown but indole-type metabolites, precursors of 4OGlc3F (4-*O*-β-D-glucosyl-indo-3-yl formamide) were proposed as substrates (Lu et al., 2015). Finally, AtABCG40 (AtPDR12) was found to be functionally related to NtPDR1 and NpPDR1 as the expression of this transporter was also found to be induced by sclareol and inhibition of wilt disease by sclareol was prevented in mutants defective in AtABCG40 (Campbell et al., 2003; Seo et al., 2012).

Insects represent a major threat to plants and several reports support the involvement of ABC transporters in the defence arsenal. In *N. tabacum*, NtABCG5 (NtPDR5) was found to be induced by wounding and *NtABCG5*-silenced plants were less resistant to the herbivore *Manduca sexta* (Bienert et al., 2012). The enzymatic and structural properties of the transporter were determined (Toussaint, Pierman, Bertin, Levy, & Boutry, 2017) but the exact substrate of this transporter still needs to be determined. In *Petunia hybrida*, PhPDR2 is predominantly expressed in multicellular glandular trichomes of leaves and stems as well as in the leaf margins. *PhPDR2*-silenced plants were found to be more susceptible to the generalist foliage feeder *Spodoptera littoralis* and their trichomes displayed reduced contents of petuniasterones and petuniolides, insecticidal steroid compounds (Sasse et al., 2016). CrABCG1 (CrTPT2) from *Catharanthus roseus* and VmABCG1 (VmTPT2) from *Vinca minor* were found to be involved in secretion of monoterpenoid indole alkaloids at the leaf surface, thereby suggesting a role in herbivore defence (Demessie, Woolfson, Yu, Qu, & De Luca, 2017; Yu & De Luca, 2013).

2.3 Surface Structures

Several plant tissues are protected by a layer of insoluble, usually polymeric, material, which is generally built from precursors synthesized in, and exported out of the cell. For instance, the cuticle covers the plant aerial surface and is made of cutin, an insoluble polymer composed of esterified hydroxy fatty acids (mainly C16 and C18) and wax, which is a complex mixture of long chain hydrocarbons. The cuticle protects the plant from biotic and abiotic stresses including desiccation. Several half-size and full-size ABCG transporters have been proposed to export cuticle precursors to the apoplasm. They were usually identified by loss-of-function mutants that showed an abnormally functioning cuticle and alterations in the levels of cutin monomers (e.g. in Arabidopsis, Bessire et al., 2011; Bird et al., 2007; Luo, Xue, Hu, Wang, & Chen, 2007; Panikashvili et al., 2007, 2010; Panikashvili, Shi, Schreiber, & Aharoni, 2011; Pighin et al., 2004; Ukitsu et al., 2007). Interestingly, AtABCG11, which is required for the cuticle formation, is also required, together with AtABCG9 and AtABCG14, for proper vascular development (Le Hir et al., 2013).

Pollen grains develop a thick wall composed of two layers. The inner layer, intine, consists primarily of cellulose and pectine. The exine, or outer layer, is composed of sporopollenin, a very stable biopolymer. As for cutin, many half-size ABCG transporters were described as being involved in exine biosynthesis, probably by mediating sporopollenin transport (Jiang, Zhang, & Cao, 2012). In Arabidopsis, analysis of mutant lines identified four transporters involved in exine synthesis: AtABCG1, AtABCG9, AtABCG16, and AtABCG26 (Choi et al., 2011, 2014; Dou et al., 2011; Kuromori, Ito, Sugimoto, & Shinozaki, 2011; Quilichini, Friedmann, Samuels, & Douglas, 2010; Yadav et al., 2014).

Another essential wax polymer found in plants is suberin. A possible role is to act as a barrier for water and solutes in the cell walls of the periderm and the endoderm, where it reinforces the role played by lignified structures known as Casparian strips. Suberin is mainly composed of aliphatic and aromatic monomers linked together by glycerol moieties. Arabidopsis half-size ABCG transporters found to be involved in the translocation of suberin precursors are AtABCG2, AtABCG6, and AtABCG20 (Yadav et al., 2014).

2.4 Detoxification

Any cell has to cope with toxic compounds, which can either be generated by the cell itself or arise from its environment. In animals and fungi, several

ABC transporters are crucial for the export of toxic molecules, especially in the ABCB and ABCG subfamilies. This activity gave the original names to some phylogenetic clusters such as, in animals, MDR (multiple drug resistance) within the ABCGB subfamily or, in fungi, PDR (pleiotropic drug resistance) within the ABCG subfamily. However, so far, it does not seem that MDR or PDR transporters are involved in this kind of activity in plants. Cellular detoxification is rather achieved by ABCC subfamily transporters, which are most often localized to the tonoplast.

ABCC transporters were the first to be identified in plants, through biochemical approaches: glutathionated compounds were found to be transported into the vacuole in the presence of ATP. Transport of glutathione conjugates was demonstrated for several plants ABCC transporters expressed in yeast (Liu, Sanchez-Fernandez, Li, & Rea, 2001; Lu, Li, & Rea, 1997; Martinoia, Grill, Tommasini, Kreuz, & Amrhein, 1993; Tommasini et al., 1998). First reports used *Saccharomyces cerevisiae* vacuolar microsomes to successively demonstrate transport of glutathione *S*-conjugates by AtABCC1 (AtMRP1) and AtABCC2 (AtMRP2) (Lu et al., 1998, 1997). Other substrates were then identified: folates and antifolates for ABCC1 (Raichaudhuri et al., 2009) and glucoronate conjugates for AtABCC2 (Liu et al., 2001). These transporters also confer tolerance to cadmium, mercury, and arsenic, which are co-transported into the vacuole with phytochelatins (Park et al., 2012; Song et al., 2014, 2010). Chlorophyll catabolites, herbicides, and organic anions are other putative substrates (Frelet-Barrand et al., 2008; Lu et al., 1998). AtABCC3 mediates transport of glutathione conjugates and chlorophyll catabolites (Tommasini et al., 1998) as well as phytochelatin–Cd complexes (Bovet et al., 2003; Brunetti et al., 2015).

Not all ABCC have been shown to be involved in detoxification. For instance, AtABCC4 was found to be involved in the regulation of stomatal aperture. Indeed, plants disrupted for *AtABCC4* displayed larger stomatal aperture than wild-type plants, resulting in increased drought susceptibility (Klein et al., 2004).

AtABCC5 (AtMRP5) was found to be involved in both stomatal opening and root development (Gaedeke et al., 2001; Klein et al., 2003; Suh et al., 2007). First indications concerning its putative substrates were obtained by working with microsomes isolated from *S. cerevisiae* cells expressing AtABCC5, which displayed ATP-dependent transport of $E_2$17G (oestradiol-17-β-D-glucuronide) (Gaedeke et al., 2001). Insertion mutants of *AtMRP5* displayed a low inositol hexa*kis*phosphate (phytic acid) phenotype in seed tissues. The heterologous expression in yeast confirmed that AtMRP5 encodes a specific and high affinity ATP-dependent inositol hexa*kis*phosphate transporter (Nagy et al., 2009).

2.5 Additional Functions

As described above, most ABC transporters localize either to the tonoplast or the plasma membrane where they are generally involved in hormone transport, response to (a)biotic stresses, maintenance of surface structures, or cellular detoxification. However, some transporters display alternative subcellular localizations and therefore fulfil more peculiar functions.

One example is that of AtABCB25/ATM3 from the small HMT/ATM cluster (half-size ABCB), which localizes to the inner mitochondrial membrane. Arabidopsis *atm3* mutants were found to be hypersensitive to an inhibitor of glutathione biosynthesis and accumulated glutathione disulphide (GSSG) in the mitochondria. Expression in *Lactococcus lactis* showed that AtABCB25 exports GSSG. Authors proposed that AtABCB25 exports glutathione polysulfide involved in iron–sulphur cluster assembly in the cytosol (Schaedler et al., 2014).

On the other hand, the Arabidopsis AtABCD1 is a full-size transporter localized to the peroxisomal membrane. It is involved in the peroxisomal import of substrates for β-oxidation and the glyoxylate cycle (Footitt et al., 2007, 2002; Hooks et al., 2007; Kunz et al., 2009; Park et al., 2013). The enzyme expressed in *S. cerevisiae* was stimulated by fatty acyl-CoAs (Nyathi et al., 2010). Still in yeast, transport of oleic acid by AtABCD1 required peroxisomal acyl-CoA synthetases, indicating that the substrate is CoA esters (De Marcos Lousa et al., 2013). Expression in *Sf9* insect cells indicated that AtABCD1 possesses fatty acyl-CoA thioesterase activity. The authors proposed that the cleavage of the CoA occurs during the transport cycle. AtABCD1 thus displays an additional enzymatic function besides its transport activity, which is unique among plant ABC transporters.

3. OPEN QUESTIONS

Several ABC transporters have hydrophobic substrates such as alkaloids and terpenoids. These compounds are expected to easily cross membranes, which raises the question as to why a transporter is needed. This question was thoroughly discussed by Widhalm, Jaini, Morgan, and Dudareva (2015) concerning the release of volatile organic compounds (VOCs) from plants. Many VOCs are small and hydrophobic and thus freely diffuse through the membranes. However, free diffusion is not an active transport and the steady-state concentration within the cell and, in particular, within membranes, might still be too high and cause toxicity. These authors considered two typical VOCs of snapdragon, methylbenzoate, a benzenoid/phenylpropanoid, as well as nerolidol, a terpenoid. They calculated that, to explain the observable emission rate by diffusion alone, the resulting VOC levels in the plasma membrane would be very high (120 and 48mM, respectively), most probably resulting in cytotoxicity. They concluded that biological mechanisms are probably involved in facilitating emission flux and reducing VOC membrane levels (Widhalm et al., 2015). This prediction was later confirmed for the emission of VOCs from petunia flowers. Indeed, a half-size ABCG transporter, PhABCG1, was found to be involved in their active export. Transgenic petunia RNAi lines with a decreased expression of *PhABCG1* displayed an accumulation of internal VOCs as well as a decrease of VOC emission as compared to wild-type lines. In addition, PhABCG1 expressed in *N. tabacum* BY-2 cells was shown to transport methyl benzoate and benzyl alcohol, two major VOCs constituents emitted by petunia flowers (Adebesin et al., 2017).

We should stress here that the identification of the substrates of many plant ABC transporters remains speculative. Unlike several other primary or secondary transporter families, the primary sequence of ABC transporters does not allow one to predict what the substrate is. Indeed, there is so far no consensus sequence known that might give some hint concerning the transporter substrate(s). An additional problem is that the substrates of some ABC transporters are secondary metabolites whose nature considerably varies according to the plant species. In many cases the substrate has therefore been deduced from the phenotype of lines modified in the expression of the transporter. For instance, if a mutated line becomes more sensitive to a heavy metal, one might deduce that this heavy metal is the substrate of the targeted ABC transporter. However, other explanations are conceivable. For instance, the substrate may be a secondary metabolite that chelates the heavy metal. Alternatively, the substrate could be a hormone whose modified concentration alters the expression of other transporters, and therefore affect metal homeostasis.

Another important question that remains unaddressed is the possibility that some plant ABC transporters have pleiotropic functions, i.e., have multiple structurally unrelated substrates like this is the case for some animal and fungal ABC transporters. In this respect, AtABCG37/PDR9 is an interesting candidate. Based on the expression in various heterologous systems, this transporter was first proposed to act in the translocation of a range of auxinic compounds to the rhizosphere (Ito & Gray, 2006; Ruzicka et al., 2010).

Later on, the same transporter was suggested to be involved in the secretion of coumarins in response to iron deficiency, and direct transport of these metabolites was demonstrated by transport assays carried out with NtABCG3/PDR3, its *N. tabacum* homologue, expressed in BY-2 cells (Lefèvre et al., 2018). Whether AtABCG37 is directly involved in the export of both coumarins and auxins still remains to be determined and competition assays involving both putative substrates will be needed.

Unambiguous identification of the substrates therefore requires transport assays, i.e., measuring the flux of the substrate across a membrane. This is best performed when the transporter of interest is expressed in a simple (unicellular) host. *Escherichia coli* is not an appropriate host for plant membrane proteins. The yeast species, *S. cerevisiae*, *Pichia pastoris*, and *Schizosaccharomyces pombe* have been successively used for some ABC transporters. Insect and mammalian cells as well as plant suspension cells such as *Nicotiana tabacum* bright yellow-2 (BY-2) are other options (Lefèvre & Boutry, 2018). Whatever the host species, transport can then be compared between wild-type cells and cells that express an ABC transporter. A more direct, but also more demanding, approach consists of measuring transport using membranes vesicles purified after cell disruption. Whatever the transport method used, it should be stressed that measuring the transport of hydrophobic substrates might turn out to be tricky because free diffusion of these molecules through the membrane generates a high background.

The physiological roles of many ABC transporters still remain to be determined. This will require at least a careful identification of the substrates. Because many of them are secondary metabolites, which largely vary between different species, determining the roles of ABC transporters of model species like Arabidopsis will not tell the whole story. In terms of application, plants are currently used to produce high value metabolites, some of which are substrates of ABC transporters. The secretion of these compounds out of plants (or plant cells) is crucial to enhance their yield as well as to prevent/decrease their potential toxicity in the plant. Successful metabolic engineering therefore depends on an efficient transport system and on a good understanding of the functions of ABC transporters.

REFERENCES

Adebesin, F., Widhalm, J. R., Boachon, B., Lefèvre, F., Pierman, B., Lynch, J. H., et al. (2017). Emission of volatile organic compounds from petunia flowers is facilitated by an ABC transporter. *Science*, *356*, 1386–1388.

Aller, S. G., Yu, J., Ward, A., Weng, Y., Chittaboina, S., Zhuo, R., et al. (2009). Structure of P-glycoprotein reveals a molecular basis for poly-specific drug binding. *Science*, *323*(5922), 1718–1722. https://doi.org/10.1126/science.1168750.

Badri, D. V., Loyola-Vargas, V. M., Broeckling, C. D., De-la-Pena, C., Jasinski, M., Santelia, D., et al. (2008). Altered profile of secondary metabolites in the root exudates of Arabidopsis ATP-binding cassette transporter mutants. *Plant Physiology, 146*(2), 762–771. https://doi.org/10.1104/pp.107.109587.

Badri, D. V., Quintana, N., El Kassis, E. G., Kim, H. K., Choi, Y. H., Sugiyama, A., et al. (2009). An ABC transporter mutation alters root exudation of phytochemicals that provoke an overhaul of natural soil microbiota. *Plant Physiology, 151*(4), 2006–2017. https://doi.org/10.1104/pp.109.147462.

Banasiak, J., Biala, W., Staszkow, A., Swarcewicz, B., Kepczynska, E., Figlerowicz, M., et al. (2013). A Medicago truncatula ABC transporter belonging to subfamily G modulates the level of isoflavonoids. *Journal of Experimental Botany, 64*(4), 1005–1015. https://doi.org/10.1093/jxb/ers380.

Bessire, M., Borel, S., Fabre, G., Carraca, L., Efremova, N., Yephremov, A., et al. (2011). A member of the pleiotropic drug resistance family of ATP binding cassette transporters is required for the formation of a functional cuticle in Arabidopsis. *Plant Cell, 23*(5), 1958–1970. https://doi.org/10.1105/tpc.111.083121.

Biala, W., Banasiak, J., Jarzyniak, K., Pawela, A., & Jasinski, M. (2017). Medicago truncatula ABCG10 is a transporter of 4-coumarate and liquiritigenin in the medicarpin biosynthetic pathway. *Journal of Experimental Botany, 68*(12), 3231–3241. https://doi.org/10.1093/jxb/erx059.

Bienert, M. D., Siegmund, S. E., Drozak, A., Trombik, T., Bultreys, A., Baldwin, I. T., et al. (2012). A pleiotropic drug resistance transporter in Nicotiana tabacum is involved in defense against the herbivore Manduca sexta. *The Plant Journal, 72*(5), 745–757. https://doi.org/10.1111/j.1365-313X.2012.05108.x.

Bird, D., Beisson, F., Brigham, A., Shin, J., Greer, S., Jetter, R., et al. (2007). Characterization of Arabidopsis ABCG11/WBC11, an ATP binding cassette (ABC) transporter that is required for cuticular lipid secretion. *The Plant Journal, 52*(3), 485–498. https://doi.org/10.1111/j.1365-313X.2007.03252.x.

Boni, R., Chauhan, H., Hensel, G., Roulin, A., Sucher, J., Kumlehn, J., et al. (2018). Pathogen-inducible Ta-Lr34res expression in heterologous barley confers disease resistance without negative pleiotropic effects. *Plant Biotechnology Journal, 16*(1), 245–253. https://doi.org/10.1111/pbi.12765.

Borghi, L., Kang, J., Ko, D., Lee, Y., & Martinoia, E. (2015). The role of ABCG-type ABC transporters in phytohormone transport. *Biochemical Society Transactions, 43*(5), 924–930. https://doi.org/10.1042/BST20150106.

Bovet, L., Eggmann, T., Meylan-Bettex, M., Polier, J., Kammer, P., Marin, E., et al. (2003). Transcript levels of *AtMRPs* after cadmium treatment: Induction of *AtMRP3*. *Plant, Cell & Environment, 26*, 371–381.

Brunetti, P., Zanella, L., De Paolis, A., Di Litta, D., Cecchetti, V., Falasca, G., et al. (2015). Cadmium-inducible expression of the ABC-type transporter AtABCC3 increases phytochelatin-mediated cadmium tolerance in Arabidopsis. *Journal of Experimental Botany, 66*(13), 3815–3829. https://doi.org/10.1093/jxb/erv185.

Campbell, E. J., Schenk, P. M., Kazan, K., Penninckx, I. A., Anderson, J. P., Maclean, D. J., et al. (2003). Pathogen-responsive expression of a putative ATP-binding cassette transporter gene conferring resistance to the diterpenoid sclareol is regulated by multiple defense signaling pathways in Arabidopsis. *Plant Physiology, 133*(3), 1272–1284. https://doi.org/10.1104/pp.103.024182.

Campe, R., Langenbach, C., Leissing, F., Popescu, G. V., Popescu, S. C., Goellner, K., et al. (2016). ABC transporter PEN3/PDR8/ABCG36 interacts with calmodulin that, like PEN3, is required for Arabidopsis nonhost resistance. *The New Phytologist, 209*(1), 294–306. https://doi.org/10.1111/nph.13582.

Cho, M., & Cho, H. T. (2013). The function of ABCB transporters in auxin transport. *Plant Signaling & Behavior, 8*(2), e22990. https://doi.org/10.4161/psb.22990.

Cho, M., Lee, S. H., & Cho, H. T. (2007). P-glycoprotein4 displays auxin efflux transporter-like action in Arabidopsis root hair cells and tobacco cells. *Plant Cell, 19*(12), 3930–3943. https://doi.org/10.1105/tpc.107.054288.

Choi, H., Jin, J. Y., Choi, S., Hwang, J. U., Kim, Y. Y., Suh, M. C., et al. (2011). An ABCG/WBC-type ABC transporter is essential for transport of sporopollenin precursors for exine formation in developing pollen. *The Plant Journal, 65*(2), 181–193. https://doi.org/10.1111/j.1365-313X.2010.04412.x.

Choi, H., Ohyama, K., Kim, Y. Y., Jin, J. Y., Lee, S. B., Yamaoka, Y., et al. (2014). The role of Arabidopsis ABCG9 and ABCG31 ATP binding cassette transporters in pollen fitness and the deposition of steryl glycosides on the pollen coat. *Plant Cell, 26*(1), 310–324. https://doi.org/10.1105/tpc.113.118935.

Clay, N. K., Adio, A. M., Denoux, C., Jander, G., & Ausubel, F. M. (2009). Glucosinolate metabolites required for an Arabidopsis innate immune response. *Science, 323*(5910), 95–101. https://doi.org/10.1126/science.1164627.

Crouzet, J., Roland, J., Peeters, E., Trombik, T., Ducos, E., Nader, J., et al. (2013). NtPDR1, a plasma membrane ABC transporter from Nicotiana tabacum, is involved in diterpene transport. *Plant Molecular Biology, 82*(1–2), 181–192. https://doi.org/10.1007/s11103-013-0053-0.

De Marcos Lousa, C., van Roermund, C. W., Postis, V. L., Dietrich, D., Kerr, I. D., Wanders, R. J., et al. (2013). Intrinsic acyl-CoA thioesterase activity of a peroxisomal ATP binding cassette transporter is required for transport and metabolism of fatty acids. *Proceedings of the National Academy of Sciences of the United States of America, 110*(4), 1279–1284. https://doi.org/10.1073/pnas.1218034110.

Demessie, Z., Woolfson, K. N., Yu, F., Qu, Y., & De Luca, V. (2017). The ATP binding cassette transporter, VmTPT2/VmABCG1, is involved in export of the monoterpenoid indole alkaloid, vincamine in Vinca minor leaves. *Phytochemistry, 140*, 118–124. https://doi.org/10.1016/j.phytochem.2017.04.019.

Do, T. H. T., Martinoia, E., & Lee, Y. (2017). Functions of ABC transporters in plant growth and development. *Current Opinion in Plant Biology, 41*, 32–38. https://doi.org/10.1016/j.pbi.2017.08.003.

Dou, X. Y., Yang, K. Z., Zhang, Y., Wang, W., Liu, X. L., Chen, L. Q., et al. (2011). WBC27, an adenosine tri-phosphate-binding cassette protein, controls pollen wall formation and patterning in Arabidopsis. *Journal of Integrative Plant Biology, 53*(1), 74–88. https://doi.org/10.1111/j.1744-7909.2010.01010.x.

Footitt, S., Dietrich, D., Fait, A., Fernie, A. R., Holdsworth, M. J., Baker, A., et al. (2007). The COMATOSE ATP-binding cassette transporter is required for full fertility in Arabidopsis. *Plant Physiology, 144*(3), 1467–1480. https://doi.org/10.1104/pp.107.099903.

Footitt, S., Slocombe, S. P., Larner, V., Kurup, S., Wu, Y., Larson, T., et al. (2002). Control of germination and lipid mobilization by *COMATOSE*, the *Arabidopsis* homologue of human ALDP. *The EMBO Journal, 21*(12), 2912–2922.

Frelet-Barrand, A., Kolukisaoglu, H. U., Plaza, S., Ruffer, M., Azevedo, L., Hortensteiner, S., et al. (2008). Comparative mutant analysis of Arabidopsis ABCC-type ABC transporters: AtMRP2 contributes to detoxification, vacuolar organic anion transport and chlorophyll degradation. *Plant & Cell Physiology, 49*(4), 557–569. https://doi.org/10.1093/pcp/pcn034.

Gaedeke, N., Klein, M., Kolukisaoglu, H. U., Forestier, C., Müller, A., Ansorge, M., et al. (2001). The *Arabidopsis thaliana* ABC transporter AtMRP5 controls root development and stomata movement. *The EMBO Journal, 20*(8), 1875–1887.

Geisler, M., Aryal, B., Di Donato, M., & Hao, P. (2017). A critical view on ABC transporters and their interacting partners in auxin transport. *Plant and Cell Physiology, 58*(10), 1601–1614.

Geisler, M., & Murphy, A. S. (2006). The ABC of auxin transport: The role of p-glycoproteins in plant development. *FEBS Letters, 580*(4), 1094–1102. https://doi.org/10.1016/j.febslet.2005.11.054.

Higgins, C. F. (2001). ABC transporters: Physiology, structure and mechanism—An overview. *Research in Microbiology, 152*, 205–210.

Hooks, M. A., Turner, J. E., Murphy, E. C., Johnston, K. A., Burr, S., & Jaroslawski, S. (2007). The Arabidopsis ALDP protein homologue COMATOSE is instrumental in peroxisomal acetate metabolism. *The Biochemical Journal, 406*(3), 399–406. https://doi.org/10.1042/BJ20070258.

Hwang, J. U., Song, W. Y., Hong, D., Ko, D., Yamaoka, Y., Jang, S., et al. (2016). Plant ABC transporters enable many unique aspects of a terrestrial plant's lifestyle. *Molecular Plant, 9*(3), 338–355. https://doi.org/10.1016/j.molp.2016.02.003.

Ito, H., & Gray, W. M. (2006). A gain-of-function mutation in the Arabidopsis pleiotropic drug resistance transporter PDR9 confers resistance to auxinic herbicides. *Plant Physiology, 142*(1), 63–74. https://doi.org/10.1104/pp.106.084533.

Jasinski, M., Stukkens, Y., Degand, H., Purnelle, B., Marchand-Brynaert, J., & Boutry, M. (2001). A plant plasma membrane ATP binding cassette-type transporter is involved in antifungal terpenoid secretion. *Plant Cell, 13*, 1095–1107.

Jiang, J., Zhang, Z., & Cao, J. (2012). Pollen wall development: The associated enzymes and metabolic pathways. *Plant Biology, 15*(2), 249–263.

Kamimoto, Y., Terasaka, K., Hamamoto, M., Takanashi, K., Fukuda, S., Shitan, N., et al. (2012). Arabidopsis ABCB21 is a facultative auxin importer/exporter regulated by cytoplasmic auxin concentration. *Plant & Cell Physiology, 53*(12), 2090–2100. https://doi.org/10.1093/pcp/pcs149.

Kang, J., Hwang, J. U., Lee, M., Kim, Y. Y., Assmann, S. M., Martinoia, E., et al. (2010). PDR-type ABC transporter mediates cellular uptake of the phytohormone abscisic acid. *Proceedings of the National Academy of Sciences of the United States of America, 107*(5), 2355–2360. https://doi.org/10.1073/pnas.0909222107.

Kang, J., Yim, S., Choi, H., Kim, A., Lee, K. P., Lopez-Molina, L., et al. (2015). Abscisic acid transporters cooperate to control seed germination. *Nature Communications, 6,* 8113. https://doi.org/10.1038/ncomms9113.

Khare, D., Choi, H., Huh, S. U., Bassin, B., Kim, J., Martinoia, E., et al. (2017). Arabidopsis ABCG34 contributes to defense against necrotrophic pathogens by mediating the secretion of camalexin. *Proceedings of the National Academy of Sciences of the United States of America, 114*(28), E5712–E5720. https://doi.org/10.1073/pnas.1702259114.

Kim, J. Y., Henrichs, S., Bailly, A., Vincenzetti, V., Sovero, V., Mancuso, S., et al. (2010). Identification of an ABCB/P-glycoprotein-specific inhibitor of auxin transport by chemical genomics. *The Journal of Biological Chemistry, 285*(30), 23309–23317. https://doi.org/10.1074/jbc.M110.105981.

Klein, M., Geisler, M., Suh, S. J., Kolukisaoglu, H. U., Azevedo, L., Plaza, S., et al. (2004). Disruption of AtMRP4, a guard cell plasma membrane ABCC-type ABC transporter, leads to deregulation of stomatal opening and increased drought susceptibility. *The Plant Journal, 39*(2), 219–236. https://doi.org/10.1111/j.1365-313X.2004.02125.x.

Klein, M., Perfus-Barbeoch, L., Frelet, A., Gaedeke, N., Reinhardt, D., Mueller-Roeber, B., et al. (2003). The plant multidrug resistance ABC transporter AtMRP5 is involved in guard cell hormonal signalling and water use. *The Plant Journal, 33*, 119–129.

Ko, D., Kang, J., Kiba, T., Park, J., Kojima, M., Do, J., et al. (2014). Arabidopsis ABCG14 is essential for the root-to-shoot translocation of cytokinin. *Proceedings of the National Academy of Sciences of the United States of America, 111*(19), 7150–7155. https://doi.org/10.1073/pnas.1321519111.

Kobae, Y., Sekino, T., Yoshioka, H., Nakagawa, T., Martinoia, E., & Maeshima, M. (2006). Loss of AtPDR8, a plasma membrane ABC transporter of Arabidopsis thaliana, causes hypersensitive cell death upon pathogen infection. *Plant & Cell Physiology*, *47*(3), 309–318. https://doi.org/10.1093/pcp/pcj001.

Krattinger, S. G., Lagudah, E. S., Spielmeyer, W., Singh, R. P., Huerta-Espino, J., McFadden, E., et al. (2009). A putative ABC transporter confers durable resistance to multiple fungal pathogens in wheat. *Science*, *323*, 1360–1363.

Kretzschmar, T., Burla, B., Lee, Y., Martinoia, E., & Nagy, R. (2011). Functions of ABC transporters in plants. *Essays in Biochemistry*, *50*, 145–160.

Kretzschmar, T., Kohlen, W., Sasse, J., Borghi, L., Schlegel, M., Bachelier, J. B., et al. (2012). A petunia ABC protein controls strigolactone-dependent symbiotic signalling and branching. *Nature*, *483*(7389), 341–344. https://doi.org/10.1038/nature10873.

Kubes, M., Yang, H., Richter, G. L., Cheng, Y., Mlodzinska, E., Wang, X., et al. (2012). The Arabidopsis concentration-dependent influx/efflux transporter ABCB4 regulates cellular auxin levels in the root epidermis. *The Plant Journal*, *69*(4), 640–654. https://doi.org/10.1111/j.1365-313X.2011.04818.x.

Kunz, H. H., Scharnewski, M., Feussner, K., Feussner, I., Flugge, U. I., Fulda, M., et al. (2009). The ABC transporter PXA1 and peroxisomal-oxidation are vital for metabolism in mature leaves of Arabidopsis during extended darkness. *The Plant Cell Online*, *21*(9), 2733–2749. https://doi.org/10.1105/tpc.108.064857.

Kuromori, T., Fujita, M., Urano, K., Tanabata, T., Sugimoto, E., & Shinozaki, K. (2016). Overexpression of AtABCG25 enhances the abscisic acid signal in guard cells and improves plant water use efficiency. *Plant Science*, *251*, 75–81. https://doi.org/10.1016/j.plantsci.2016.02.019.

Kuromori, T., Ito, T., Sugimoto, E., & Shinozaki, K. (2011). Arabidopsis mutant of AtABCG26, an ABC transporter gene, is defective in pollen maturation. *Journal of Plant Physiology*, *168*(16), 2001–2005. https://doi.org/10.1016/j.jplph.2011.05.014.

Kuromori, T., Miyaji, T., Yabuuchi, H., Shimizu, H., Sugimoto, E., Kamiya, A., et al. (2010). ABC transporter AtABCG25 is involved in abscisic acid transport and responses. *Proceedings of the National Academy of Sciences of the United States of America*, *107*(5), 2361–2366. https://doi.org/10.1073/pnas.0912516107.

Kuromori, T., Sugimoto, E., & Shinozaki, K. (2014). Intertissue signal transfer of abscisic acid from vascular cells to guard cells. *Plant Physiology*, *164*(4), 1587–1592. https://doi.org/10.1104/pp.114.235556.

Lane, T. S., Rempe, C. S., Davitt, J., Staton, M. E., Peng, Y., Soltis, D. E., et al. (2016). Diversity of ABC transporter genes across the plant kingdom and their potential utility in biotechnology. *BMC Biotechnology*, *16*(1), 47. https://doi.org/10.1186/s12896-016-0277-6.

Le Hir, R., Sorin, C., Chakraborti, D., Moritz, T., Schaller, H., Tellier, F., et al. (2013). ABCG9, ABCG11 and ABCG14 ABC transporters are required for vascular development in Arabidopsis. *The Plant Journal*, *76*(5), 811–824. https://doi.org/10.1111/tpj.12334.

Lee, J.-Y., Ninch, L. N., Borek, D. M., Wang, J., Wang, J., Urbatsch, I. L., et al. (2016). Crystal structure of the human sterol transporter ABCG5/ABCG8. *Nature*, *533*, 561–564.

Lefèvre, F., & Boutry, M. (2018). Towards identification of the substrates of ATP-binding cassette transporters. *Plant Physiology*, *178*, 18–39.

Lefèvre, F., Fourmeau, J., Pottier, M., Baijot, A., Cornet, T., Abadia, J., et al. (2018). The *Nicotiana tabacum* ABC transporter NtPDR3 secretes *O*-methylated coumarins in response to iron deficiency. *Journal of Experimental Botany*, *69*(18), 4419–4431.

Li, Q., Zheng, J., Li, S., Huang, S., Huang, G., Skilling, S. J., et al. (2017). Transporter-mediated nuclear entry of jasmonoyl-isoleucine is essential for jasmonate signaling. *Molecular Plant*, *10*, 695–708.

Licht, A., & Schneider, E. (2011). ATP binding cassette systems: Structures, mechanisms, and functions. *Central European Journal of Biology*, *6*(5), 785–801. https://doi.org/10.2478/s11535-011-0054-4.

Liu, G., Sanchez-Fernandez, R., Li, Z. S., & Rea, P. A. (2001). Enhanced multispecificity of arabidopsis vacuolar multidrug resistance-associated protein-type ATP-binding cassette transporter, AtMRP2. *The Journal of Biological Chemistry*, *276*(12), 8648–8656. https://doi.org/10.1074/jbc.M009690200.

Liu, F., Zhang, Z., Csanady, L., Gadsby, D. C., & Chen, J. (2017). Molecular structure of the human CFTR ion channel. *Cell*, *169*, 85–95.

Lu, X., Dittgen, J., Pislewska-Bednarek, M., Molina, A., Schneider, B., Svatos, A., et al. (2015). Mutant allele-specific uncoupling of PENETRATION3 functions reveals engagement of the ATP-binding cassette transporter in distinct tryptophan metabolic pathways. *Plant Physiology*, *168*(3), 814–827. https://doi.org/10.1104/pp.15.00182.

Lu, Y.-P., Li, Z.-S., Drozdowicz, Y. M., Hörtensteiner, S., Martinoia, E., & Rea, P. A. (1998). AtMRP2, an Arabidopsis ATP binding cassette transporter able to transport glutathione S-conjugates and chlorophyll catabolites: Functional comparisons with AtMRP1. *Plant Cell*, *10*, 267–282.

Lu, Y.-P., Li, Z.-S., & Rea, P. A. (1997). AtMRP1 gene of *Arabidopsis* encodes a glutathione S-conjugate pump isolation and functional definition of a plant ATP-binding cassette transporter gene. *Proceedings of the National Academy of Sciences of the United States of America*, *94*, 8243–8248.

Luo, B., Xue, X. Y., Hu, W. L., Wang, L. J., & Chen, X. Y. (2007). An ABC transporter gene of Arabidopsis thaliana, AtWBC11, is involved in cuticle development and prevention of organ fusion. *Plant & Cell Physiology*, *48*(12), 1790–1802. https://doi.org/10.1093/pcp/pcm152.

Martinoia, E., Grill, E., Tommasini, R., Kreuz, K., & Amrhein, N. (1993). ATP-dependant glutathione *S*-cojugate 'export' pump in the vacuolar membrane of plants. *Nature*, *364*, 247–249.

Nagy, R., Grob, H., Weder, B., Green, P., Klein, M., Frelet-Barrand, A., et al. (2009). The Arabidopsis ATP-binding cassette protein AtMRP5/AtABCC5 is a high affinity inositol hexakisphosphate transporter involved in guard cell signaling and phytate storage. *The Journal of Biological Chemistry*, *284*(48), 33614–33622. https://doi.org/10.1074/jbc.M109.030247.

Nyathi, Y., De Marcos Lousa, C., van Roermund, C. W., Wanders, R. J., Johnson, B., Baldwin, S. A., et al. (2010). The Arabidopsis peroxisomal ABC transporter, comatose, complements the Saccharomyces cerevisiae pxa1 pxa2Delta mutant for metabolism of long-chain fatty acids and exhibits fatty acyl-CoA-stimulated ATPase activity. *The Journal of Biological Chemistry*, *285*(39), 29892–29902. https://doi.org/10.1074/jbc.M110.151225.

Panikashvili, D., Savaldi-Goldstein, S., Mandel, T., Yifhar, T., Franke, R. B., Hofer, R., et al. (2007). The Arabidopsis DESPERADO/AtWBC11 transporter is required for cutin and wax secretion. *Plant Physiology*, *145*(4), 1345–1360. https://doi.org/10.1104/pp.107.105676.

Panikashvili, D., Shi, J. X., Bocobza, S., Franke, R. B., Schreiber, L., & Aharoni, A. (2010). The Arabidopsis DSO/ABCG11 transporter affects cutin metabolism in reproductive organs and suberin in roots. *Molecular Plant*, *3*(3), 563–575. https://doi.org/10.1093/mp/ssp103.

Panikashvili, D., Shi, J. X., Schreiber, L., & Aharoni, A. (2011). The Arabidopsis ABCG13 transporter is required for flower cuticle secretion and patterning of the petal epidermis. *The New Phytologist*, *190*(1), 113–124. https://doi.org/10.1111/j.1469-8137.2010.03608.x.

Park, S., Gidda, S. K., James, C. N., Horn, P. J., Khuu, N., Seay, D. C., et al. (2013). The alpha/beta hydrolase CGI-58 and peroxisomal transport protein PXA1 coregulate lipid homeostasis and signaling in Arabidopsis. *Plant Cell, 25*(5), 1726–1739. https://doi.org/10.1105/tpc.113.111898.

Park, J., Lee, Y., Martinoia, E., & Geisler, M. (2017). Plant hormone transporters: What we know and what we would like to know. *BMC Biology, 15*(1), 93. https://doi.org/10.1186/s12915-017-0443-x.

Park, J., Song, W. Y., Ko, D., Eom, Y., Hansen, T. H., Schiller, M., et al. (2012). The phytochelatin transporters AtABCC1 and AtABCC2 mediate tolerance to cadmium and mercury. *The Plant Journal, 69*(2), 278–288. https://doi.org/10.1111/j.1365-313X.2011.04789.x.

Pierman, B., Toussaint, F., Bertin, A., Levy, D., Smargiasso, N., De Pauw, E., et al. (2017). Activity of the purified plant ABC transporter NtPDR1 is stimulated by diterpenes and sesquiterpenes involved in constitutive and induced defenses. *The Journal of Biological Chemistry, 292*(47), 19491–19502. https://doi.org/10.1074/jbc.M117.811935.

Pighin, J. A., Zheng, H., Balakshin, L. J., Goodman, I. P., Western, T. L., Jetter, R., et al. (2004). Plant cuticular lipid export requires an ABC transporter. *Science, 306*, 702–704.

Quilichini, T. D., Friedmann, M. C., Samuels, A. L., & Douglas, C. J. (2010). ATP-binding cassette transporter G26 is required for male fertility and pollen exine formation in Arabidopsis. *Plant Physiology, 154*(2), 678–690. https://doi.org/10.1104/pp.110.161968.

Raichaudhuri, A., Peng, M., Naponelli, V., Chen, S., Sanchez-Fernandez, R., Gu, H., et al. (2009). Plant vacuolar ATP-binding cassette transporters that translocate folates and antifolates in vitro and contribute to Antifolate tolerance in vivo. *The Journal of Biological Chemistry, 284*(13), 8449–8460. https://doi.org/10.1074/jbc.M808632200.

Rea, P. A. (2007). Plant ATP-binding cassette transporters. *Annual Review of Plant Biology, 58*, 347–375. https://doi.org/10.1146/annurev.arplant.57.032905.105406.

Rinaldo, A., Gilbert, B., Boni, R., Krattinger, S. G., Singh, D., Park, R. F., et al. (2017). The Lr34 adult plant rust resistance gene provides seedling resistance in durum wheat without senescence. *Plant Biotechnology Journal, 15*(7), 894–905. https://doi.org/10.1111/pbi.12684.

Risk, J. M., Selter, L. L., Chauhan, H., Krattinger, S. G., Kumlehn, J., Hensel, G., et al. (2013). The wheat Lr34 gene provides resistance against multiple fungal pathogens in barley. *Plant Biotechnology Journal, 11*(7), 847–854. https://doi.org/10.1111/pbi.12077.

Ruzicka, K., Strader, L. C., Bailly, A., Yang, H., Blakeslee, J., Langowski, L., et al. (2010). Arabidopsis PIS1 encodes the ABCG37 transporter of auxinic compounds including the auxin precursor indole-3-butyric acid. *Proceedings of the National Academy of Sciences of the United States of America, 107*(23), 10749–10753. https://doi.org/10.1073/pnas.1005878107.

Sanchez-Fernandez, R., Davies, T. G., Coleman, J. O., & Rea, P. A. (2001). The Arabidopsis thaliana ABC protein superfamily, a complete inventory. *The Journal of Biological Chemistry, 276*(32), 30231–30244. https://doi.org/10.1074/jbc.M103104200.

Sasse, J., Schlegel, M., Borghi, L., Ullrich, F., Lee, M., Liu, G. W., et al. (2016). Petunia hybrida PDR2 is involved in herbivore defense by controlling steroidal contents in trichomes. *Plant, Cell & Environment, 39*(12), 2725–2739. https://doi.org/10.1111/pce.12828.

Sasse, J., Simon, S., Gubeli, C., Liu, G. W., Cheng, X., Friml, J., et al. (2015). Asymmetric localizations of the ABC transporter PaPDR1 trace paths of directional strigolactone transport. *Current Biology, 25*(5), 647–655. https://doi.org/10.1016/j.cub.2015.01.015.

Schaedler, T. A., Thornton, J. D., Kruse, I., Schwarzlander, M., Meyer, A. J., van Veen, H. W., et al. (2014). A conserved mitochondrial ATP-binding cassette transporter exports glutathione polysulfide for cytosolic metal cofactor assembly. *The Journal of Biological Chemistry, 289*(34), 23264–23274. https://doi.org/10.1074/jbc.M114.553438.

Seo, S., Gomi, K., Kaku, H., Abe, H., Seto, H., Nakatsu, S., et al. (2012). Identification of natural diterpenes that inhibit bacterial wilt disease in tobacco, tomato and Arabidopsis. *Plant & Cell Physiology, 53*(8), 1432–1444. https://doi.org/10.1093/pcp/pcs085.

Shibata, Y., Ojika, M., Sugiyama, A., Yazaki, K., Jones, D. A., Kawakita, K., et al. (2016). The full-size ABCG transporters Nb-ABCG1 and NbABCG2 function in pre- and postinvasion defense against *Phytophthora infestans in Nicotiana benthamiana. The Plant Cell, 28*(5), 1163–1181.

Shitan, N., Bazin, I., Dan, K., Obata, K., Kigawa, K., Ueda, K., et al. (2003). Involvement of CjMDR1, a plant multidrug-resistance-type ATP-binding cassette protein, in alkaloid transport in *Coptis japonica. Proceedings of the National Academy of Sciences of the United States of America, 100*(2), 751–756.

Song, W. Y., Mendoza-Cozatl, D. G., Lee, Y., Schroeder, J. I., Ahn, S. N., Lee, H. S., et al. (2014). Phytochelatin-metal(loid) transport into vacuoles shows different substrate preferences in barley and Arabidopsis. *Plant, Cell & Environment, 37*(5), 1192–1201. https://doi.org/10.1111/pce.12227.

Song, W.-Y., Park, J., Mendoza-Cozatl, D. G., Suter-Grotemeyer, M., Shim, D., Hörtenseiner, S., et al. (2010). Arsenic tolerence in *Arabidopsis* is mediated by two ABCC-type phytochelatin transporters. *Proceedings of the National Academy of Sciences of the United States of America, 107*(49), 21187–21192.

Stein, M., Dittgen, J., Sanchez-Rodriguez, C., Hou, B. H., Molina, A., Schulze-Lefert, P., et al. (2006). Arabidopsis PEN3/PDR8, an ATP binding cassette transporter, contributes to nonhost resistance to inappropriate pathogens that enter by direct penetration. *Plant Cell, 18*(3), 731–746. https://doi.org/10.1105/tpc.105.038372.

Strader, L. C., & Bartel, B. (2009). The Arabidopsis pleiotropic drug resistance8/ABCG36 ATP binding cassette transporter modulates sensitivity to the auxin precursor indole-3-butyric acid. *Plant Cell, 21*(7), 1992–2007. https://doi.org/10.1105/tpc.109.065821.

Stukkens, Y., Bultreys, A., Grec, S., Trombik, T., Vanham, D., & Boutry, M. (2005). NpPDR1, a pleiotropic drug resistance-type ATP-binding cassette transporter from Nicotiana plumbaginifolia, plays a major role in plant pathogen defense. *Plant Physiology, 139*(1), 341–352. https://doi.org/10.1104/pp.105.062372.

Sucher, J., Boni, R., Yang, P., Rogowsky, P., Buchner, H., Kastner, C., et al. (2017). The durable wheat disease resistance gene Lr34 confers common rust and northern corn leaf blight resistance in maize. *Plant Biotechnology Journal, 15*(4), 489–496. https://doi.org/10.1111/pbi.12647.

Sugiyama, A., Fukuda, S., Takanashi, K., Yoshioka, M., Yoshioka, H., Narusaka, Y., et al. (2015). Molecular characterization of LjABCG1, an ATP-binding cassette protein in Lotus japonicus. *PLoS One, 10*(9), e0139127. https://doi.org/10.1371/journal.pone.0139127.

Suh, S. J., Wang, Y. F., Frelet, A., Leonhardt, N., Klein, M., Forestier, C., et al. (2007). The ATP binding cassette transporter AtMRP5 modulates anion and calcium channel activities in Arabidopsis guard cells. *The Journal of Biological Chemistry, 282*(3), 1916–1924. https://doi.org/10.1074/jbc.M607926200.

Taylor, N. M. I., Manolaridis, I., Jackson, S. M., Kowal, J., Stahlberg, H., & Locher, K. P. (2017). Structure of the human multidrug transporter ABCG2. *Nature, 546*, 504–509.

Theodoulou, F. L., Job, K., Slocombe, S. P., Footitt, S., Holdsworth, M., Baker, A., et al. (2005). Jasmonic acid levels are reduced in COMATOSE ATP-binding cassette transporter mutants. Implications for transport of jasmonate precursors into peroxisomes. *Plant Physiology, 137*(3), 835–840. https://doi.org/10.1104/pp.105.059352.

Tommasini, R., Vogt, E., Fromenteau, M., Hörtenseiner, S., Matile, P., Amrhein, N., et al. (1998). An ABC-transporter of *Arabidopsis thaliana* has both glutathione-conjugate and chlorophyll catabolite transport activity. *The Plant Journal*, *13*(6), 773–780.

Toussaint, F., Pierman, B., Bertin, A., Levy, D., & Boutry, M. (2017). Purification and biochemical characterization of NpABCG5/NpPDR5, a plant pleiotropic drug resistance transporter expressed in Nicotiana tabacum BY-2 suspension cells. *The Biochemical Journal*, *474*(10), 1689–1703. https://doi.org/10.1042/BCJ20170108.

Ukitsu, H., Kuromori, T., Toyooka, K., Goto, Y., Matsuoka, K., Sakuradani, E., et al. (2007). Cytological and biochemical analysis of COF1, an Arabidopsis mutant of an ABC transporter gene. *Plant & Cell Physiology*, *48*(11), 1524–1533. https://doi.org/10.1093/pcp/pcm139.

Underwood, W., & Somerville, S. C. (2013). Perception of conserved pathogen elicitors at the plasma membrane leads to relocalization of the Arabidopsis PEN3 transporter. *Proceedings of the National Academy of Sciences of the United States of America*, *110*(30), 12492–12497. https://doi.org/10.1073/pnas.1218701110.

Vasiliou, V., Vasiliou, K., & Nebert, D. W. (2009). Human ATP-binding cassette (ABC) transporter family. *Human Genomics*, *3*(3), 281–290. https://doi.org/10.1186/1479-7364-3-3-281.

Verrier, P. J., Bird, D., Burla, B., Dassa, E., Forestier, C., Geisler, M., et al. (2008). Plant ABC proteins—A unified nomenclature and updated inventory. *Trends in Plant Science*, *13*(4), 151–159. https://doi.org/10.1016/j.tplants.2008.02.001.

Widhalm, J. R., Jaini, R., Morgan, J. A., & Dudareva, N. (2015). Rethinking how volatiles are released from plant cells. *Trends in Plant Science*, *20*(9), 545–550. https://doi.org/10.1016/j.tplants.2015.06.009.

Yadav, V., Molina, I., Ranathunge, K., Castillo, I. Q., Rothstein, S. J., & Reed, J. W. (2014). ABCG transporters are required for suberin and pollen wall extracellular barriers in Arabidopsis. *Plant Cell*, *26*(9), 3569–3588. https://doi.org/10.1105/tpc.114.129049.

Yang, H., & Murphy, A. S. (2009). Functional expression and characterization of Arabidopsis ABCB, AUX 1 and PIN auxin transporters in Schizosaccharomyces pombe. *The Plant Journal*, *59*(1), 179–191. https://doi.org/10.1111/j.1365-313X.2009.03856.x.

Yu, F., & De Luca, V. (2013). ATP-binding cassette transporter controls leaf surface secretion of anticancer drug components in Catharanthus roseus. *Proceedings of the National Academy of Sciences of the United States of America*, *110*(39), 15830–15835. https://doi.org/10.1073/pnas.1307504110.

Zhang, K., Novak, O., Wei, Z., Gou, M., Zhang, X., Yu, Y., et al. (2014). Arabidopsis ABCG14 protein controls the acropetal translocation of root-synthesized cytokinins. *Nature Communications*, *5*, 3274. https://doi.org/10.1038/ncomms4274.

CHAPTER TWO

Plant Aquaporins

Lionel Verdoucq[1], Christophe Maurel
Biochimie et Physiologie Moléculaire des Plantes (BPMP), Université de Montpellier, CNRS, INRA, SupAgro, Montpellier, France
[1]Corresponding author: e-mail address: lionel.verdoucq@cnrs.fr

Contents

Abstract

Aquaporins are membrane channels that facilitate the transport of water and small neutral molecules across biological membranes of most living organisms. Plant aquaporins constitute a large family of evolutionarily related channels showing a high diversity of cellular localisations, transport selectivity, and regulation properties. Plant aquaporins can transport various small solutes in addition to water, such as hydrogen peroxide, ammonia, urea, metalloids, gases and even ions, indicating a wide range of cellular functions. At the cellular level, plant aquaporins are subjected to various regulation mechanisms leading to active/inactive channels in their target membranes. Structure–function studies have been instrumental to address the molecular and cellular mechanisms of plant aquaporin gating and subcellular trafficking. We present our current knowledge on their co- and post-translational modifications and more specifically discuss the importance of lipidic environment in aquaporin function. Overall, the present chapter shows how the 25 years of research performed since the discovery of aquaporins has established their crucial role in plants.

Advances in Botanical Research, Volume 87
ISSN 0065-2296
https://doi.org/10.1016/bs.abr.2018.09.011

1. INTRODUCTION

Terrestrial plants must cope with an ever-fluctuating environment to achieve proper growth and development. For this, they need to continuously sense and respond to these environmental variations. In addition, plants are sessile and autotrophic organisms requiring light, carbon dioxide (CO_2), water and mineral nutrients for their life cycle. Thus, they have to be particularly adaptable to variations in water and nutrient availability. This is especially true for roots, which show a remarkable ability to explore the soil for taking up these essential resources.

With respect to their water status, terrestrial plants display specific adaptations. During photosynthesis, assimilation of CO_2 within chloroplasts creates a net diffusion flow of CO_2 from the atmosphere to inner leaf tissues. In parallel, water diffuses from the leaf interior to the drier air. Plants have addressed this trade-off by differentiating aerial parts covered with an impermeable barrier (the cuticule) perforated by microscopic pores (the stomata). Thus, transpirational water loss and CO_2 uptake share the same diffusional path across tightly regulated stomatal apertures.

At the tissue level, cell plasma membranes provide an additional barrier that limits the exchange of water, solutes and gases between the metabolically active cells and their environment. These exchanges are controlled by dedicated channel or transporter proteins that are embedded in the phospholipidic bilayer. For decades, water diffusion across the lipid phase of membranes was thought to be sufficient to support water exchanges within living cells and tissues (Verkman, 1992). The pioneering work of Preston et al. revealed in the early 1990s that a proteinaceous pore of the red blood cell membrane, named Aquaporin1 (AQP1), functions as a water channel (Preston & Agre, 1991). AQP1 belongs to the ancient superfamily of major intrinsic proteins (MIPs), with homologs in nearly all living organisms. At present, and this will be the case in the present chapter, the general term "aquaporin" (AQP) often refers to all MIP family members, whatever their transport selectivity.

Since the early description of AQP1 function, huge progress has been made in the characterisation of plant AQPs, providing deep insights into their role in plant–water relations (Maurel et al., 2015). In particular, the function and regulation of AQPs have been quite extensively studied to explain plant hydraulic responses to environmental challenges. In the past 10 years, many studies have also revealed that plant AQPs are not only water channels but allow the diffusion across cell membranes of small unpolar substrates of great physiological significance, such as carbon dioxide (CO_2), hydrogen peroxide (H_2O_2), boron or silicon. In the present review, we cover the many facets of AQP functions in plants. We describe the diversity of plant AQP isoforms, their substrate selectivity and cellular localisations. In addition, we emphasise their integrated physiological functions with respect to their regulation in response to various environmental stresses.

2. PLANT AQUAPORIN DIVERSITY

2.1 Evolution and Diversity of Plant Aquaporins

MIPs form an ancient family of proteins, found in all taxonomic kingdoms of life (Eukarya, Bacteria, and Archaea), suggesting a key role in basal life functions, with an exception for thermophilic Archaea and intracellular bacteria (Abascal, Irisarri, & Zardoya, 2014). MIPs consist of small proteins (21–35 kDa), with a reduced (~25%) sequence conservation and a typical symmetrical organisation due to the early tandem intragenic duplication of a three transmembrane domain protein (Park & Saier, 1996; Preston & Agre, 1991). Thus, MIPs comprise six transmembrane α-helices domains connected by five loops with the N- and C-terminal ends of the protein located in the cytoplasm. Two additional short α-helices, containing two highly conserved Asn-Pro-Ala (NPA) motifs, coming from loops B and E, fold halfway back into the membrane from opposite sides to form a pore. Biochemical, structural and functional studies have shown that MIPs associate as tetramers in membranes, each monomer forming an independent channel. In plants, AQPs assemble in homotetramers or heterotetramers, as was described in several plant species (see later). The raising number of plant genome sequences available has revealed that the MIPs form a large protein family in plants that has expanded and diversified from ancestors due to several rounds of horizontal gene transfer (HGT) and gene and/or genome duplications events. It is now commonly accepted that an early gene duplication event gave rise to two distinct water channel and glycerol channel subfamilies (Finn, Chauvigne, Hlidberg, Cutler, & Cerda, 2014; Laloux et al., 2018; Soto, Alleva, Amodeo, Muschietti, & Ayub, 2012; Zardoya, 2005). The latter is present in most eukaryotes, including green algae and mosses, but not in vascular plants. In contrast, the former is present in all eukaryotes.

While mammals usually possess 12–15 AQP isoforms, higher plants possess far more isoforms, in part due to the greater duplication rate of their genomes. To date, 33 isoforms were found in rice (*Oryza sativa*) (Sakurai, Ishikawa, Yamaguchi, Uemura, & Maeshima, 2005), 35 in the model species *Arabidopsis thaliana* (Johanson et al., 2001; Quigley, Rosenberg, Shachar-Hill, & Bohnert, 2001), 38 in *Zea mays* (Chaumont, Barrieu, Wojcik, Chrispeels, & Jung, 2001), 55 in poplar (*Populus trichocarpa*) (Gupta & Sankararamakrishnan, 2009) and 71 in upland cotton (*Gossypium hirsutum*) (Park, Scheffler, Bauer, & Campbell, 2010). The record seems to be held by oil seed rape *Brassica napus*, which genome displays 121 *AQP* genes (Yuan et al., 2017). Plants MIPs are classified into seven subfamilies with five of these subfamilies found in seed plants: plasma membrane intrinsic proteins (PIPs), tonoplast intrinsic proteins (TIPs), NOD26-like intrinsic proteins (NIPs), small basic intrinsic proteins (SIPs), and X (for unrecognised) intrinsic proteins (XIP). This latter subfamily is absent in Brassicaceae and monocots (Danielson & Johanson, 2008; Lopez et al., 2012). Mosses have up to two additional subfamilies: spike mosses (e.g., *Selaginella moellendorffii*) possess hybrid intrinsic proteins (HIPs) while *Physcomitrella patens* has, in addition to HIPs, GLpF-like intrinsic proteins (GIPs), which probably originate through HGT from a bacterial ancestral gene (Gustavsson, Lebrun, Norden, Chaumont, & Johanson, 2005). These differences suggest that loss of these two subfamilies has occurred during evolution between ancestor and vascular and seed plants (Abascal et al., 2014). Finally, a unique AQP family was recently discovered in diatoms. These so-called large intrinsic proteins (LIPs) are characterised by a specific NPM motif in place of the second NPA (Khabudaev, Petrova, Grachev, & Likhoshway, 2014).

Deeper knowledge on the diversity of AQPs in plants can be found in several excellent recent reviews (Fox, Maistriaux, & Chaumont, 2017; Laloux et al., 2018).

2.2 Cellular and Subcellular Localisation

2.2.1 Plant AQPs Exhibit Membrane Specialisation

The phylogenetic diversity of plant AQPs somehow reflects specific membrane localisations. Among the NIPs, GmNOD26, one of the first identified AQPs, was found to be exclusively expressed in the peribacteroid membrane of nitrogen-fixing symbiotic nodules of legume roots (Fortin, Morrison, & Verma, 1987; Rivers et al., 1997). Fusions with fluorescent protein reporters have indicated that NIPs can also be expressed at the plasma membrane of non-legume plant species (Ma et al., 2006; Takano et al., 2006; Wang et al., 2017) or at the endoplasmic reticulum (ER) (Mizutani, Watanabe, Nakagawa, & Maeshima, 2006). SIPs are also localised at the

ER (Ishikawa, Suga, Uemura, Sato, & Maeshima, 2005) while XIPs sit at the plasma membrane (Bienert, Bienert, Jahn, Boutry, & Chaumont, 2011). PIPs generally localise to the plasma membrane, but recent evidence shows that some isoforms could also localise, at least in minute amounts, to the chloroplast envelope (Ferro et al., 2010; Simm et al., 2013; Zybailov et al., 2008). Finally, TIPs represent the only subclass of AQPs that is found in vacuolar membranes. Note that some TIPs were also found in the plasma, chloroplast and thylakoid membranes (Beebo, Mathai, Schoefs, & Spetea, 2013; Ferro et al., 2010). Thus, seed-specific *At*TIP3;1 and *At*TIP3;2, which predominantly localise in *Arabidopsis* protein storage vacuoles, show a transient expression at the plasma membrane during the early stages of seed germination and maturation (Gattolin, Sorieul, & Frigerio, 2011). The surprising co-localisation of two classes of highly permeable AQPs (PIPs and TIPs) in the chloroplast and thylakoid membranes may reflect a crucial role in supplying water and CO_2 for photosynthetic reactions, as discussed later (Beebo et al., 2013).

2.2.2 Increasing Evidence of Finite Subcellular Localisation

While most of plant AQPs seem to be evenly expressed in their final destination membrane, some isoforms can be confined to membrane subdomains. For instance, the use of GFP-fused proteins or immunolocalisations revealed that two NIP homologs, *Arabidopsis* *At*NIP5;1 (Takano et al., 2010) and rice *Os*NIP2;1 (Lsi1) (Ma et al., 2006; Yamaji, Mitatni, & Ma, 2008), are specifically expressed on the outer (distal) side of root cell plasma membranes where they mediate a passive cellular efflux of B and Si, respectively. Interestingly, influx transporters of the same solutes (BOR1 and Lsi2, respectively) show an opposite polar localisation in the same cell types. This organisation creates an efficient path for transcellular transport and centripetal transfer of B and Si into xylem vessels. In addition, a polar expression was described for some rice and maize PIP isoforms, with preferential accumulation on the exofacial and endofacial sides of root cells, respectively (Hachez, Heinen, Draye, & Chaumont, 2008; Sakurai-Ishikawa et al., 2011). TIPs can also show a non-uniform expression: accumulation of TIPs fused to GFP was observed in facing tonoplast regions of two adjacent vacuoles (Beebo et al., 2013). This localisation may allow privileged exchange between neighbouring vacuoles. TIPs fused to GFP were also observed in intravacuolar invaginations named bulbs (Saito et al., 2002). Recent studies suggest, however, that the bulb formation itself is due to the accumulation of membrane-anchored GFPs with an antiparallel dimerisation of GFP expressed in two facing membranes (Segami, Makino, Miyake, Asaoka, & Maeshima, 2014). Finally, PIPs were found to be enriched in detergent-resistant membrane fractions tentatively associated with membrane lipid rafts (Kierszniowska, Seiwert, & Schulze, 2009; Mongrand et al., 2004; Morel et al., 2006). Recent evidence based on super-resolution microscopy confirmed the confinement of *At*PIP2;1 in membrane microdomains associated with specific sterol compositions (Li, Wang, et al., 2011). The biological significance of this membrane partitioning is still unknown. It will have to be revisited, especially in the context of AQP–lipid interactions, as discussed later in the text.

2.3 Specialised Substrate Specificities

Functional expression of plant AQPs in heterologous systems (*Xenopus* oocytes or yeast cells) revealed that AQPs can transport, in addition to water, a great diversity of small solutes, as summarised in Table 1. Plant AQP substrates include solutes transported by animal and bacterial homologs (urea, glycerol, H_2O_2), metalloid species [boric acid: $B(OH)_3$, silicic acid: $Si(OH)_4$, arsenious acid: $As(OH)_3$, antimonite acid: $Sb(OH)_3$] or related (Aluminium-malate), lactic acid, sodium ions, or dissolved gas molecules (O_2, CO_2, ammonia: NH_3) (Maurel et al., 2015). By facilitating the diffusion of water and solutes along their chemical potential gradient, AQPs play key roles in diverse physiological functions as described in the following paragraphs.

Table 1 Functional Properties of Plant Aquaporins (AQPs)

	Transport Selectivity													
AQPs	**H_2O**	**CO_2**	**O_2**	**H_2O_2**	**Glycerol**	**Urea**	**NH_3**	**Lactic Acid**	**Ions**	**Si**	**B**	**Se**	**As**	**Sb**
PIPs	+	+	+	+	–	–	–	?	+	?	+	?	±	?
TIPs	+	–	?	+	+	+	+	?	?	–	–	?	±	–
NIPs	+	+	?	–	+	+	–	+	+	+	+	+	+	+
SIPs	+	–	?	+	–	–	–	?	?	–	–	?	–	–
XIPs	+	–	?	+	+	+	+	?	?	–	+	?	+	–

The transport selectivity of the different plant AQP subfamilies are shown. Substrates: H_2O, water; CO_2, carbon dioxide; O_2, oxygen; H_2O_2, hydrogen peroxide; glycerol; urea; NH_3, ammonia; lactic acid; ions; Si, silicon; B, boric acid; Se, selenium; As, arsenic; Sb, antimony; PIPs, plasma membrane intrinsic proteins; TIPs, tonoplast intrinsic proteins; NIPs, noduline26-like intrinsic proteins; SIPs, small basic intrinsic proteins and XIPs, uncharacterised X intrinsic proteins. Question marks indicate that the transport selectivity was not tested for the mentioned AQP subfamilies.

2.3.1 Water

AQPs of vascular plants are key players of water uptake, from the soil into the root, and of its distribution (water homeostasis) at the whole plant level. This review will not attempt to draw a complete overview of each plant AQP expression profile and their different roles in plant water homeostasis, as recent and extensive reviews already addressed this point (Chaumont & Tyerman, 2014; Fox et al., 2017; Maurel et al., 2015). Water is transported across tissues down their water potential gradients ($\Delta\Psi$), in other words, from high Ψ to low Ψ. For instance, during transpiration, water is taken up by the roots (high Ψ) and transported across the plant to the leaves, where it is evaporated in the stomatal chamber (low Ψ). This long-distance axial transport occurs across the xylem vessels consisting mostly in low-resistance dead cells. In addition, short-distance radial water transport across the root and out of the leaf vasculature is mediated through living tissues. During this process, water can flow along concurrent paths: the apoplastic path (through the cell wall continuum), the symplastic path (across the plasmodesmata and cytoplasmic continuities), and the transcellular path (across cell membranes) (Steudle & Peterson, 1998). In the latter path, AQP abundance and activity determines most of the membrane water permeability (Chaumont & Tyerman, 2014; Heinen, Ye, & Chaumont, 2009; Maurel et al., 2015). More generally, pharmacological studies have shown that PIPs can contribute to up of 85% of root hydraulic conductivity (Lp_r) (Maurel et al., 2015). Reverse genetic analyses also indicated that the single AQP isoform, *At*PIP2;2, can account for 14% of *Arabidopsis* Lp_r (Javot et al., 2003). PIPs also contribute to inner water transport in leaves (leaf hydraulic conductance) (Postaire et al., 2010; Prado et al., 2013; Shatil-Cohen, Attia, & Moshelion, 2011), but to a much lesser extent than in the roots, indicating that leaf vasculature represents an important hydraulic limitation.

Trans-cellular water transport is also required to maintain and regulate cell turgor, a key element in several physiological processes such as stomata movements and cell expansion. For instance, *Arabidopsis* *At*PIP2;1 was recently shown to be essential for stomata closure in response to the drought-induced hormone abscisic acid (ABA) (Grondin et al., 2015). Several recent reports pointed to roles of PIPs and TIPs in plant cell growth and organ development, and lateral root emergence in particular (Peret et al., 2012; Reinhardt et al., 2016). Cell turgor regulation is also essential in plant reproduction. Here again, a role of AQPs in anther dehydration (Bots et al., 2005), pollen development and germination (Di Giorgio et al., 2016; Wudick, Luu, Tournaire-Roux, Sakamoto, & Maurel, 2014) and seed germination (Gao, Young, Bonham-Smith, & Gusta, 1999; Vander Willigen, Postaire, Tournaire-Roux, Boursiac, & Maurel, 2006) has been described.

2.3.2 Hydrogen Peroxide

When present at low concentration, hydrogen peroxide (H_2O_2) can serve as a signalling molecule involved in the regulation of a wide range of metabolic processes (Neill, Desikan, & Hancock, 2002). At higher (mM) concentrations, H_2O_2 can however be toxic as it can oxidise various cellular targets and cause oxidative damage, eventually leading to apoptosis. Significant progresses were recently made in understanding the physiological contribution of PIPs to H_2O_2 transport and homeostasis. While early evidence established that various plant AQPs (including PIPs, TIPs and XIPs) can mediate H_2O_2 transmembrane diffusion in heterologous systems (Bienert et al., 2007), several of these AQPs were recently shown to be truly involved in plant signal transduction. For instance, *At*PIP1;4 was shown to facilitate the cellular entry of apoplastic H_2O_2 produced in mesophyll cells after infection with the pathogenic *Pseudomonas syringae* pv. *tomato* (*Pst* DC3000 strain) (Tian et al., 2016). In this study, *Atpip1;4* knockout mutants showed reduced accumulation of H_2O_2 into cells whereas they were not affected in the primary, pathogen-induced generation of H_2O_2, which accumulated as high apoplastic concentrations. The defective cytoplasmic accumulation of H_2O_2 in mutant plants compromised plant defense responses and caused a hypersensitivity to *Pst* DC3000. In another recent study, *Arabidopsis At*PIP2;1 was shown to facilitate H_2O_2 entry into guard cells, leading to stomatal closure in response to ABA or the bacterial elicitor flg22 (Rodrigues et al., 2017). Here, authors showed that accumulation of H_2O into guard cells in response to either flg22 and ABA was abolished in *Atpip2;1* knockout mutants. They also showed that the intrinsic transport activity of *At*PIP2;1 strongly relies on *At*PIP2;1 phosphorylation by the guard cell SnRK2;6 (OST1) protein kinase. These works are reminiscent of observation made in mammals, where AQP3, AQP8, and AQP9 are also capable of facilitating the transmembrane diffusion of H_2O_2 (Hara-Chikuma et al., 2015; Miller, Dickinson, & Chang, 2010; Vieceli Dalla Sega et al., 2014; Watanabe, Moniaga, Nielsen, & Hara-Chikuma, 2016). For instance, AQP3 facilitates H_2O_2 diffusion and epidermal growth factor (EGF) signaling (Miller et al., 2010), an observation recently extended to NF-κB signaling in keratinocytes (Hara-Chikuma et al., 2015) or to the response of colonic epithelia to environmental stresses (Thiagarajah, Chang, Goettel, Verkman, & Lencer, 2017). Similarly, AQP8 facilitates cellular accumulation of H_2O_2 after VEGF stimulation, thereby enhancing PI3K activity and phosphorylation of MAPKs, two essential processes for leukemia cell proliferation (Vieceli Dalla Sega et al., 2014).

2.3.3 Ammonia and Urea

Nitrogen is one of essential constituents of plants. As an illustration, various nitrogen forms such as nitrate (NO_3^-), ammonium/ammonia (NH_4^+/NH_3) and urea ($CO(NH_2)_2$) are broadly used as fertilizers to enhance plant growth. Several studies using heterologous expression systems such as yeast or *Xenopus* oocytes showed that tonoplast AQPs can facilitate the transmembrane diffusion of NH_3 (Holm et al., 2005; Jahn et al., 2004; Loque, Ludewig, Yuan, & von Wiren, 2005) or urea (Gerbeau, Guclu, Ripoche, & Maurel, 1999; Liu, Ludewig, Gassert, Frommer, & von Wiren, 2003). While these results may suggest a role for these TIPs in vacuolar storage and remobilisation of nitrogen metabolites, overexpression of *At*TIP2;1 or *At*TIP2;3 in *Arabidopsis* failed to enhance whole plant NH_4^+/NH_3 accumulation (Loque et al., 2005). In addition, a few NIPs and PIPs were also reported to facilitate transmembrane diffusion of urea in plants (Gaspar et al., 2003; Gu, Chen, Zhou, & Yuan, 2012). As plant membranes are equipped with ammonium transporters (AMT/Rh) providing a major route for ammonium diffusion (Ludewig, Neuhauser, & Dynowski, 2007), the role of TIPs in vacuolar partitioning of nitrogen metabolites may apply to specific contexts. The symbiotic interaction of plants with soil microorganisms may be one of them. Here, the interplay between AMTs and AQPs may provide a basis to explain the fungus-based nitrogen nutrition of plants in symbiotic roots. Indeed, mycorrhizal symbiosis causes significant changes in AQP expression in host plants (Aroca, Porcel, & Ruiz-Lozano, 2007; Marjanovic et al., 2005; Uehlein et al., 2007) while some specific AQPs of legumes (e.g. *Gm*NOD26, the first AQP discovered in plants) are expressed in the peribacteroid membrane of nitrogen-fixing root nodules (Fortin et al., 1987). Considering that *Gm*NOD26 functions as NH_3-transporting AQP and binds cytosolic glutamine synthase, it may possibly enhance ammonia assimilation efficiency by creating a metabolite funnel towards this enzyme (Masalkar, Wallace, Hwang, & Roberts, 2010).

2.3.4 Metalloids

Some metalloids (boron, silicon, selenium) represent essential or at least beneficial elements for plant growth or human health while others (arsenic and antimony) are toxic. The capacity of several plant and mammalian AQPs to transport trivalent metalloids, which are structurally related to glycerol, another common substrate of MIPs, seems to be an ancestral feature of these proteins (Maciaszczyk-Dziubinska, Wawrzycka, & Wysocki, 2012). It is noteworthy that high concentrations of boron (B) can be toxic to plants, and therefore, transport of this element has to be tightly balanced. Thus, *At*NIP5;1 gene expression is strongly induced in response to B deficiency and allows efficient boric acid transport within the plant root, as revealed using *nip5;1* knock-out plants (Takano et al., 2006). Other plant AQPs were also identified as boric acid channels, such as *At*NIP6;1 and *At*NIP7;1, which mediate its allocation to shoots and floral anthers, respectively (Li, Choi, Wallace, Baudry, & Roberts, 2011 Tanaka, Wallace, Takano, Roberts, & Fujiwara, 2008). XIPs and PIPs were also recently proposed to act as boric acid channels in *Solanaceae* and rice, respectively (Bienert et al., 2011; Mosa et al., 2016). A role for TIPs in B homeostasis was also recently uncovered. Overexpression in *Arabidopsis* of *Bv*COLD1, a TIP1 homolog from sugar beet only present in the Chenopodioideae subfamily of the Amaranthaceae family, conferred enhanced tolerance to various abiotic stresses including B deficiency (Porcel, Bustamante, Ros, Serrano, & Mulet, 2018). Finally, genetic studies in barley showed that a quantitative trait locus for high B concentration toxicity was associated to expression of the *HvNIP2;1* gene (Schnurbusch et al., 2010).

Silicon (Si) is an abundant soil mineral which exerts beneficial effects on plant growth, especially in cereals where it can account for up to 10% of shoot dry weight and provides protection from numerous biotic and abiotic stresses (Epstein, 1994; Fauteux, Remus-Borel, Menzies, & Belanger, 2005). Rice *Os*NIP2;1 (Lsi1) was identified by an elegant forward genetic approach as the first Si transport protein of plants (Ma et al., 2006). A similar approach also showed *Os*NIP2;2 (Lsi6) to play a role in Si xylem unloading in rice panicles (Yamaji et al., 2008). Plant Si transporters belonging to the NIP family have been reported in several other plants, such as barley (Chiba, Mitani, Yamaji, & Ma, 2009) and maize (Mitani, Yamaji, & Ma, 2009), but *Arabidopsis* does not possess NIPs similar to the Si channels of cereals (Liu & Zhu, 2010).

Several AQPs were also shown to facilitate the absorption of critically important metalloids (arsenic (As) or antimonite (Sb)), which can be highly toxic for humans once in the food chain, or for plants when present at high levels. In contrast, selenium (Se) can be beneficial for human health. Heterologous expression in *Saccharomyces cerevisiae* was used to show that *A. thaliana At*NIP5;1, *At*NIP6;1 and *At*NIP7;1, *O. sativa Os*NIP2;1,

*Os*NIP2;2 and *Os*NIP3;2, *Lotus japonicus Lj*NIP5;1 and *Lj*NIP6;1 all can mediate As and Sb transport (Bienert et al., 2008). Direct genetic studies complemented these observations and pointed to a central role for NIP1;1 in As sensitivity of *Arabidopsis* (Kamiya et al., 2009) and rice (Ma et al., 2008). In the latter plant species, the Si transporter *Os*NIP2;1 (Lsi1) can serve for Se and As uptake in seeds. Interestingly, it was shown that toxic accumulation of As can be counteracted by providing an excess of Si to rice plants (Zhao, Mitani, Yamaji, Shen, & Ma, 2010). In addition to NIPs, rice *Os*PIP2;4, *Os*PIP2;6, and *Os*PIP2;7 were reported to mediate As transport in oocytes (Mosa et al., 2012). When these AQPs were overexpressed in *Arabidopsis*, the influx or efflux of As (depending on the outer metalloid concentration) was facilitated, and, more importantly, an overall increase in As resistance was observed. Very recently, disturbance of As radial transport in rice roots through overexpression of *Os*NIP1;1 and *Os*NIP3;3 was found to decrease overall plant As accumulation (Sun et al., 2018).

2.3.5 Gases

The first evidence for AQP-mediated diffusion of CO_2 across membranes was obtained by monitoring pH changes in *Xenopus* oocytes expressing human AQP1 together with a carbonic anhydrase (Nakhoul, Davis, Romero, & Boron, 1998). The subsequent demonstration that plant AQPs can also increase membrane permeability to CO_2 was made by a similar approach, by expressing *Nt*AQP1, a tobacco PIP1 homologue, in *Xenopus* oocytes (Uehlein, Lovisolo, Siefritz, & Kaldenhoff, 2003). In addition to *Nt*AQP1, *At*PIP1;2 was also shown to facilitate CO_2 transmembrane transport after heterologous expression in yeast cells (Heckwolf, Pater, Hanson, & Kaldenhoff, 2011). Genetic alteration of their function in tobacco or *Arabidopsis* plants, using either antisense suppression or overexpression, revealed a positive correlation between their expression and the rate of CO_2 assimilation (Heckwolf et al., 2011). To date, several AQPs from several plant species, all belonging to the PIP1 and PIP2 subclasses, have been shown to facilitate CO_2 membrane diffusion (Heinen et al., 2014; Mori et al., 2014; Navarro-Rodenas et al., 2013; Otto et al., 2010; Wang et al., 2016). While the commonly accepted idea is that these PIPs facilitate CO_2 transfer from the atmosphere to mesophyll cell chloroplasts (Groszmann, Osborn, & Evans, 2017), there have been concerns whether AQPs can significantly enhance this path (Boron, Endeward, Gros, Musa-Aziz, & Pohl, 2011). Indeed, mesophyll conductance to CO_2 (g_m) is possibly dominated by other components such as cell walls and carbonic anhydrases (Evans,

Kaldenhoff, Genty, & Terashima, 2009). In addition, plant membranes have a high content in sterols, which are known to reduce membrane permeability to CO_2 (Itel et al., 2012; Kai & Kaldenhoff, 2014; Tsiavaliaris et al., 2015). Thus, the abundance of aquaporins in membranes could impact the overall membrane CO_2 permeability by reducing the amount of lipids contributing to gas diffusion (Endeward, Al-Samir, Itel, & Gros, 2014). Finally, g_m is linked to stomatal conductance and the plant water status by as yet unknown physiological mechanisms. Thus, the effects of AQP genetic deregulation on g_m may be an indirect consequence of hydraulic defects in plant aquaporin mutants (Maurel et al., 2015).

In addition to CO_2, two AQPs (human AQP1 and tobacco *Nt*PIP1;3) were recently shown to facilitate oxygen (O_2) import when expressed in yeast protoplasts, and likely act as O_2 channels in plants (Zwiazek, Xu, Tan, Navarro-Rodenas, & Morte, 2017). These findings point to a putative role of AQPs in plant response to hypoxia, as typically induced by submergence stress. Finally, human AQP1 and AQP4 were also reported to facilitate transmembrane diffusion of nitric oxide (NO) (Herrera, Hong, & Garvin, 2006), and further studies are required to determine if plant AQPs may transport this plant signalling molecule as well.

2.3.6 Ions

Initial reconstitution experiments indicated that *Gm*NOD26, a soybean NIP, can act as a voltage-gated ion channel (Weaver, Shomer, Louis, & Roberts, 1994). Several animal and plant AQPs, including mammalian AQP1 (Anthony et al., 2000) and AQP6 (Yool & Campbell, 2012), *Drosophila Dm*BIB (Yanochko & Yool, 2002), and *Arabidopsis At*PIP2;1 (Byrt et al., 2017) and *At*PIP2;2 (Kourghi et al., 2017), were also shown to function as ion channels when expressed in *Xenopus* oocytes. Whether this ion transport activity has a physiological relevance remains an open question. This issue is more extensively discussed in chapter "Xylem ion loading and its implication for plant abiotic stress tolerance" of this volume by Ishikawa et al.

3. MOLECULAR FUNCTION AND REGULATION

3.1 Structural Conformation and Specificity Determinants

Molecular structures of AQPs at atomic resolution were first determined for animal (human AQP1) (Murata et al., 2000) and microbial isoforms (*Escherichia coli* GlpF) (Tajkhorshid et al., 2002). At present, 50 atomic structures, from 11 different AQP isoforms including two from plants (Kirscht et al., 2016; Törnroth-Horsefield et al., 2006), have been released. All these studies have revealed a conserved aquaporin fold whereby six transmembrane helices surround a narrow water-conducting channel in which water molecules align in a single file (Tornroth-Horsefield, Hedfalk, Fischer, Lindkvist-Petersson, & Neutze, 2010).

3.1.1 A Conserved Overall Structural Conformation

X-ray crystallography of spinach *So*PIP2;1 represented a crucial advance as it showed the structure of this plant AQP in its closed or open conformations (Nyblom et al., 2009; Törnroth-Horsefield et al., 2006), thereby giving crucial insights into the gating mechanisms of AQPs (detailed later). These structures indicated how each monomer can form an individual transmembrane pore (Fig. 1A). The water molecules passing the channel are forced

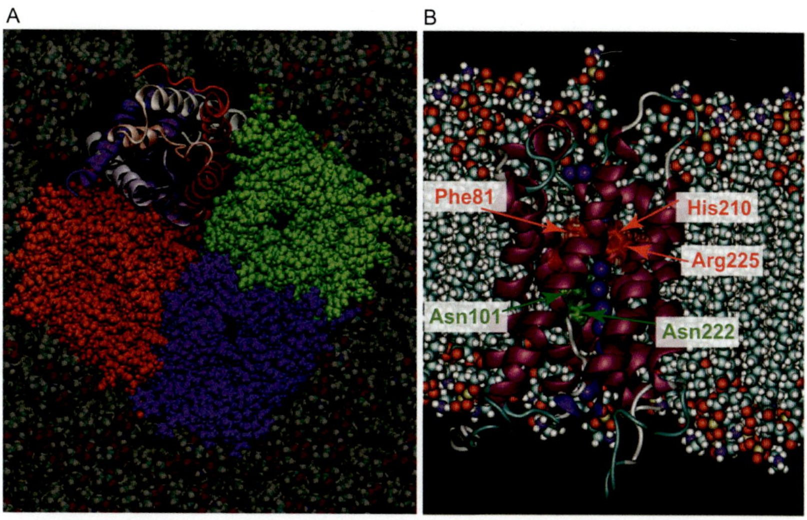

Fig. 1 Representative atomic structure of a plant aquaporin (A) and general molecular mechanisms of transport selectivity (B). (A) Structure of *So*PIP2;1 in its open conformation (PDB code 2B5F, Törnroth-Horsefield et al., 2006) showing a typical tetrameric arrangement. Each monomer is constituted by six tilted transmembrane helices (Nter: red, Cter: blue) as represented on the top left monomer. The pores of individual monomers are emphasised by the space filling representation of the three other monomers. (B) The figure shows how the two highly conserved Asn-Pro-Ala (NPA) motifs (represented by Asn101 and Asn222, in green) are positioned in close vicinity to form one of main pore constrictions. Another constriction, represented in red and called Ar/R, is formed, on the extra-cytoplasmic side of the membrane, by a spatial arrangement of aromatic (Ar) residues, such as Phe81 and His210, facing an Arg (R) residue, here Arg225. Protons transport is blocked by electrostatic repulsion in the Ar/R constriction and dipole orientation of the water molecule by the two Asn residues of the NPA motifs. This results in a transient isolation of the water molecule within the single file which fills the pore (in orange).

by the positive electrostatic potential formed by the two NPA motifs to flip at the centre of the channel, this mechanism preventing proton (H^+) permeation (de Groot, Frigato, Helms, & Grubmuller, 2003). Furthermore, an aromatic/arginine (Ar/R) motif consisting of a strictly conserved arginine and three additional amino acids, two of which being conserved in water-specific AQPs, forms a typical constriction (Fig. 1B). This constriction, which corresponds to the narrowest part of the channel, serves as selectivity filter. The Ar/R motif was also proposed to restrict H^+ permeation, based on mutational analyses in human AQP1 (Beitz, Wu, Holm, Schultz, & Zeuthen, 2006). The recent crystal structure of *Pichia pastoris* AQY1 at 0.88 Å (Eriksson et al., 2013) provided a more direct explanation, showing how the positive charge of Arg227 side-chain, within the Ar/R region, can create a repulsion barrier against H^+.

3.1.2 Selectivity Filters

The release of AQP atomic structures also allowed homology modelling to explore the transport selectivity of plant AQPs (Wallace & Roberts, 2004). Conserved residues were identified in the selectivity filter of each AQP subfamily. The NIP subfamily was further divided in three subclasses, corresponding to distinct predicted selectivities (Wallace & Roberts, 2005). For instance, NIPs of subclass III have a highly hydrophilic and widely open selectivity filter allowing the passage of large diameter solutes such as silicic acid (Mitani, Yamaji, & Ma, 2008). PIPs and TIPs also display specific selectivity filter configurations. Here, the Ile254 residue of barley *Hv*PIP2;3 is crucial for CO_2 transport, as its substitution by a corresponding Met254 residue found in the *Hv*PIP2;4 isoform impairs CO_2 but not water transport (Mori et al., 2014). In addition, a Val-to-Ile substitution at a critical position in the selectivity filter of tulip *Tg*TIP1;1 and *Tg*TIP1;2 was shown to determine the range of transported solutes (Azad, Yoshikawa, Ishikawa, Sawa, & Shibata, 2012). While the individual pore of each AQP monomer unambiguously represents a main path for water and solute transport, a putative transporting role of the central cavity formed by the four monomers has been proposed. For instance, diffusion of CO_2, O_2 and ions through the highly hydrophobic central cavity of AQP1 tetramers could be observed by molecular dynamic simulations (Wang & Tajkhorshid, 2007). This alternative path may play a physiological role either in membranes with low intrinsic gas permeability or when a major fraction of the membrane is occupied by AQPs. It was also proposed, based on a steric considerations, that the central region of spinach *So*PIP2;1 tetramers could mediates CO_2 transport, since the molecular structure showed that a detergent molecule can penetrate into this region (Frick, Jarva, Ekvall, et al., 2013). However, the presence of a lipid originating from the native membrane in the central pore of human AQP5 tetramers suggests that, in this case, the lipid would prevent the passage of CO_2 (or ions) through the centre of the tetramer, because of steric hindrance (Horsefield et al., 2008).

3.2 Various Levels of Regulation

3.2.1 Cellular Trafficking and Aquaporin Interactions

Altering AQP density within a membrane can provide an efficient means for controlling its water permeability. This mechanism which operates in animals and plants relies on a very precise control of AQP subcellular distribution. For recent reviews in plants, see Hachez, Besserer, Chevalier, and Chaumont (2013), Luu and Maurel (2013), Chaumont and Tyerman (2014), and Chevalier and Chaumont (2015b). Briefly, most of our understanding of AQP trafficking was obtained on *Arabidopsis* and maize PIPs. Diacidic motifs (Asp/Glu-X-Asp/Glu) located in the N-terminal tail of maize and *Arabidopsis* PIP2s likely interact with the COPII sorting machinery and direct their trafficking out of the ER into the secretory pathway (Sorieul, Santoni, Maurel, & Luu, 2011; Zelazny, Miecielica, Borst, Hemminga, & Chaumont, 2009). In addition, a Leu-XXX-Ala motif in the third transmembrane domain of PIP2s (Chevalier, Bienert, & Chaumont, 2014; Chevalier & Chaumont, 2015a), together with a phosphorylation site located in their C-terminal tail (Prak et al., 2008), assists their exit out of the ER. Other export or retention signals must exist since these two motifs, alone or in combination, do not allow *Zm*PIP1 translocation to the plasma membrane. For instance, a F220A mutation within the fifth transmembrane domain of *Zm*PIP1;2 was found to significantly increase its plasma membrane localisation in oocytes (Berny, Gilis, Rooman, & Chaumont, 2016).

Post-Golgi transport and insertion of PIPs into the plasma membrane were shown to rely on Q-SNAREs of the syntaxin family (Besserer et al., 2012; Tyrrell et al., 2007). In particular, two syntaxins (SYP121 and SYP61) physically interact with *At*PIP2;7 and *Zm*PIP2;5 and ensure their proper delivery to the plasma membrane (Besserer et al., 2012; Hachez, Laloux, et al., 2014).

While some PIP1s reach the plasma membrane alone, as exemplified with tobacco *Nt*AQP1 (Uehlein et al., 2003), maize *Zm*PIP1s are retained in the ER and require hetero-oligomerisation with *Zm*PIP2s to exit this compartment (Bienert, Heinen, Berny, & Chaumont, 2014; Fetter, Van Wilder, Moshelion, & Chaumont, 2004; Zelazny et al., 2007). A similar requirement of PIP2s for PIP1 targeting to the plasma membrane was described in many other species (Jozefkowicz et al., 2016; Mori et al., 2014; Yaneff et al., 2014). In contrast to PIPs, motifs that determine TIP routing to the vacuole remain largely unknown. While TIP1 targeting to vacuole is sensitive to Brefeldin A, trafficking of TIP2s and TIP3s does not pass through the same Golgi-dependent route (Rivera-Serrano, Rodriguez-Welsh, Hicks, & Rojas-Pierce, 2012).

Stimulus-induced subcellular trafficking of plant aquaporins was first characterised in ice plant *Mesembryanthemum crystallinum* where *Mc*TIP1;2 was redistributed from tonoplast to intracellular vesicles in response to an osmotic stress (Vera-Estrella, Barkla, Bohnert, & Pantoja, 2004). *Arabidopsis* PIPs and TIPs can also be rapidly removed from their respective membranes during root response to salt or oxidative stresses (Hachez et al., 2013; Luu & Maurel, 2013). Here, roots treated with 0.1 M NaCl for 4 h or with 2 mM H_2O_2 for 15 min showed an intracellular relocalisation of several highly expressed AQP isoforms, with a concomitant inhibition of Lp_r by up to 70% (Boursiac et al., 2008, 2005). Ser283 phosphorylation of *At*PIP2;1 was also shown to partly govern the redirection of this AQP to the prevacuolar compartment, this process being enhanced by both salt or oxidative stresses (Prak et al., 2008). Super-resolution microscopic approaches have also revealed that the low lateral membrane mobility of*At*PIP2;1 at the cell surface (Martiniere et al., 2012), as compared to that of 10-fold more diffusive TIPs (Hosy, Martiniere, Choquet, Maurel, & Luu, 2015; Li, Wang, et al., 2011; Luu, Martiniere, Sorieul, Runions, & Maurel, 2012), was increased by twofold during salt stress, while *At*PIP2;1 density at the plasma membrane was concomitantly decreased by 46% (Li, Wang, et al., 2011).

AQPs physically interact with much more proteins than other AQP isoforms (to form heterotetramers) or syntaxins. Thus, it has recently been reported that almost 400 proteins can directly or indirectly interact with *At*PIP2;1 and *At*PIP1;2 (Bellati et al., 2016), with about 80% of the interactants being shared between the two isoforms. For instance, *At*PIP2;1 can interact with receptor-like kinases or with the carbonic anhydrase enzyme, βCA4. The *At*PIP2;1–βCA4 complex contributes to a transport metabolon regulating stomatal closure in response to internal leaf CO_2 concentrations (Wang et al., 2016). In addition, *At*PIP1;4 was recently reported to interact with the rice bacterial blight pathogen protein harpin Hpa1 (Li et al., 2015). This interaction, which requires the N-terminal part of Hpa1 and apoplastic domains of *At*PIP1;4, increased leaf photosynthesis rates and plant growth and was proposed to be the result of increased CO_2 transport activity of *At*PIP1;4 (Li et al., 2015). *At*PIP2;7 can interact with the tryptophan-rich sensory protein/translocator (TSPO) multi-stress regulator, which balances *At*PIP2;7 accumulation in the plasma membrane and its vacuolar degradation through the autophagic pathway (Hachez, Veljanovski, et al., 2014). Interestingly, this interaction is stimulated by ABA. This example points to the physiological relevance of key AQP-protein interactions, of which many remain to be elucidated.

3.2.2 Gating

Rapid adjustments of membrane water permeability, as observed in plants under hormonal stimulation or environmental stress, can also be achieved through regulated opening and closing of the AQP water channel pore. First evidence for plant AQP gating was obtained by *in vitro* studies using purified plant membranes or proteoliposomes (Gerbeau et al., 2002; Guenther et al., 2003; Verdoucq, Grondin, & Maurel, 2008). These studies revealed that divalent cations, H^+, and phosphorylation can control plant AQP intrinsic activity. High resolution structures of spinach *So*PIP2;1 in open and closed conformations allowed to refine a model mechanism for plant AQP gating (Törnroth-Horsefield et al., 2006). In the closed conformational state, the cytosolic loop D (His193) of *So*PIP2;1 is linked to the N-terminal tail (Asp28 and Glu31) and the cytosolic loop B (Ser115 and Arg118) through a network of ionic interactions and hydrogen bonds. This results in the displacement of the hydrophobic Leu197 residue into the pore mouth and in its occlusion. Protonation of His193 tightens the closed conformation, partly through ionic interactions with the acidic residues (Asp28 and Glu31) of the N-terminal tail (Frick, Jarva, & Tornroth-Horsefield, 2013). Molecular dynamics simulations based on *So*PIP2;1 structures have provided a strong support to previous functional and biochemical data on PIP gating (Fig. 2). In particular, functional assays in oocytes showed that the pH-dependent inhibition of PIPs relies on protonation of a perfectly conserved His residue of loop D (Tournaire-Roux et al., 2003), such as His193 in *So*PIP2;1. In plants, cytosolic acidification is typically induced by anoxia, as observed under flooding conditions. Thus, the mechanism above provides molecular

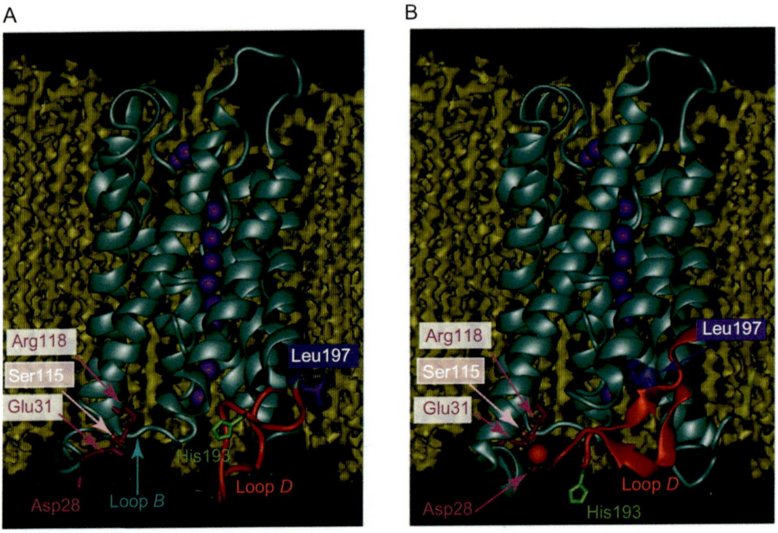

Fig. 2 Molecular mechanisms of PIP gating. The *So*PIP2;1 structure was solved in an open (A: PDB code 2B5F) and in a closed (B: PDB code:1Z98) conformation (Törnroth-Horsefield et al., 2006). The His193 residue (green) is perfectly conserved in all PIPs (loop D in red) and is distal from the cytoplasmic loop B (in cyan) in the open conformation (A). When protonated, His193 interacts with an acidic residue (Asp28, in purple) of the N-terminal tail. This induces a displacement of loop D (B) and of the hydrophobic side chain of Leu197 (in yellow) which occludes the pore. Binding of divalent cations (Cd^{2+} in the atomic structure, in red) would also involve Asp28 and an adjacent acidic residue (Glu31, in purple). A network of H-bond and hydrostatic interactions, involving Arg118 (in purple), maintains Loop D in the closed pore conformation. In addition, molecular dynamics simulation showed that phosphorylation of loop B at Ser115 (in pink) favours the open pore conformation by disrupting the interactions linking the N-terminal tail and the loop D.

details to explain how these stress conditions induce a rapid and reversible decrease in Lp_r (Maurel et al., 2015; Tournaire-Roux et al., 2003). In contrast to H^+-dependent gating, the mode of action of phosphorylation in AQP gating is not fully clear. For instance, phosphorylation of Ser115 and Ser188 in the loop D and Ser274 in the C-terminal of *So*PIP2;1 was proposed to prevent the channel closure (Törnroth-Horsefield et al., 2006). However, *So*PIP2;1 mutants carrying phosphomimetic mutations at Ser115 and Ser274 did not show enhanced water transport activity after reconstitution in proteoliposomes and displayed a closed structural conformation (Nyblom et al., 2009). Finally, the inhibition of PIP water transport by divalent cations (Gerbeau et al., 2002) can be explained by a direct binding of the cation (probably Ca^{2+} *in vivo*) to N-terminal Asp28 and Glu31 which, through an interaction network involving His193 in loop D and the Ser115 in loop B, stabilises the closed pore conformation. A second binding site was recently identified between loop D and the C-terminal tail of *So*PIP2;1 (Frick, Jarva, Ekvall, et al., 2013) which supports studies showing a biphasic dose–response curve for Ca^{2+} inhibition of water transport in *Beta vulgaris* plasma membrane vesicles (Alleva et al., 2006). Heteromerisation of PIPs can also affect their intrinsic permeability (Yaneff et al., 2014). Molecular dynamics simulations of a ZmPIP1;2/ZmPIP2;5 heterotetramer identified a key role of the interface between transmembrane domains 4 and 5 in modulating the water transport activity of ZmPIP1;2 monomers (Vajpai, Mukherjee, & Sankararamakrishnan, 2018).

Although the atomic structure of a TIP homolog, *At*TIP2;1, is available (Kirscht et al., 2016), the structural basis of TIP gating remains largely unknown. Yet, there are several examples of H^+-dependent gating of TIPs. For instance, the water permeability of tonoplast vesicles from *B. vulgaris* roots showed a 42% decrease when exposed to acidic pH (Sutka, Alleva, Parisi, & Amodeo, 2005). Moreover, an unusual His residue found in the cytoplasmic loop D of *Vitis vinifera Vn*TIP2;1 was shown to be required for its gating in response to cytoplasmic pH (Leitao, Prista, Moura, Loureiro-Dias, & Soveral, 2012). This property may provide a clue for coordinated regulation of TIPs and PIPs in grapevine tissues under anoxic stress. At variance with this mechanism, *At*TIP5;1 activity was shown to be dependent on extracellular pH, and this regulation relied on a His residue located in the second extra-cytoplasmic loop (loop C) (Soto et al., 2010).

Finally, the role of several cytosol-exposed phosphorylation sites in controlling water transport of TIPs, PIPs, and NIPs was dissected in structure–function analyses using *Xenopus* oocyte expression (Guenther et al., 2003; Johansson et al., 1998; Maurel, Kado, Guern, & Chrispeels, 1995). These studies will be described in the following section.

3.2.3 Cotranslational and Posttranslational Modification

While transcriptional regulation of AQPs is associated to developmental processes and long-term responses (>day) to the environment, posttranslational modifications mediate most of the short-term responses (minutes to hours). Mass spectrometry approaches have revealed that aquaporins carry numerous co- and post-translational modifications (di Pietro et al., 2013; Kline, Barrett-Wilt, & Sussman, 2010; Prak et al., 2008; Santoni et al., 2006; Santoni, Vinh, Pflieger, Sommerer, & Maurel, 2003; Van Wilder et al., 2008). These modifications, which have mainly been reported for PIPs, include N-terminal acetylation, phosphorylation, methylation, ubiquitination and/or deamidation. As described in previous paragraphs, AQP phosphorylation is implicated in both AQP gating and the regulation of their subcellular localisation. These regulations can occur in various physiological contexts. One typical example is regulation of *Arabidopsis* rosette hydraulic conductivity as a function of light regime, this property being strictly dependent on *At*PIP2;1 function (Prado et al., 2013). A prolonged darkness was found to enhance both *Arabidopsis* rosette hydraulic conductivity and phosphorylation of *At*PIP2;1 at two C-terminal sites, whereas the abundance of the protein was unchanged. When introduced in an *Atpip2;1* knock-out background, a phosphomimetic but not a phosphodeficient form *At*PIP2;1 was able to restore the dark-induced rosette hydraulic conductivity, demonstrating that C-terminal diphosphorylation of *At*PIP2;1 is necessary for this regulation.

Ubiquitination was shown to play a role in retention of *At*PIP2;1 in the ER, with subsequent degradation by the proteasome (Lee et al., 2009), while the N-glycosylation of *M. crystallinum* *Mc*TIP1;2 affects its targeting to endosomal compartments (Vera-Estrella et al., 2004). However, the significance of other post-translational modifications like methylation, acetylation, and deamidation still remains elusive. Methylation of *At*PIP2;1 on two adjacent residues, Lys3 and Asp6, was proposed to influence its subcellular trafficking (Santoni et al., 2006) as these residues overlap with a diacidic motif involved in ER export (see previous section).

3.2.4 Importance of the Lipidic Environment

The nature of phospholipids is known to affect the physical properties of a membrane (fluidity, charges, thickness, etc.), which in turn can induce conformational changes in intrinsic membrane proteins through hydrophobic matching (Jensen & Mouritsen, 2004). We also note that membrane lipid composition can vary locally, by asymmetric distribution between leaflets, or in the membrane plane through lipid microdomains. A role for membrane lipids in AQP gating is emerging in animals (Tong, Briggs, & McIntosh, 2012; Tong, Canty, Briggs, & McIntosh, 2013) but was scarcely explored in plants. For instance, *Nt*AQP1 activity was shown to be strongly dependent on the phospholipid/sterol molar percentage of the membrane (Kai & Kaldenhoff, 2014). Cold, drought and salt stress can affect the lipid composition of plasma membranes and simultaneously influence AQP activity (López-Pérez, Martínez-Ballesta, Maurel, & Carvajal, 2009; Minami et al., 2009). The lipid composition of membranes may also influence the ability of a given PIP to interact with other proteins and/or traffic through the secretory pathway. Thus, barley *Hv*PIP1;2 needed to be co-expressed with PIP2s to increase the water permeability of *Xenopus* oocytes (Horie et al., 2011) whereas such co-expression was not required in yeast (Besse et al., 2011). Discrepancies in substrate specificities can also be observed between different membrane systems. For instance, *Nt*PIP2;1 was shown to facilitate CO_2 diffusion in artificial membranes (Uehlein et al., 2012) but not in yeast cells (Otto et al., 2010).

It is noteworthy that proteomic analyses in several plant species have revealed the presence of PIPs in detergent-resistant membranes (Minami et al., 2009; Mongrand et al., 2004), or their interaction with resident proteins of these microdomains, such as Flotillin2 (Junkova et al., 2018). Membrane microdomains, which are typically enriched in sphingolipids and sterols, can provide a particular membrane environment for formation of specific protein complexes (reviewed in Minami et al., 2009; Sezgin, Levental, Mayor, & Eggeling, 2017). In line with these ideas, the distribution and trafficking of *At*PIP2;1 were dependent on the sterol and sphingolipid content of the plasma membrane (Li, Wang, et al., 2011). This AQP exhibited a highly confined diffusion mode in contrast to *At*TIP1;1 which displayed a much higher velocity in the tonoplast, a membrane with a higher membrane lipid fluidity (Hosy et al., 2015).

4. CONCLUSION AND PERSPECTIVES

Tremendous progresses have been made over the past 20 years in understanding the role and functioning of AQPs in plants. These progresses have been made possible thanks to elaborate structure/function analyses, relying on a large variety of analytical tools (for review: Deshmukh, Sonah, & Belanger, 2016), in combination with in depth genetic and physiological studies (Maurel et al., 2015). In line with their high isoform multiplicity, and diversity of subcellular localisations and transport selectivities, AQPs can fulfill an amazing range of functions related to water relations but also mineral nutrition and growth control. By providing an enhanced knowledge of AQP regulation mechanisms, the coming years will undoubtedly shed more light onto these functions, in particular during stress responses. One promising direction is to explore the function of the multiple protein kinases or interacting partners acting on AQP trafficking and gating. Resolution at single-cell level of the mechanisms described so far in vitro or in whole tissue will also be crucial. Finally, the functional interactions between AQPs and their lipid environment remain largely unexplored.

REFERENCES

Abascal, F., Irisarri, I., & Zardoya, R. (2014). Diversity and evolution of membrane intrinsic proteins. *Biochimica et Biophysica Acta, 1840*, 1468–1481.

Alleva, K., Niemietz, C. M., Sutka, M., Maurel, C., Parisi, M., Tyerman, S. D., et al. (2006). Plasma membrane of *Beta vulgaris* storage root shows high water channel activity regulated by cytoplasmic pH and a dual range of calcium concentrations. *Journal of Experimental Botany, 57*, 609–621.

Anthony, T. L., Brooks, H. L., Boassa, D., Leonov, S., Yanochko, G. M., Regan, J. W., et al. (2000). Cloned human aquaporin-1 is a cyclic GMP-gated ion channel. *Molecular Pharmacology, 57*, 576–588.

Aroca, R., Porcel, R., & Ruiz-Lozano, J. M. (2007). How does arbuscular mycorrhizal symbiosis regulate root hydraulic properties and plasma membrane aquaporins in *Phaseolus vulgaris* under drought, cold or salinity stresses? *The New Phytologist, 173*, 808–816.

Azad, A. K., Yoshikawa, N., Ishikawa, T., Sawa, Y., & Shibata, H. (2012). Substitution of a single amino acid residue in the aromatic/arginine selectivity filter alters the transport profiles of tonoplast aquaporin homologs. *Biochimica et Biophysica Acta, 1818*, 1–11.

Beebo, A., Mathai, J. C., Schoefs, B., & Spetea, C. (2013). Assessment of the requirement for aquaporins in the thylakoid membrane of plant chloroplasts to sustain photosynthetic water oxidation. *FEBS Letters, 587*, 2083–2089.

Beitz, E., Wu, B., Holm, L., Schultz, J., & Zeuthen, T. (2006). Point mutations in the aromatic/arginine region in aquaporin 1 allow passage of urea, glycerol, ammonia, and protons. *Proceedings of the National Academy of Sciences of the United States of America, 103*, 269–274.

Bellati, J., Champeyroux, C., Hem, S., Rofidal, V., Krouk, G., Maurel, C., et al. (2016). Novel aquaporin regulatory mechanisms revealed by interactomics. *Molecular & Cellular Proteomics, 15*, 3473–3487.

Berny, M. C., Gilis, D., Rooman, M., & Chaumont, F. (2016). Single mutations in the transmembrane domains of maize plasma membrane aquaporins affect the activity of monomers within a heterotetramer. *Molecular Plant, 9*, 986–1003.

Besse, M., Knipfer, T., Miller, A. J., Verdeil, J. L., Jahn, T. P., & Fricke, W. (2011). Developmental pattern of aquaporin expression in barley (*Hordeum vulgare* L.) leaves. *Journal of Experimental Botany, 62*, 4127–4142.

Besserer, A., Burnotte, E., Bienert, G. P., Chevalier, A. S., Errachid, A., Grefen, C., et al. (2012). Selective regulation of maize plasma membrane aquaporin trafficking and activity by the SNARE SYP121. *The Plant Cell, 24*, 3463–3481.

Bienert, G. P., Bienert, M. D., Jahn, T. P., Boutry, M., & Chaumont, F. (2011). Solanaceae XIPs are plasma membrane aquaporins that facilitate the transport of many uncharged substrates. *The Plant Journal, 66*, 306–317.

Bienert, G. P., Heinen, R. B., Berny, M. C., & Chaumont, F. (2014). Maize plasma membrane aquaporin ZmPIP2;5, but not ZmPIP1;2, facilitates transmembrane diffusion of hydrogen peroxide. *Biochimica et Biophysica Acta, 1838*, 216–222.

Bienert, G. P., Moller, A. L., Kristiansen, K. A., Schulz, A., Moller, I. M., Schjoerring, J. K., et al. (2007). Specific aquaporins facilitate the diffusion of hydrogen peroxide across membranes. *The Journal of Biological Chemistry, 282*, 1183–1192.

Bienert, G. P., Thorsen, M., Schüssler, M. D., Nilsson, H. R., Wagner, A., Tamás, M. J., et al. (2008). A subgroup of plant aquaporins facilitate the bi-directional diffusion of $As(OH)_3$ and $Sb(OH)_3$ across membranes. *BMC Biology, 6*, 26.

Boron, W. F., Endeward, V., Gros, G., Musa-Aziz, R., & Pohl, P. (2011). Intrinsic CO_2 permeability of cell membranes and potential biological relevance of CO_2 channels. *Chemphyschem, 12*, 1017–1019.

Bots, M., Vergeldt, F., Wolters-Arts, M., Weterings, K., van As, H., & Mariani, C. (2005). Aquaporins of the PIP2 class are required for efficient anther dehiscence in tobacco. *Plant Physiology*, *137*, 1049–1056.

Boursiac, Y., Boudet, J., Postaire, O., Luu, D. T., Tournaire-Roux, C., & Maurel, C. (2008). Stimulus-induced downregulation of root water transport involves reactive oxygen species-activated cell signalling and plasma membrane intrinsic protein internalization. *The Plant Journal*, *56*, 207–218.

Boursiac, Y., Chen, S., Luu, D. T., Sorieul, M., van den Dries, N., & Maurel, C. (2005). Early effects of salinity on water transport in Arabidopsis roots. Molecular and cellular features of aquaporin expression. *Plant Physiology*, *139*, 790–805.

Byrt, C. S., Zhao, M., Kourghi, M., Bose, J., Henderson, S. W., Qiu, J., et al. (2017). Non-selective cation channel activity of aquaporin AtPIP2;1 regulated by Ca^{2+} and pH. *Plant, Cell & Environment*, *40*, 802–815.

Chaumont, F., Barrieu, F., Wojcik, E., Chrispeels, M. J., & Jung, R. (2001). Aquaporins constitute a large and highly divergent protein family in maize. *Plant Physiology*, *125*, 1206–1215.

Chaumont, F., & Tyerman, S. D. (2014). Aquaporins: Highly regulated channels controlling plant water relations. *Plant Physiology*, *164*, 1600–1618.

Chevalier, A. S., Bienert, G. P., & Chaumont, F. (2014). A new LxxxA motif in the transmembrane Helix3 of maize aquaporins belonging to the plasma membrane intrinsic protein PIP2 group is required for their trafficking to the plasma membrane. *Plant Physiology*, *166*, 125–138.

Chevalier, A. S., & Chaumont, F. (2015a). The LxxxA motif in the third transmembrane helix of the maize aquaporin ZmPIP2;5 acts as an ER export signal. *Plant Signaling & Behavior*, *10*, e990845.

Chevalier, A. S., & Chaumont, F. (2015b). Trafficking of plant plasma membrane aquaporins: Multiple regulation levels and complex sorting signals. *Plant & Cell Physiology*, *56*, 819–829.

Chiba, Y., Mitani, N., Yamaji, N., & Ma, J. F. (2009). HvLsi1 is a silicon influx transporter in barley. *The Plant Journal*, *57*, 810–818.

Danielson, J. A., & Johanson, U. (2008). Unexpected complexity of the aquaporin gene family in the moss *Physcomitrella patens*. *BMC Plant Biology*, *8*, 45.

de Groot, B. L., Frigato, T., Helms, V., & Grubmuller, H. (2003). The mechanism of proton exclusion in the aquaporin-1 water channel. *Journal of Molecular Biology*, *333*, 279–293.

Deshmukh, R. K., Sonah, H., & Belanger, R. R. (2016). Plant aquaporins: Genome-wide identification, transcriptomics, proteomics, and advanced analytical tools. *Frontiers in Plant Science*, 7, 1896.

Di Giorgio, J. A., Bienert, G. P., Ayub, N. D., Yaneff, A., Barberini, M. L., Mecchia, M. A., et al. (2016). Pollen-specific aquaporins NIP4;1 and NIP4;2 are required for pollen development and pollination in Arabidopsis thaliana. *The Plant Cell*, *28*, 1053–1077.

di Pietro, M., Vialaret, J., Li, G. W., Hem, S., Prado, K., Rossignol, M., et al. (2013). Coordinated post-translational responses of aquaporins to abiotic and nutritional stimuli in Arabidopsis roots. *Molecular & Cellular Proteomics*, *12*, 3886–3897.

Endeward, V., Al-Samir, S., Itel, F., & Gros, G. (2014). How does carbon dioxide permeate cell membranes? A discussion of concepts, results and methods. *Frontiers in Physiology*, *4*, 382.

Epstein, E. (1994). The anomaly of silicon in plant biology. *Proceedings of the National Academy of Sciences of the United States of America*, *91*, 11–17.

Eriksson, U. K., Fischer, G., Friemann, R., Enkavi, G., Tajkhorshid, E., & Neutze, R. (2013). Subangstrom resolution X-ray structure details aquaporin-water interactions. *Science*, *340*, 1346–1349.

Evans, J. R., Kaldenhoff, R., Genty, B., & Terashima, I. (2009). Resistances along the CO_2 diffusion pathway inside leaves. *Journal of Experimental Botany, 60*, 2235–2248.

Fauteux, F., Remus-Borel, W., Menzies, J. G., & Belanger, R. R. (2005). Silicon and plant disease resistance against pathogenic fungi. *FEMS Microbiology Letters, 249*, 1–6.

Ferro, M., Brugiere, S., Salvi, D., Seigneurin-Berny, D., Court, M., Moyet, L., et al. (2010). AT_CHLORO, a comprehensive chloroplast proteome database with subplastidial localization and curated information on envelope proteins. *Molecular & Cellular Proteomics, 9*, 1063–1084.

Fetter, K., Van Wilder, V., Moshelion, M., & Chaumont, F. (2004). Interactions between plasma membrane aquaporins modulate their water channel activity. *The Plant Cell, 16*, 215–228.

Finn, R. N., Chauvigne, F., Hlidberg, J. B., Cutler, C. P., & Cerda, J. (2014). The lineage-specific evolution of aquaporin gene clusters facilitated tetrapod terrestrial adaptation. *PLoS One, 9*, e113686.

Fortin, M., Morrison, N., & Verma, D. (1987). Nodulin-26, a peribacteroid membrane nodulin is expressed independently of the development of the peribacteroid compartment. *Nucleic Acids Research, 15*, 813–824.

Fox, A. R., Maistriaux, L. C., & Chaumont, F. (2017). Toward understanding of the high number of plant aquaporin isoforms and multiple regulation mechanisms. *Plant Science, 264*, 179–187.

Frick, A., Jarva, M., Ekvall, M., Uzdavinys, P., Nyblom, M., & Tornroth-Horsefield, S. (2013). Mercury increases water permeability of a plant aquaporin through a non-cysteine-related mechanism. *The Biochemical Journal, 454*, 491–499.

Frick, A., Jarva, M., & Tornroth-Horsefield, S. (2013). Structural basis for pH gating of plant aquaporins. *FEBS Letters, 587*, 989–993.

Gao, Y. P., Young, L., Bonham-Smith, P., & Gusta, L. V. (1999). Characterization and expression of plasma and tonoplast membrane aquaporins in primed seed of *Brassica napus* during germination under stress conditions. *Plant Molecular Biology, 40*, 635–644.

Gaspar, M., Bousser, A., Sissoëff, I., Roche, O., Hoarau, J., & Mahe, A. (2003). Cloning and characterization of *ZmPIP1-5b*, an aquaporin transporting water and urea. *Plant Science, 165*, 21–31.

Gattolin, S., Sorieul, M., & Frigerio, L. (2011). Mapping of tonoplast intrinsic proteins in maturing and germinating Arabidopsis seeds reveals dual localization of embryonic TIPs to the tonoplast and plasma membrane. *Molecular Plant, 4*, 180–189.

Gerbeau, P., Amodeo, G., Henzler, T., Santoni, V., Ripoche, P., & Maurel, C. (2002). The water permeability of *Arabidopsis* plasma membrane is regulated by divalent cations and pH. *The Plant Journal, 30*, 71–81.

Gerbeau, P., Guclu, J., Ripoche, P., & Maurel, C. (1999). Aquaporin Nt-TIPa can account for the high permeability of tobacco cell vacuolar membrane to small neutral solutes. *The Plant Journal, 18*, 577–587.

Grondin, A., Rodrigues, O., Verdoucq, L., Merlot, S., Leonhardt, N., & Maurel, C. (2015). Aquaporins contribute to ABA-triggered stomatal closure through OST1-mediated phosphorylation. *The Plant Cell, 27*, 1945–1954.

Groszmann, M., Osborn, H. L., & Evans, J. R. (2017). Carbon dioxide and water transport through plant aquaporins. *Plant, Cell & Environment, 40*, 938–961.

Gu, R., Chen, X., Zhou, Y., & Yuan, L. (2012). Isolation and characterization of three maize aquaporin genes, ZmNIP2;1, ZmNIP2;4 and ZmTIP4;4 involved in urea transport. *BMB Reports, 45*, 96–101.

Guenther, J. F., Chanmanivone, N., Galetovic, M. P., Wallace, I. S., Cobb, J. A., & Roberts, D. M. (2003). Phosphorylation of soybean nodulin 26 on serine 262 enhances water permeability and is regulated developmentally and by osmotic signals. *The Plant Cell, 15*, 981–991.

Gupta, A. B., & Sankararamakrishnan, R. (2009). Genome-wide analysis of major intrinsic proteins in the tree plant *Populus trichocarpa*: Characterization of XIP subfamily of aquaporins from evolutionary perspective. *BMC Plant Biology, 9*, 134.

Gustavsson, S., Lebrun, A., Norden, K., Chaumont, F., & Johanson, U. (2005). A novel plant major intrinsic protein in *Physcomitrella patens* most similar to bacterial glycerol channels. *Plant Physiology, 139*, 287–295.

Hachez, C., Besserer, A., Chevalier, A. S., & Chaumont, F. (2013). Insights into plant plasma membrane aquaporin trafficking. *Trends in Plant Science, 18*, 344–352.

Hachez, C., Heinen, R. B., Draye, X., & Chaumont, F. (2008). The expression pattern of plasma membrane aquaporins in maize leaf highlights their role in hydraulic regulation. *Plant Molecular Biology, 68*, 337–353.

Hachez, C., Laloux, T., Reinhardt, H., Cavez, D., Degand, H., Grefen, C., et al. (2014). Arabidopsis SNAREs SYP61 and SYP121 coordinate the trafficking of plasma membrane aquaporin PIP2;7 to modulate the cell membrane water permeability. *The Plant Cell, 26*, 3132–3147.

Hachez, C., Veljanovski, V., Reinhardt, H., Guillaumot, D., Vanhee, C., Chaumont, F., et al. (2014). The Arabidopsis abiotic stress-induced TSPO-related protein reduces cell-surface expression of the aquaporin PIP2;7 through protein-protein interactions and autophagic degradation. *The Plant Cell, 26*, 4974–4990.

Hara-Chikuma, M., Satooka, H., Watanabe, S., Honda, T., Miyachi, Y., Watanabe, T., et al. (2015). Aquaporin-3-mediated hydrogen peroxide transport is required for NF-kappaB signalling in keratinocytes and development of psoriasis. *Nature Communications, 6*, 7454.

Heckwolf, M., Pater, D., Hanson, D. T., & Kaldenhoff, R. (2011). The *Arabidopsis thaliana* aquaporin *At*PIP1;2 is a physiologically relevant CO_2 transport facilitator. *The Plant Journal, 67*, 795–804.

Heinen, R. B., Bienert, G. P., Cohen, D., Chevalier, A. S., Uehlein, N., Hachez, C., et al. (2014). Expression and characterization of plasma membrane aquaporins in stomatal complexes of Zea mays. *Plant Molecular Biology, 86*, 335–350.

Heinen, R. B., Ye, Q., & Chaumont, F. (2009). Role of aquaporins in leaf physiology. *Journal of Experimental Botany, 60*, 2971–2985.

Herrera, M., Hong, N. J., & Garvin, J. L. (2006). Aquaporin-1 transports NO across cell membranes. *Hypertension, 48*, 157–164.

Holm, L. M., Jahn, T. P., Moller, A. L., Schjoerring, J. K., Ferri, D., Klaerke, D. A., et al. (2005). NH_3 and NH_4^+ permeability in aquaporin-expressing *Xenopus* oocytes. *Pflugers Archiv, 450*, 415–428.

Horie, T., Kaneko, T., Sugimoto, G., Sasano, S., Panda, S. K., Shibasaka, M., et al. (2011). Mechanisms of water transport mediated by PIP aquaporins and their regulation via phosphorylation events under salinity stress in barley roots. *Plant & Cell Physiology, 52*, 663–675.

Horsefield, R., Norden, K., Fellert, M., Backmark, A., Tornroth-Horsefield, S., Terwisscha van Scheltinga, A. C., et al. (2008). High-resolution x-ray structure of human aquaporin 5. *Proceedings of the National Academy of Sciences of the United States of America, 105*, 13327–13332.

Hosy, E., Martiniere, A., Choquet, D., Maurel, C., & Luu, D. T. (2015). Super-resolved and dynamic imaging of membrane proteins in plant cells reveal contrasting kinetic profiles and multiple confinement mechanisms. *Molecular Plant, 8*, 339–342.

Ishikawa, F., Suga, S., Uemura, T., Sato, M. H., & Maeshima, M. (2005). Novel type aquaporin SIPs are mainly localized to the ER membrane and show cell-specific expression in *Arabidopsis thaliana*. *FEBS Letters, 579*, 5814–5820.

Itel, F., Al-Samir, S., Oberg, F., Chami, M., Kumar, M., Supuran, C. T., et al. (2012). CO_2 permeability of cell membranes is regulated by membrane cholesterol and protein gas channels. *FASEB Journal, 26*, 5182–5191.

Jahn, T. P., Moller, A. L., Zeuthen, T., Holm, L. M., Klaerke, D. A., Mohsin, B., et al. (2004). Aquaporin homologues in plants and mammals transport ammonia. *FEBS Letters*, *574*, 31–36.

Javot, H., Lauvergeat, V., Santoni, V., Martin-Laurent, F., Guclu, J., Vinh, J., et al. (2003). Role of a single aquaporin isoform in root water uptake. *The Plant Cell*, *15*, 509–522.

Jensen, M. O., & Mouritsen, O. G. (2004). Lipids do influence protein function-the hydrophobic matching hypothesis revisited. *Biochimica et Biophysica Acta*, *1666*, 205–226.

Johanson, U., Karlsson, M., Gustavsson, S., Sjovall, S., Fraysse, L., Weig, A. R., et al. (2001). The complete set of genes encoding major intrinsic proteins in *Arabidopsis* provides a framework for a new nomenclature for major intrinsic proteins in plants. *Plant Physiology*, *126*, 1358–1369.

Johansson, I., Karlsson, M., Shukla, V. K., Chrispeels, M. J., Larsson, C., & Kjellbom, P. (1998). Water transport activity of the plasma membrane aquaporin PM28A is regulated by phosphorylation. *The Plant Cell*, *10*, 451–460.

Jozefkowicz, C., Sigaut, L., Scochera, F., Soto, G., Ayub, N., Pietrasanta, L. I., et al. (2016). PIP water transport and its pH dependence are regulated by tetramer stoichiometry. *Biophysical Journal*, *110*, 1312–1321.

Junkova, P., Danek, M., Kocourkova, D., Brouzdova, J., Kroumanova, K., Zelazny, E., et al. (2018). Mapping of plasma membrane proteins interacting with Arabidopsis thaliana flotillin 2. *Frontiers in Plant Science*, *9*, 991.

Kai, L., & Kaldenhoff, R. (2014). A refined model of water and CO(2) membrane diffusion: Effects and contribution of sterols and proteins. *Scientific Reports*, *4*, 6665.

Kamiya, T., Tanaka, M., Mitani, N., Ma, J. F., Maeshima, M., & Fujiwara, T. (2009). NIP1;1, an aquaporin homolog, determines the arsenite sensitivity of *Arabidopsis thaliana*. *The Journal of Biological Chemistry*, *284*, 2114–2120.

Khabudaev, K. V., Petrova, D. P., Grachev, M. A., & Likhoshway, Y. V. (2014). A new subfamily LIP of the major intrinsic proteins. *BMC Genomics*, *15*, 173.

Kierszniowska, S., Seiwert, B., & Schulze, W. X. (2009). Definition of Arabidopsis sterol-rich membrane microdomains by differential treatment with methyl-beta-cyclodextrin and quantitative proteomics. *Molecular & Cellular Proteomics*, *8*, 612–623.

Kirscht, A., Kaptan, S. S., Bienert, G. P., Chaumont, F., Nissen, P., de Groot, B. L., et al. (2016). Crystal structure of an ammonia-permeable aquaporin. *PLoS Biology*, *14*, e1002411.

Kline, K. G., Barrett-Wilt, G. A., & Sussman, M. R. (2010). *In planta* changes in protein phosphorylation induced by the plant hormone abscisic acid. *Proceedings of the National Academy of Sciences of the United States of America*, *107*, 15986–15991.

Kourghi, M., Nourmohammadi, S., Pei, J. V., Qiu, J., McGaughey, S., Tyerman, S. D., et al. (2017). Divalent cations regulate the ion conductance properties of diverse classes of aquaporins. *International Journal of Molecular Sciences*, *18*, e2323.

Laloux, T., Junqueira, B., Maistriaux, L. C., Ahmed, J., Jurkiewicz, A., & Chaumont, F. (2018). Plant and mammal aquaporins: Same but different. *International Journal of Molecular Sciences*, *19*, e521.

Lee, H. K., Cho, S. K., Son, O., Xu, Z., Hwang, I., & Kim, W. T. (2009). Drought stress-induced Rma1H1, a RING membrane-anchor E3 ubiquitin ligase homolog, regulates aquaporin levels via ubiquitination in transgenic *Arabidopsis* plants. *Plant Cell*, *21*, 622–641.

Leitao, L., Prista, C., Moura, T. F., Loureiro-Dias, M. C., & Soveral, G. (2012). Grapevine aquaporins: Gating of a tonoplast intrinsic protein (TIP2;1) by cytosolic pH. *PLoS One*, *7*, e33219.

Li, T., Choi, W. G., Wallace, I. S., Baudry, J., & Roberts, D. M. (2011). Arabidopsis thaliana NIP7;1: An anther-specific boric acid transporter of the aquaporin superfamily regulated by an unusual tyrosine in helix 2 of the transport pore. *Biochemistry*, *50*, 6633–6641.

Li, L., Wang, H., Gago, J., Cui, H., Qian, Z., Kodama, N., et al. (2015). Harpin Hpa1 interacts with aquaporin PIP1;4 to promote the substrate transport and photosynthesis in Arabidopsis. *Scientific Reports, 5*, 17207.

Li, X., Wang, X., Yang, Y., Li, R., He, Q., Fang, X., et al. (2011). Single-molecule analysis of PIP2;1 dynamics and partitioning reveals multiple modes of Arabidopsis plasma membrane aquaporin regulation. *Plant Cell, 23*, 3780–3797.

Liu, L. H., Ludewig, U., Gassert, B., Frommer, W. B., & von Wiren, N. (2003). Urea transport by nitrogen-regulated tonoplast intrinsic proteins in *Arabidopsis*. *Plant Physiology, 133*, 1220–1228.

Liu, Q., & Zhu, Z. (2010). Functional divergence of the NIP III subgroup proteins involved altered selective constraints and positive selection. *BMC Plant Biology, 10*, 256.

Lopez, D., Bronner, G., Brunel, N., Auguin, D., Bourgerie, S., Brignolas, F., et al. (2012). Insights into Populus XIP aquaporins: evolutionary expansion, protein functionality, and environmental regulation. *Journal of Experimental Botany, 63*, 2217–2230.

López-Pérez, L., Martínez-Ballesta, M. C., Maurel, C., & Carvajal, M. (2009). Changes in plasma membrane lipids, aquaporins and proton pump of broccoli roots, as an adaptation mechanism to salinity. *Phytochemistry, 70*, 492–500.

Loque, D., Ludewig, U., Yuan, L., & von Wiren, N. (2005). Tonoplast intrinsic proteins AtTIP2;1 and AtTIP2;3 facilitate NH_3 transport into the vacuole. *Plant Physiology, 137*, 671–680.

Ludewig, U., Neuhauser, B., & Dynowski, M. (2007). Molecular mechanisms of ammonium transport and accumulation in plants. *FEBS Letters, 581*, 2301–2308.

Luu, D. T., Martiniere, A., Sorieul, M., Runions, J., & Maurel, C. (2012). Fluorescence recovery after photobleaching reveals high cycling dynamics of plasma membrane aquaporins in Arabidopsis roots under salt stress. *The Plant Journal, 69*, 894–905.

Luu, D. T., & Maurel, C. (2013). Aquaporin trafficking in plant cells: An emerging membrane-protein model. *Traffic, 14*, 629–635.

Ma, J. F., Tamai, K., Yamaji, N., Mitani, N., Konishi, S., Katsuhara, M., et al. (2006). A silicon transporter in rice. *Nature, 440*, 688–691.

Ma, J. F., Yamaji, N., Mitani, N., Xu, X. Y., Su, Y. H., McGrath, S. P., et al. (2008). Transporters of arsenite in rice and their role in arsenic accumulation in rice grain. *Proceedings of the National Academy of Sciences of the United States of America, 105*, 9931–9935.

Maciaszczyk-Dziubinska, E., Wawrzycka, D., & Wysocki, R. (2012). Arsenic and antimony transporters in eukaryotes. *International Journal of Molecular Sciences, 13*, 3527–3548.

Marjanovic, Z., Uehlein, N., Kaldenhoff, R., Zwiazek, J. J., Weiss, M., Hampp, R., et al. (2005). Aquaporins in poplar: What a difference a symbiont makes! *Planta, 222*, 258–268.

Martiniere, A., Lavagi, I., Nageswaran, G., Rolfe, D. J., Maneta-Peyret, L., Luu, D. T., et al. (2012). Cell wall constrains lateral diffusion of plant plasma-membrane proteins. *Proceedings of the National Academy of Sciences of the United States of America, 109*, 12805–12810.

Masalkar, P., Wallace, I. S., Hwang, J. H., & Roberts, D. M. (2010). Interaction of cytosolic glutamine synthetase of soybean root nodules with the C-terminal domain of the symbiosome membrane nodulin 26 aquaglyceroporin. *The Journal of Biological Chemistry, 285*, 23880–23888.

Maurel, C., Boursiac, Y., Luu, D. T., Santoni, V., Shahzad, Z., & Verdoucq, L. (2015). Aquaporins in plants. *Physiological Reviews, 95*, 1321–1358.

Maurel, C., Kado, R. T., Guern, J., & Chrispeels, M. J. (1995). Phosphorylation regulates the water channel activity of the seed-specific aquaporin a-TIP. *The EMBO Journal, 14*, 3028–3035.

Miller, E. W., Dickinson, B. C., & Chang, C. J. (2010). Aquaporin-3 mediates hydrogen peroxide uptake to regulate downstream intracellular signaling. *Proceedings of the National Academy of Sciences of the United States of America, 107*, 15681–15686.

Minami, A., Fujiwara, M., Furuto, A., Fukao, Y., Yamashita, T., Kamo, M., et al. (2009). Alterations in detergent-resistant plasma membrane microdomains in *Arabidopsis thaliana* during cold acclimation. *Plant & Cell Physiology*, *50*, 341–359.

Mitani, N., Yamaji, N., & Ma, J. F. (2008). Characterization of substrate specificity of a rice silicon transporter, Lsi1. *Pflugers Archiv*, *456*, 679–686.

Mitani, N., Yamaji, N., & Ma, J. F. (2009). Identification of maize silicon influx transporters. *Plant & Cell Physiology*, *50*, 5–12.

Mizutani, M., Watanabe, S., Nakagawa, T., & Maeshima, M. (2006). Aquaporin NIP2;1 is mainly localized to the ER membrane and shows root-specific accumulation in *Arabidopsis thaliana*. *Plant & Cell Physiology*, *47*, 1420–1426.

Mongrand, S., Morel, J., Laroche, J., Claverol, S., Carde, J.-P., Hartmann, M. A., et al. (2004). Lipid rafts in higher plant cells: Purification and characterization of Triton X-100-insoluble microdomains from tobacco plasma membrane. *The Journal of Biological Chemistry*, *279*, 36277–36286.

Morel, J., Claverol, S., Mongrand, S., Furt, F., Fromentin, J., Bessoule, J. J., et al. (2006). Proteomics of plant detergent-resistant membranes. *Molecular & Cellular Proteomics*, *5*, 1396–1411.

Mori, I. C., Rhee, J., Shibasaka, M., Sasano, S., Kaneko, T., Horie, T., et al. (2014). CO_2 transport by PIP2 aquaporins of barley. *Plant & Cell Physiology*, *55*, 251–257.

Mosa, K. A., Kumar, K., Chhikara, S., McDermott, J., Liu, Z., Musante, C., et al. (2012). Members of rice plasma membrane intrinsic proteins subfamily are involved in arsenite permeability and tolerance in plants. *Transgenic Research*, *21*, 1265–1277.

Mosa, K. A., Kumar, K., Chhikara, S., Musante, C., White, J. C., & Dhankher, O. P. (2016). Enhanced boron tolerance in plants mediated by bidirectional transport through plasma membrane intrinsic proteins. *Scientific Reports*, *6*, 21640.

Murata, K., Mitsuoka, K., Hirai, T., Walz, T., Agre, P., Heymann, J. B., et al. (2000). Structural determinants of water permeation through aquaporin-1. *Nature*, *407*, 599–605.

Nakhoul, N. L., Davis, B. A., Romero, M. F., & Boron, W. F. (1998). Effect of expressing the water channel aquaporin-1 on the CO_2 permeability of Xenopus oocytes. *The American Journal of Physiology*, *43*, C543–C548.

Navarro-Rodenas, A., Barzana, G., Nicolas, E., Carra, A., Schubert, A., & Morte, A. (2013). Expression analysis of aquaporins from desert truffle mycorrhizal symbiosis reveals a fine-tuned regulation under drought. *Molecular Plant-Microbe Interactions*, *26*, 1068–1078.

Neill, S., Desikan, R., & Hancock, J. (2002). Hydrogen peroxide signalling. *Current Opinion in Plant Biology*, *5*, 388–395.

Nyblom, M., Frick, A., Wang, Y., Ekvall, M., Hallgren, K., Hedfalk, K., et al. (2009). Structural and functional analysis of *So*PIP2;1 mutants add insight into plant aquaporin gating. *Journal of Molecular Biology*, *387*, 653–668.

Otto, B., Uehlein, N., Sdorra, S., Fischer, M., Ayaz, M., Belastegui-Macadam, X., et al. (2010). Aquaporin tetramer composition modifies the function of tobacco aquaporins. *The Journal of Biological Chemistry*, *285*, 31253–31260.

Park, J. H., & Saier, M. H. (1996). Phylogenetic characterization of the MIP family of transmembrane channel proteins. *Journal of Membrane Biology*, *153*, 171–180.

Park, W., Scheffler, B. E., Bauer, P. J., & Campbell, B. T. (2010). Identification of the family of aquaporin genes and their expression in upland cotton (Gossypium hirsutum L.). *BMC Plant Biology*, *10*, 142.

Peret, B., Li, G., Zhao, J., Band, L. R., Voss, U., Postaire, O., et al. (2012). Auxin regulates aquaporin function to facilitate lateral root emergence. *Nature Cell Biology*, *14*, 991–998.

Porcel, R., Bustamante, A., Ros, R., Serrano, R., & Mulet, J. M. (2018). BvCOLD1: A novel aquaporin from sugar beet (Beta vulgaris L.) involved in boron homeostasis and abiotic stress. *Plant, Cell & Environment*. https://doi.org/10.1111/pce.13416. [Epub ahead of print].

Postaire, O., Tournaire-Roux, C., Grondin, A., Boursiac, Y., Morillon, R., Schäffner, T., et al. (2010). A PIP1 aquaporin contributes to hydrostatic pressure-induced water transport in both the root and rosette of Arabidopsis. *Plant Physiology*, *152*, 1418–1430.

Prado, K., Boursiac, Y., Tournaire-Roux, C., Monneuse, J. M., Postaire, O., Da Ines, O., et al. (2013). Regulation of Arabidopsis leaf hydraulics involves light-dependent phosphorylation of aquaporins in veins. *Plant Cell*, *25*, 1029–1039.

Prak, S., Hem, S., Boudet, J., Viennois, G., Sommerer, N., Rossignol, M., et al. (2008). Multiple phosphorylations in the C-terminal tail of plant plasma membrane aquaporins: Role in subcellular trafficking of AtPIP2;1 in response to salt stress. *Molecular & Cellular Proteomics*, 7, 1019–1030.

Preston, G., & Agre, P. (1991). Isolation of the cDNA for erythrocyte integral membrane protein of 28 kilodaltons: Member of an ancient channel family. *Proceedings of the National Academy of Sciences of the United States of America*, *88*, 11110–11114.

Quigley, F., Rosenberg, J. M., Shachar-Hill, Y., & Bohnert, H. J. (2001). From genome to function: The *Arabidopsis* aquaporins. *Genome Biology*, *3*, 1–17.

Reinhardt, H., Hachez, C., Bienert, M. D., Beebo, A., Swarup, K., Voss, U., et al. (2016). Tonoplast aquaporins facilitate lateral root emergence. *Plant Physiology*, *170*, 1640–1654.

Rivera-Serrano, E. E., Rodriguez-Welsh, M. F., Hicks, G. R., & Rojas-Pierce, M. (2012). A small molecule inhibitor partitions two distinct pathways for trafficking of tonoplast intrinsic proteins in Arabidopsis. *PLoS One*, 7, e44735.

Rivers, R. L., Dean, R. M., Chandy, G., Hall, J. E., Roberts, D. M., & Zeidel, M. L. (1997). Functional analysis of nodulin 26, an aquaporin in soybean root symbiosomes. *The Journal of Biological Chemistry*, *272*, 16256–16261.

Rodrigues, O., Reshetnyak, G., Grondin, A., Saijo, Y., Leonhardt, N., Maurel, C., et al. (2017). Aquaporins facilitate hydrogen peroxide entry into guard cells to mediate ABA- and pathogen-triggered stomatal closure. *Proceedings of the National Academy of Sciences of the United States of America*, *114*, 9200–9205.

Saito, C., Ueda, T., Abe, H., Wada, Y., Kuroiwa, T., Hisada, A., et al. (2002). A complex and mobile structure forms a distinct subregion within the continuous vacuolar membrane in young cotyledons of *Arabidopsis*. *The Plant Journal*, *29*, 245–255.

Sakurai, J., Ishikawa, F., Yamaguchi, T., Uemura, M., & Maeshima, M. (2005). Identification of 33 rice aquaporin genes and analysis of their expression and function. *Plant & Cell Physiology*, *46*, 1568–1577.

Sakurai-Ishikawa, J., Murai-Hatano, M., Hayashi, H., Ahamed, A., Fukushi, K., Matsumoto, T., et al. (2011). Transpiration from shoots triggers diurnal changes in root aquaporin expression. *Plant, Cell & Environment*, *34*, 1150–1163.

Santoni, V., Verdoucq, L., Sommerer, N., Vinh, J., Pflieger, D., & Maurel, C. (2006). Methylation of aquaporins in plant plasma membrane. *The Biochemical Journal*, *400*, 189–197.

Santoni, V., Vinh, J., Pflieger, D., Sommerer, N., & Maurel, C. (2003). A proteomic study reveals novel insights into the diversity of aquaporin forms expressed in the plasma membrane of plant roots. *The Biochemical Journal*, *372*, 289–296.

Schnurbusch, T., Hayes, J., Hrmova, M., Baumann, U., Ramesh, S. A., Tyerman, S. D., et al. (2010). Boron toxicity tolerance in barley through reduced expression of the multifunctional aquaporin HvNIP2;1. *Plant Physiology*, *153*, 1706–1715.

Segami, S., Makino, S., Miyake, A., Asaoka, M., & Maeshima, M. (2014). Dynamics of vacuoles and H+-pyrophosphatase visualized by monomeric green fluorescent protein in Arabidopsis: Artifactual bulbs and native intravacuolar spherical structures. *The Plant Cell*, *26*, 3416–3434.

Sezgin, E., Levental, I., Mayor, S., & Eggeling, C. (2017). The mystery of membrane organization: Composition, regulation and roles of lipid rafts. *Nature Reviews. Molecular Cell Biology*, *18*, 361–374.

Shatil-Cohen, A., Attia, Z., & Moshelion, M. (2011). Bundle-sheath cell regulation of xylem-mesophyll water transport via aquaporins under drought stress: A target of xylem-borne ABA? *The Plant Journal*, *67*, 72–80.

Simm, S., Papasotiriou, D. G., Ibrahim, M., Leisegang, M. S., Muller, B., Schorge, T., et al. (2013). Defining the core proteome of the chloroplast envelope membranes. *Frontiers in Plant Science*, *4*, 11.

Sorieul, M., Santoni, V., Maurel, C., & Luu, D. T. (2011). Mechanisms and effects of retention of over-expressed aquaporin AtPIP2;1 in the endoplasmic reticulum. *Traffic*, *12*, 473–482.

Soto, G., Alleva, K., Amodeo, G., Muschietti, J., & Ayub, N. D. (2012). New insight into the evolution of aquaporins from flowering plants and vertebrates: Orthologous identification and functional transfer is possible. *Gene*, *503*, 165–176.

Soto, G., Fox, R., Ayub, N., Alleva, K., Guaimas, F., Erijman, E., et al. (2010). TIP5;1 is an aquaporin specifically targeted to pollen mitochondria and is probably involved in nitrogen remobilization in *Arabidopsis thaliana*. *The Plant Journal*, *64*, 1038–1047.

Steudle, E., & Peterson, C. A. (1998). How does water get through roots? *Journal of Experimental Botany*, *49*, 775–788.

Sun, S. K., Chen, Y., Che, J., Konishi, N., Tang, Z., Miller, A. J., et al. (2018). Decreasing arsenic accumulation in rice by overexpressing OsNIP1;1 and OsNIP3;3 through disrupting arsenite radial transport in roots. *The New Phytologist*, *219*, 641–653.

Sutka, M., Alleva, K., Parisi, M., & Amodeo, G. (2005). Tonoplast vesicles of *Beta vulgaris* storage root show functional aquaporins regulated by protons. *Biology of the Cell*, *97*, 837–846.

Tajkhorshid, E., Nollert, P., Jensen, M., Miercke, L., O'Connell, J., Stroud, R., et al. (2002). Control of the selectivity of the aquaporin water channel family by global orientational tuning. *Science*, *296*, 525–530.

Takano, J., Tanaka, M., Toyoda, A., Miwa, K., Kasai, K., Fuji, K., et al. (2010). Polar localization and degradation of Arabidopsis boron transporters through distinct trafficking pathways. *Proceedings of the National Academy of Sciences of the United States of America*, *107*, 5220–5225.

Takano, J., Wada, M., Ludewig, U., Schaaf, G., von Wiren, N., & Fujiwara, T. (2006). The *Arabidopsis* major intrinsic protein NIP5;1 is essential for efficient boron uptake and plant development under boron limitation. *Plant Cell*, *18*, 1498–1509.

Tanaka, M., Wallace, I. S., Takano, J., Roberts, D. M., & Fujiwara, T. (2008). NIP6;1 is a boric acid channel for preferential transport of boron to growing shoot tissues in Arabidopsis. *Plant Cell*, *20*, 2860–2875.

Thiagarajah, J. R., Chang, J., Goettel, J. A., Verkman, A. S., & Lencer, W. I. (2017). Aquaporin-3 mediates hydrogen peroxide-dependent responses to environmental stress in colonic epithelia. *Proceedings of the National Academy of Sciences of the United States of America*, *114*, 568–573.

Tian, S., Wang, X., Li, P., Wang, H., Ji, H., Xie, J., et al. (2016). Plant aquaporin AtPIP1;4 links apoplastic H_2O_2 induction to disease immunity pathways. *Plant Physiology*, *171*, 1635–1650.

Tong, J., Briggs, M. M., & McIntosh, T. J. (2012). Water permeability of aquaporin-4 channel depends on bilayer composition, thickness, and elasticity. *Biophysical Journal*, *103*, 1899–1908.

Tong, J., Canty, J. T., Briggs, M. M., & McIntosh, T. J. (2013). The water permeability of lens aquaporin-0 depends on its lipid bilayer environment. *Experimental Eye Research*, *113*, 32–40.

Tornroth-Horsefield, S., Hedfalk, K., Fischer, G., Lindkvist-Petersson, K., & Neutze, R. (2010). Structural insights into eukaryotic aquaporin regulation. *FEBS Letters*, *584*, 2580–2588.

Törnroth-Horsefield, S., Wang, Y., Hedfalk, K., Johanson, U., Karlsson, M., Tajkhorshid, E., et al. (2006). Structural mechanism of plant aquaporin gating. *Nature*, *439*, 688–694.
Tournaire-Roux, C., Sutka, M., Javot, H., Gout, E., Gerbeau, P., Luu, D. T., et al. (2003). Cytosolic pH regulates root water transport during anoxic stress through gating of aquaporins. *Nature*, *425*, 393–397.
Tsiavaliaris, G., Itel, F., Hedfalk, K., Al-Samir, S., Meier, W., Gros, G., et al. (2015). Low CO2 permeability of cholesterol-containing liposomes detected by stopped-flow fluorescence spectroscopy. *The FASEB Journal*, *29*, 1780–1793.
Tyrrell, M., Campanoni, P., Sutter, J. U., Pratelli, R., Paneque, M., Sokolovski, S., et al. (2007). Selective targeting of plasma membrane and tonoplast traffic by inhibitory (dominant-negative) SNARE fragments. *The Plant Journal*, *51*, 1099–1115.
Uehlein, N., Fileschi, K., Eckert, M., Bienert, G. P., Bertl, A., & Kaldenhoff, R. (2007). Arbuscular mycorrhizal symbiosis and plant aquaporin expression. *Phytochemistry*, *68*, 122–129.
Uehlein, N., Lovisolo, C., Siefritz, F., & Kaldenhoff, R. (2003). The tobacco aquaporin NtAQP1 is a membrane CO_2 pore with physiological functions. *Nature*, *425*, 734–737.
Uehlein, N., Otto, B., Eilingsfeld, A., Itel, F., Meier, W., & Kaldenhoff, R. (2012). Gas-tight triblock-copolymer membranes are converted to CO(2) permeable by insertion of plant aquaporins. *Scientific Reports*, *2*, 538.
Vajpai, M., Mukherjee, M., & Sankararamakrishnan, R. (2018). Cooperativity in plant plasma membrane intrinsic proteins (PIPs): Mechanism of increased water transport in maize PIP1 channels in hetero-tetramers. *Scientific Reports*, *8*, 12055.
Van Wilder, V., Miecielica, U., Degand, H., Derua, R., Waelkens, E., & Chaumont, F. (2008). Maize plasma membrane aquaporins belonging to the PIP1 and PIP2 subgroups are in vivo phosphorylated. *Plant & Cell Physiology*, *49*, 1364–1377.
Vander Willigen, C., Postaire, O., Tournaire-Roux, C., Boursiac, Y., & Maurel, C. (2006). Expression and inhibition of aquaporins in germinating *Arabidopsis* seeds. *Plant & Cell Physiology*, *47*, 1241–1250.
Vera-Estrella, R., Barkla, B. J., Bohnert, H. J., & Pantoja, O. (2004). Novel regulation of aquaporins during osmotic stress. *Plant Physiology*, *135*, 2318–2329.
Verdoucq, L., Grondin, A., & Maurel, C. (2008). Structure-function analysis of plant aquaporin AtPIP2;1 gating by divalent cations and protons. *The Biochemical Journal*, *415*, 409–416.
Verkman, A. S. (1992). Water channels in cell membranes. *Annual Review of Physiology*, *54*, 97–108.
Vieceli Dalla Sega, F., Zambonin, L., Fiorentini, D., Rizzo, B., Caliceti, C., Landi, L., et al. (2014). Specific aquaporins facilitate Nox-produced hydrogen peroxide transport through plasma membrane in leukaemia cells. *Biochimica et Biophysica Acta*, *1843*, 806–814.
Wallace, I. S., & Roberts, D. M. (2004). Homology modeling of representative subfamilies of *Arabidopsis* major intrinsic proteins. Classification based on the aromatic/arginine selectivity filter. *Plant Physiology*, *135*, 1059–1068.
Wallace, I. S., & Roberts, D. M. (2005). Distinct transport selectivity of two structural subclasses of the nodulin-like intrinsic protein family of plant aquaglyceroporin channels. *Biochemistry*, *44*, 16826–16834.
Wang, C., Hu, H., Qin, X., Zeise, B., Xu, D., Rappel, W. J., et al. (2016). Reconstitution of CO_2 regulation of SLAC1 anion channel and function of CO_2-permeable PIP2;1 aquaporin as CARBONIC ANHYDRASE4 interactor. *Plant Cell*, *28*, 568–582.
Wang, Y., Li, R., Li, D., Jia, X., Zhou, D., Li, J., et al. (2017). NIP1;2 is a plasma membrane-localized transporter mediating aluminum uptake, translocation, and tolerance in Arabidopsis. *Proceedings of the National Academy of Sciences of the United States of America*, *114*, 5047–5052.

Wang, Y., & Tajkhorshid, E. (2007). Molecular mechanisms of conduction and selectivity in aquaporin water channels. *The Journal of Nutrition, 137*, 1509S–1515S. discussion 1516S–1517S.

Watanabe, S., Moniaga, C. S., Nielsen, S., & Hara-Chikuma, M. (2016). Aquaporin-9 facilitates membrane transport of hydrogen peroxide in mammalian cells. *Biochemical and Biophysical Research Communications, 471*, 191–197.

Weaver, C. D., Shomer, N. H., Louis, C. F., & Roberts, D. M. (1994). Nodulin 26, a nodule-specific symbiosome membrane protein from soybean, is an ion channel. *The Journal of Biological Chemistry, 269*, 17858–17862.

Wudick, M. M., Luu, D. T., Tournaire-Roux, C., Sakamoto, W., & Maurel, C. (2014). Vegetative and sperm cell-specific aquaporins of Arabidopsis highlight the vacuolar equipment of pollen and contribute to plant reproduction. *Plant Physiology, 164*, 1697–1706.

Yamaji, N., Mitatni, N., & Ma, J. F. (2008). A transporter regulating silicon distribution in rice shoots. *Plant Cell, 20*, 1381–1389.

Yaneff, A., Sigaut, L., Marquez, M., Alleva, K., Pietrasanta, L. I., & Amodeo, G. (2014). Heteromerization of PIP aquaporins affects their intrinsic permeability. *Proceedings of the National Academy of Sciences of the United States of America, 111*, 231–236.

Yanochko, G. M., & Yool, A. J. (2002). Regulated cationic channel function in Xenopus oocytes expressing Drosophila big brain. *The Journal of Neuroscience, 22*, 2530–2540.

Yool, A. J., & Campbell, E. M. (2012). Structure, function and translational relevance of aquaporin dual water and ion channels. *Molecular Aspects of Medicine, 33*, 553–561.

Yuan, D., Li, W., Hua, Y., King, G. J., Xu, F., & Shi, L. (2017). Genome-wide identification and characterization of the aquaporin gene family and transcriptional responses to boron deficiency in Brassica napus. *Frontiers in Plant Science, 8*, 1336.

Zardoya, R. (2005). Phylogeny and evolution of the major intrinsic protein family. *Biology of the Cell, 97*, 397–414.

Zelazny, E., Borst, J. W., Muylaert, M., Batoko, H., Hemminga, M. A., & Chaumont, F. (2007). FRET imaging in living maize cells reveals that plasma membrane aquaporins interact to regulate their subcellular localization. *Proceedings of the National Academy of Sciences of the United States of America, 104*, 12359–12364.

Zelazny, E., Miecielica, U., Borst, J. W., Hemminga, M. A., & Chaumont, F. (2009). An N-terminal diacidic motif is required for the trafficking of maize aquaporins *Zm*PIP2;4 and *Zm*PIP2;5 to the plasma membrane. *The Plant Journal, 57*, 346–355.

Zhao, X. Q., Mitani, N., Yamaji, N., Shen, R. F., & Ma, J. F. (2010). Involvement of silicon influx transporter OsNIP2;1 in selenite uptake in rice. *Plant Physiology, 153*, 1871–1877.

Zwiazek, J. J., Xu, H., Tan, X., Navarro-Rodenas, A., & Morte, A. (2017). Significance of oxygen transport through aquaporins. *Scientific Reports, 7*, 40411.

Zybailov, B., Rutschow, H., Friso, G., Rudella, A., Emanuelsson, O., Sun, Q., et al. (2008). Sorting signals, N-terminal modifications and abundance of the chloroplast proteome. *PLoS One, 3*, e1994.

FURTHER READING

Biela, A., Grote, K., Otto, B., Hoth, S., Hedrich, R., & Kaldenhoff, R. (1999). The *Nicotiana tabacum* plasma membrane aquaporin NtAQP1 is mercury-insensitive and permeable for glycerol. *The Plant Journal, 18*, 565–570.

Li, Q., Shen, W., Zheng, Q., Fowler, D. B., & Zou, J. (2016). Adjustments of lipid pathways in plant adaptation to temperature stress. *Plant Signaling & Behavior, 11*, e1058461.

Preston, G. M., Carroll, T. P., Guggino, W. B., & Agre, P. (1992). Appearance of water channels in *Xenopus* oocytes expressing red cell CHIP28 protein. *Science, 256*, 385–387.

CHAPTER THREE

Heavy Metal Pumps in Plants: Structure, Function and Origin

Jeppe Thulin Østerberg, Michael Palmgren[1]
Department of Plant and Environmental Sciences, University of Copenhagen, Frederiksberg, Denmark
[1]Corresponding author: e-mail address: palmgren@plen.ku.dk

Contents

Abstract

In all forms of life, primary active transport of heavy metal ions is mediated by members of a subfamily of P-type ATPase pumps, the P1B ATPases. All organisms require heavy metals such as copper and zinc for essential life processes, but too much heavy metal is hazardous for cells, so control of cellular heavy metal homeostasis is essential. Heavy metal exporting P1B ATPases evolved in prokaryotes where they are omnipresent. In this review, we discuss the structure, function and evolution of plant P1B ATPases. These pumps control heavy metal homeostasis in the plant both at the cellular and organismal levels. They operate by a mechanism similar to other P-type ATPase but have structural features that make them specific for copper or zinc, although a subset of pumps have less specificity. Remarkably, zinc pumps are absent from animals, which employ secondary active transport systems for the export of zinc. A subset of copper pumps in plants share an evolutionary origin with animal copper pumps, but chloroplastic copper pumps evolved from endosymbiotic gene transfer from cyanobacteria. In plants and other photosynthetic eukaryotes, two types of zinc pumps may both have a cyanobacterial origin.

Advances in Botanical Research, Volume 87
ISSN 0065-2296
https://doi.org/10.1016/bs.abr.2018.09.004

1. COPPER AND ZINC HOMEOSTASIS IN EUKARYOTES

Copper and zinc are essential micronutrients which are used as cofactors in multiple proteins. However, in higher doses both elements easily become toxic. Therefore an intricate system is required to maintain heavy metal homeostasis of any organism (for reviews, see Andresen, Peiter, & Kupper, 2018; Kupper & Andresen, 2016; Palmer & Guerinot, 2009).

Copper, due to its redox properties, is a cofactor in proteins that are involved in electron transfer reactions (Linder & Goode, 1991; Lippard & Berg, 1994). In all eukaryotic cells, copper is required in the mitochondrial electron transport chain for the function of cytochrome oxidase, which accommodates two bound copper ions, and within the cytosol where it acts as a cofactor of superoxide dismutase. In plants, copper plays an essential role in photosynthesis as it is an essential cofactor in the electron carrier plastocyanin, but is also required here for the function of superoxide dismutase and polyphenol oxidase. Many other plant proteins, including the ethylene receptor, depend on copper for their function (Marschner, 1995; Rodriguez et al., 1999). Therefore, plants depend on systems that can control copper homeostasis in the cytosol, mitochondria and chloroplasts (Aguirre & Pilon, 2015; Burkhead, Reynolds, Abdel-Ghany, Cohu, & Pilon, 2009; Garcia, Welchen, & Gonzalez, 2014).

Zinc is not redox active but acts in biological systems as a Lewis acid, i.e. it can accept lone electron pairs from donor groups with strong binding as a result (Williams, 1987). Due to this property, it can play catalytical roles, by coordinating a substrate molecule in the active site of enzymes, or structural roles, by stabilizing proteins (Bertini, Luchinat, & Monnanni, 1985; McCall, Huang, & Fierke, 2000). For example, a large number of transcription factors and zinc-finger containing proteins require zinc as a structural component. In the model plant *Arabidopsis thaliana*, >1.200 proteins are predicted to contain, bind, or transport zinc (Kramer & Clemens, 2005).

In plants, multiple transport systems are involved in the uptake, redistribution, sequestration and secretion of copper and zinc (Clemens, Palmgren, & Kramer, 2002; Grotz & Guerinot, 2006). All living organisms have a negative inside membrane potential, which acts as a strong driving force for the entry through channel proteins or symport carriers of divalent cations such as copper and zinc. However, the same membrane potential prohibits passive efflux of cations, which poses a problem if excess heavy metals are to be exported.

This requires some form of active transport, where the energy source can be either direct (primary), i.e. using metabolic chemical energy, or indirect (secondary), i.e. utilizing potential energy stored in an electrochemical gradient. In both plants and animals, excess copper is actively pumped out by primary active transporters, i.e. pumps. As zinc is equally important for plants and animals, one would also expect related mechanisms for handling this element. This is however not the case. Whereas plants utilize primary active transporters, i.e. pumps, for export of zinc, animals use secondary active transporters, i.e. antiporters, for the same purpose (Lichten & Cousins, 2009).

This review will focus on primary active transport of copper and zinc across cell membranes, which is always driven by the hydrolysis of adenosine-triphosphate (ATP), and how it evolved in eukaryotic cells. The reader is referred to other recent reviews for coverage of how animals (Lichten & Cousins, 2009; Navarro & Schneuwly, 2017) and plants (Sinclair & Kramer, 2012) in general regulate copper and zinc homeostasis, how copper and zinc travels over long distances in the plant body (Clemens et al., 2002; Olsen & Palmgren, 2014), and how hyperaccumulator plants handle zinc (Leitenmaier & Kupper, 2013).

2. P-Type ATPases ARE PRIMARY ACTIVE PUMPS FOUND IN ALL CELLS

In plants, primary active transport of copper and zinc is carried out by P-type ATPases (Clemens et al., 2002; Grotz & Guerinot, 2006). All cells are surrounded by a lipid bilayer membrane that restricts the free in- or outflow of solutes. This barrier is essential to maintain a constant intracellular environment. In eukaryotic cells, cellular compartments such as organelles are likewise surrounded by one or more biological membranes. P-type ATPases constitute a large family of biological pumps that by splitting ATP get energy to actively transport charged ions and molecules across the lipid bilayer membrane (Palmgren & Nissen, 2011). Such primary active transport, primary because it directly utilizes metabolic energy, can move a charged solute not only against a concentration gradient but also against an electrical gradient such as the membrane potential. In this way, P-type pumps establish and maintain electrochemical gradients across membranes. Electrochemical gradients constitute stored potential energy that can be used to energize a myriad of secondary active transport processes in the cell.

P-type ATPases from all organisms are divided into five major subfamilies named P1–P5, which again are divided into a number of subclades (Axelsen & Palmgren, 1998). Heavy metals are pumped by a subclade of P1 ATPases. The different subclades of P2 ATPases each transport ions of softer metals such as Ca^{2+}, Na^{+} and K^{+}. The biggest subclade of P3 ATPases, P3A ATPases, only transport the smallest of all ions, namely, H^{+} (which also can be used by other P-type ATPases as a counterion). P4 ATPases, on the contrary, transport a giant substrate compared to the size of the pump, namely, a phospholipid, the hydrophilic headgroup of which is bound and taken through the membrane with the hydrophobic tails following after. P5 ATPases are unique to eukaryotes where they are found in all cells but their biochemical function remains to be elucidated.

P1 ATPases are divided into P1A and P1B ATPases. P1A ATPases (or KdpB ATPases) transport K^{+} whereas P1B ATPases transport heavy metal ions, such as copper and zinc, and serve to detoxify and maintain homeostasis of heavy metals. P1A ATPases are only found in eubacteria whereas P1B ATPases are found in all life forms. In contrast, however, P1B ATPases from eubacteria to eukaryotes transport heavy metals and only heavy metals. This consistency would suggest that substrate specificities were fixed very early in evolution and have remained constant since then (Axelsen & Palmgren, 1998).

In *A. thaliana*, the first plant to have its genome fully sequenced, representatives for all five P-type ATPase families can be found (Axelsen & Palmgren, 2001). A total of 46 P-type ATPases are present in *A. thaliana* and P-type ATPases can be found in most cell membrane systems (Fig. 1). Most other plants have a setup of P-type ATPases much like *A. thaliana*. However, two major differences exist between plants and animals: First, the bioenergetics circuits at their plasma membranes depend on pumps with different ion specificities. Thus, animals have a plasma membrane sodium/potassium exchange pump (the Na^{+}/K^{+}-ATPase) whereas plants do not. They have a plasma membrane H^{+}-ATPase instead. For a discussion as to why plant do not have a Na^{+}/K^{+}-ATPase, and whether they would potentially benefit from having one, the reader is referred to a recent review on this subject (Pedersen & Palmgren, 2017). The second major difference is that plants have zinc pumps but animals do not. As zinc pumps evolved already in eubacteria, it is interesting why these pumps do not occur in animals. Possible reasons could be that zinc pumps were lost in an early animal ancestor, or that zinc pumps in plants were acquired by horizontal rather than vertical gene transfer.

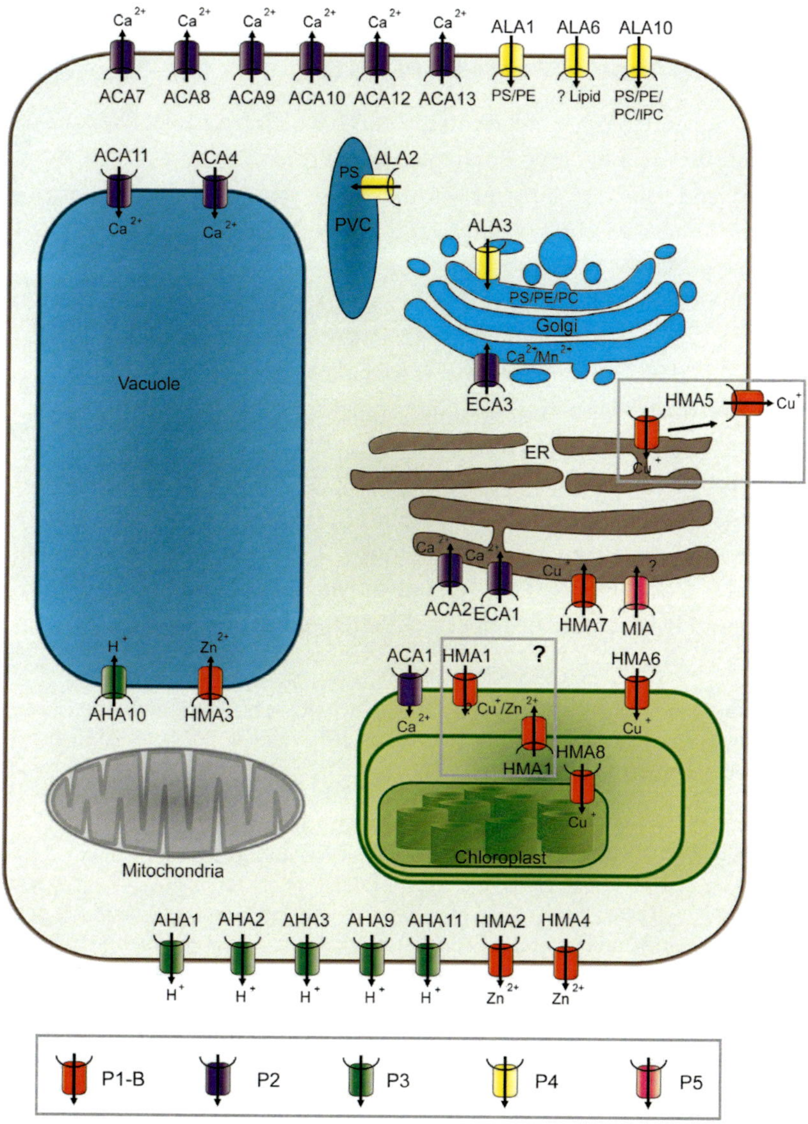

Fig. 1 Schematic overview of subcellular localization of P-type ATPase in *Arabidopsis thaliana*. Subcellular localization of P-type ATPases in Arabidopsis is only shown when experimentally demonstrated. Transport direction and transported substrate are indicated. P1B type ATPases are shown in red, P2 ATPases in blue, P3A ATPases in green, P4 ATPases in yellow and the P5 ATPase in purple. Boxes indicate, for AtHMA1, the uncertain location and transport direction within the chloroplast envelope and for AtHMA5, the change in localization upon increased copper levels. References for localization of P-type ATPases are: P1B-ATPases: AtHMA1, chloroplast membrane

(Continued)

3. P1B-Type ATPases IN PLANTS

Compared to other eukaryotes, plants have a large array of heavy metal ATPases. Where only two such pumps are present in, e.g. the human genome, and two are present in the yeast genome, the genomes of *A. thaliana* and rice (*Oryza sativa*) contain eight and nine P1B ATPases, respectively (Baxter et al., 2003).

In *A. thaliana*, Heavy Metal ATPase5 (HMA5) to HMA8 transport copper. Among these pumps, HMA5 is the sole copper P1B ATPase to be located in the plasma membrane. It is mainly expressed in the root, where it serves a detoxifying function by exporting copper from the cytosol (Andres-Colas et al., 2006; Kobayashi et al., 2008). Long-distance transport of cations that involves crossing apoplastic gaps is likely to be mediated by active transporters (Olsen & Palmgren, 2014). The rice homologue OsHMA5 loads the root xylem with copper to enable long-distance transport to the shoot (Deng, Yamaji, Xia, & Ma, 2013), and a similar role of Arabidopsis HMA5 could be expected. Whether HMA5 also exports copper

Fig. 1—cont'd (Seigneurin-Berny et al., 2006); AtHMA2/4, plasma membrane (Hussain et al., 2004; Verret et al., 2004); AtHMA3, tonoplast (Gravot et al., 2004); AtHMA5, plasma membrane/endoplasmic reticulum (Li et al., 2017a); AtHMA6, outer chloroplast membrane (Abdel-Ghany, Muller-Moule, Niyogi, Pilon, & Shikanai, 2005; Catty et al., 2011); AtHMA7, endoplasmic reticulum (Li et al., 2017b). AtHMA8, thylakoid membrane (Abdel-Ghany et al., 2005). The P2 ATPases: AtECA1, endoplasmic reticulum (Liang, Cunningham, Harper, & Sze, 1997) AtECA3, Golgi (Mills et al., 2008); AtACA1, chloroplast envelope (Evans, 1994; Huang, Berkelman, Franklin, & Hoffman, 1993); AtACA2, endoplasmic reticulum (Dunkley et al., 2006); AtACA4, vacuole (Geisler, Frangne, Gomes, Martinoia, & Palmgren, 2000); AtACA7, plasma membrane (Lucca & Leon, 2012); AtACA8, plasma membrane (Bonza et al., 2000); AtACA9, plasma membrane (Schiott et al., 2004); AtACA10, plasma membrane (Dunkley et al., 2006); AtACA11, vacuole (Lee et al., 2007); AtACA12, plasma membrane (Limonta et al., 2014); AtACA13, plasma membrane (Iwano et al., 2014). The P3A ATPases: AtAHA1, plasma membrane (Hashimoto-Sugimoto et al., 2013); AtAHA2, plasma membrane (DeWitt, Hong, Sussman, & Harper, 1996); AtAHA3, plasma membrane (DeWitt et al., 1996); AtAHA9, plasma membrane (Rodrigues et al., 2014); AtAHA10, tonoplast (Appelhagen et al., 2015); AtAHA11, plasma membrane (Rodrigues et al., 2014). The P4 ATPases: AtALA1, plasma membrane (Lopez-Marques, Poulsen, & Palmgren, 2012); AtALA2, prevacuolar compartment (Lopez-Marques et al., 2010); AtALA3, Golgi (Poulsen et al., 2008); AtALA6, plasma membrane (McDowell et al., 2015); AtALA10, plasma membrane (Poulsen et al., 2015). P5 ATPase: MIA, endoplasmic reticulum (Jakobsen et al., 2005). No subcellular localization has to our knowledge been confirmed for AtECA2, and AtECA4, AtAHA4, 5, 6, 7 and 8, AtALA4, 5, 7, 8, 9, 11 and 12.

from the mother plant to the developing seed, in analogy with the function in long distance transport of zinc by HMA2 and HMA4 (see below), remains to be shown (See also Chapter 4 by Eroglu). HMA5 only migrates to the plasma membrane when plants are subjected to high external copper concentrations and otherwise reside in the endoplasmic reticulum (Li, Iqbal, et al., 2017). This mechanism is similar to the migration in response to copper of the human ATP7A and ATP7B copper pumps (Petris et al., 1996; Roelofsen et al., 2000). HMA6 to HMA8 are all involved in the delivery of copper to intracellular compartments for incorporation into copper containing proteins. HMA7 pumps copper into the endoplasmic reticulum (Li, Lacey, et al., 2017), delivering copper for incorporation into newly synthesized proteins. An example is the ethylene receptor that requires copper as a cofactor (Rodriguez et al., 1999). HMA7 was initially identified in a screen for mutants that were positive for the ethylene triple response in the presense of an ethylene antagonist and hence named *Responsive to Antagonist* (*RAN1*) (Hirayama et al., 1999; Woeste & Kieber, 2000). HMA6 and HMA8 deliver copper into the chloroplastic stroma and thylakoid lumen, respectively (Abdel-Ghany et al., 2005; Shikanai, Muller-Moule, Munekage, Niyogi, & Pilon, 2003).

In *A. thaliana*, HMA1 to HMA4 transport zinc. HMA1 is not specific for zinc and transports a variety of other biologically relevant metals such as cobalt, copper and calcium (Moreno et al., 2008; Seigneurin-Berny et al., 2006). HMA1 is located in the chloroplast envelope but its orientation is uncertain and it has been suggested both as a copper importer into chloroplasts (Boutigny et al., 2014; Seigneurin-Berny et al., 2006), and as an exporter of excess zinc from this organelle (Kim et al., 2009). In barley, HvHMA1 was demonstrated to mediate export of copper and zinc from chloroplasts (Mikkelsen et al., 2012). In isolated chloroplasts, a negative inside membrane potential of around $-100\,mV$ has been detected across the envelope membrane (Demmig & Gimmler, 1983). Thus, active transport is likely to be required for a cation to leave the chloroplast from the stroma whereas the need for a pump for import purposes is more difficult to understand. This calls for a reinvestigation of the exact membrane localizations of HMA1 and HMA6.

HMA2 and HMA4 are zinc pumps located in the plasma membrane. Knockout mutants of either of these pumps display weak zinc deficiency phenotypes. However, when both are knocked out, a severe zinc deficiency phenotype is evident, indicating a redundant function of these two pumps (Hussain et al., 2004). Indeed, rice seems to have only one pump serving this function (Baxter et al., 2003). HMA2 and HMA4 are expressed in the root pericycle cells, where they load zinc into the root xylem for long distance transport to the shoot (Hussain et al., 2004) and in seeds, where they are involved in seed loading of zinc across the apoplastic gaps between parent and progeny (Olsen et al., 2016). A similar distribution has been demonstrated in monocot species, and in addition, a high expression in the nodes of rice shoots has been ascribed to the direction of zinc towards the seeds (Yamaji, Xia, Mitani-Ueno, Yokosho, & Feng Ma, 2013). In the zinc hyperaccumulating relative *Arabidopsis halleri*, a triplication of the *HMA4* gene is responsible for the hyperaccumulation trait (Kramer, 2010). HMA3 is located in the vacuole membrane and involved in sequestration of excess zinc and of the toxic zinc analogue cadmium into the vacuolar lumen (Morel et al., 2009). Arabidopsis HMA3 was demonstrated to be expressed in guard cells, vascular tissue and root apex. Its functional HMA3 homologues in rice are mainly expressed in root tissues (Miyadate et al., 2011). The homologue in *Thlaspi arvense* is expressed in leaves and roots and contributes to the cadmium hyperaccumulation trait of this species (Ueno et al., 2011).

4. MECHANISM OF PUMPING BY P-Type ATPases

In order to understand the structure-function relationships in P1B ATPases and their implications for the evolution of these pumps, we need first to take a look at the general structure and mechanism of P-type ATPases. Although P-type ATPases can have accessory subunits, the catalytic subunit is always a single polypeptide, typically around 1000 amino acid residues (Palmgren & Nissen, 2011). This simplicity is in stark contrast to other primary active transport systems, e.g. the ATP synthase of chloroplasts, which can operate in the reverse direction and pump protons at the expense of ATP. The ATP synthase is an assembly of 26 subunits that operates like a nanomachine with a rotating turbine and many parts (Hahn, Vonck, Mills, Meier, & Kuhlbrandt, 2018).

P-type ATPases are also nanomachines exhibiting large motions during each step of the transport cycle, but here the individual functional parts are joined together like pearls on a string (Palmgren & Nissen, 2011). The basic machinery in all P-type ATPases characterized so far is essentially the same. The catalytic subunit consists of a membrane-embedded domain (the M domain), which in its middle contains the binding site(s) for the ligand to be transported. Very little of the protein is exposed to the extracytoplasmic side of the membrane. In contrast, the major mass of the protein protrudes into the cytoplasm. The cytoplasmic portion of the pump is divided into three globular domains, the P, N and A domains (Fig. 2). The P (phosphorylation) domain is phosphorylated (and subsequently dephosphorylated) at a conserved aspartyl residue once during each catalytic cycle. This phosphorylation event is not to be mistaken for regulatory phosphorylation, which takes place in many proteins and involves phosphorylation at threonine, serine or tyrosine residues. P-type ATPases derive their name (P for phosphorylation) from this unusual phosphorylation reaction, unusual because it intuitively is not a straightforward process for ATP to phosphorylate an aspartate residue as both are negatively charged.

The N (nucleotide binding) domain is inserted in the P domain and the connecting two strands are so close together that the N domain can move up and down as if it is connected to the P domain by a hinge. ATP binds to the N domain via its adenosine group with its γ-phosphate protruding to the medium. Once the ligand(s) to be transported is bound in the middle of the M domain, this event is signaled by intramolecular motions to the

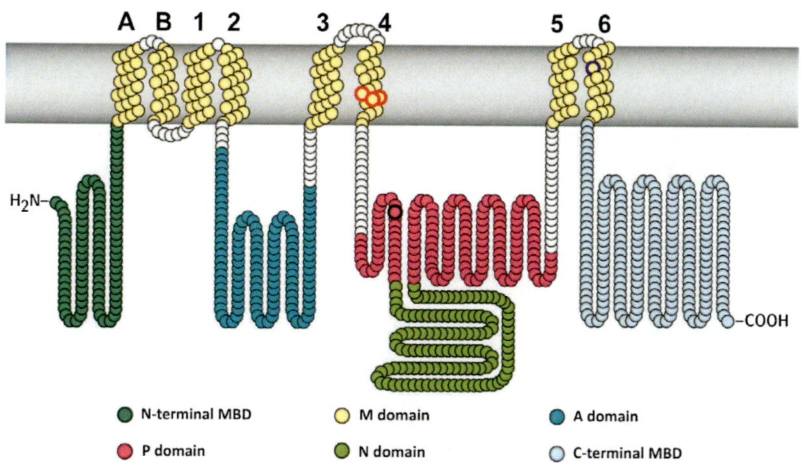

Fig. 2 Schematic overview of the structural organization of P1B ATPases. The M domain (yellow) includes eight transmembrane segments, MA, MB and M1–M6. The conserved cytosolic domains of P-type ATPases comprises the actuator (A) domain (blue) between M2 and M3, the phosphorylation (P) domain (purple) and the nucleotide binding (N) domain (green) between M4 and M5. Coloured circles in M4 and M6 indicate motifs important for metal coordination. Red circles in M4 mark a conserved Cys motif, whereas a blue circle in M6 marks a conserved Met, Asp or His in P1B-1, P1B-2 and P1B-4 ATPases, respectively. In addition to the core domains, the N-terminal (dark green) and C-terminal (light blue) metal binding domains (MBDs) are shown. The topology map was modified from a model generated by Protter (Omasits, Ahrens, Muller, & Wollscheid, 2014).

P domain, where a magnesium-binding site is now formed adjacent to the conserved aspartate. As magnesium is a divalent cation it neutralizes the negative charge of the aspartyl residue and the N domain can move down to phosphorylate the P domain.

Phosphorylation of the P domain introduces energy into the molecule and results in a major conformational change that is transmitted down to the M domain. Prior to phosphorylation, the ligand binding site(s) of the M domain is (are) exposed to the cytoplasm and freely accessible for an intracellular ligand(s) all the way down to the binding site in the middle of the membrane. Following phosphorylation, however, the entrance gate closes and an exit gate opens up towards the extracytoplasmic site. The unphosphorylated major conformation, called Enzyme 1 (E1), has high affinity for its ligand(s) in contrast to the phosphorylated conformation, called Enzyme 2 (E2). By this so-called alternating access mechanism the transported ligand is actually not pumped from one position to another, but rather diffuses to and from a single ion binding site that changes affinities when it alternates between being exposed to either side of the membrane.

Simultaneously with the transition to the E2 conformation, the A (actuator) domain moves towards the N domain and takes over its position at the top of the P domain while the N domain moves up. Whereas the N domain can be characterized as a built-in protein kinase, the A domain is a built-in protein phosphatase. The job of this domain is to dephosphorylate the phosphorylated P domain. Once this mission is accomplished, the structure of the P-type ATPase relaxes and it returns to the high-affinity E1 conformation. For more details on this intricate mechanism, that has been elucidated by more than thousand mutation mutagenesis experiments and interpreted in the light of more than hundred 3D crystal structures of P-type ATPases in the major and multiple intermediate conformations, the reader is referred to recent reviews (Moller, Olesen, Winther, & Nissen, 2010; Palmgren & Nissen, 2011; Toyoshima, 2009).

5. STRUCTURE AND MECHANISM OF P1B ATPases

At the time of this writing, two crystal structures of P1B ATPases have been solved, and both are of eubacterial origin (Gourdon et al., 2011; Wang et al., 2014). The first is from a copper pump, LpCopA from *Legionella pneumophila* (Gourdon et al., 2011), and the second from a zinc pump, ZntA from *Shigella sonnei* (Wang et al., 2014). This has made it possible to directly compare copper and zinc pumps structurally. In the absence of structures of eukaryotic origin, these two structures can serve as guidelines for understanding structure-function relationships in P1B ATPases of plants. It should be noted, that both of the solved structures are in the E2 conformation, which implies that the binding sites for copper and zinc, respectively, are in the low affinity states with no ions bound. Therefore, the likely ion binding sites have been hypothesized from modeling the protein structures into the E1 conformation. For a detailed description of both structures, the reader is referred to recent reviews (Gronberg, Sitsel, Lindahl, Gourdon, & Andersson, 2016; Sitsel et al., 2015).

At a first glance, both structures retain the functional schematic described for other cation-transporting P-type ATPases, such as the sarcoplasmic reticulum Ca^{2+}-ATPase (SERCA) and the Na^{+}/K^{+}-ATPase with a large cytosolic portion divided into three domains: A, P and N. There are six transmembrane segments (M1–6) in the core M domain but two extra transmembrane segments preceding the basal six transmembrane segments. These additional segments, which are unique for P1B ATPases, are named MA and MB (Fig. 3).

The substrate binding and release mechanisms of P1B ATPases resemble those of other well-characterized P-type ATPases. The heavy metals enter the M domain via a half-channel exposed to the cytoplasm. Binding involves coordination of the heavy metal ion from side chains in two transmembrane segments, M4 and M6 (Fig. 3B). As high levels of any heavy metal are toxic to the cell, heavy metals are maintained at picomolar to nanomolar levels in the cytosol. Not surprisingly, therefore, P1B ATPases that have been characterized biochemically have very high affinity for their transported substrate with activation optima in the range of 0.1–1 μM (Hung, Layton, Voskoboinik, Mercer, & Camakaris, 2007; Leitenmaier et al., 2011).

Copper and zinc are both Lewis acids, i.e. they are capable of accepting an electron pair from another molecule or atom (a Lewis base). Copper is a soft Lewis acid with a low charge state and can be coordinated by soft Lewis bases such as methionine and cysteine, which both have sulphur as a donor atom. Zinc is a hard Lewis acid with a high charge state and can also be coordinated by harder Lewis bases, which typically have donor atoms such as oxygen or nitrogen, such as aspartate and glutamate.

In the copper pump LpCopA, copper is likely coordinated by three sulphur atoms. These are present within two conserved cysteines (C) in M4 (in a CPC motif) and in a conserved methionine in M6 (in an MXXS motif). Coordination of Cu^{+} by these three ligands leads to a characteristic trigonal coordination geometry of the binding site (Mattle et al., 2015; Sitsel et al., 2015) (Fig. 3B). In the zinc pump ZntA, zinc is likely

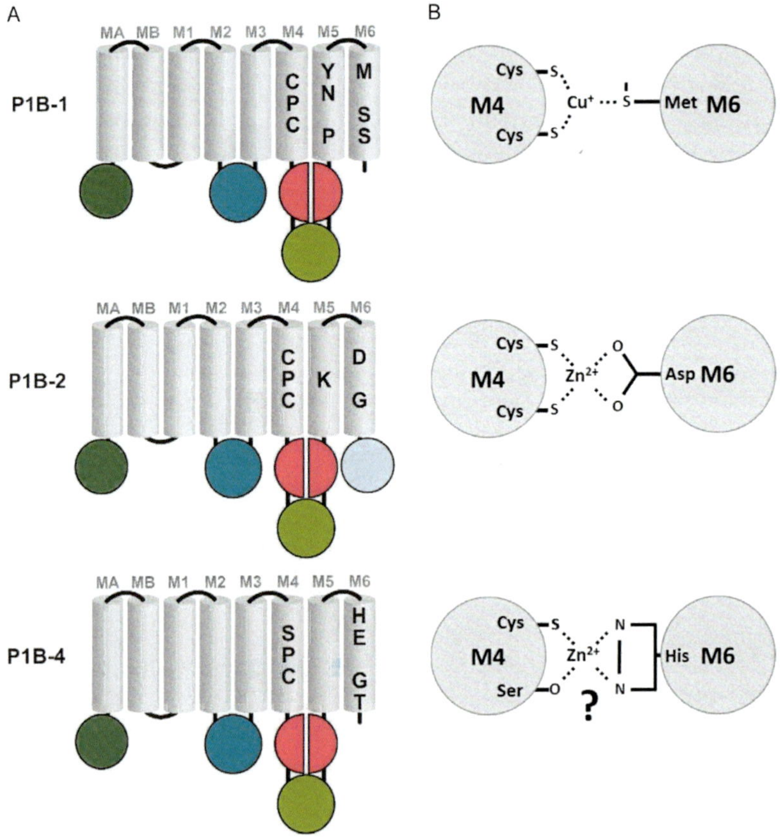

Fig. 3 Structural organization and signature sequences for metal coordination in P1B-1, P1B-2 and P1B-4 ATPases. P1B-1 ATPases pump copper, P1B-2 ATPases pump zinc, whereas P1B-4 ATPases have a broad metal binding specificity. Signature amino acid residues for metal coordination are present in transmembrane segments. P1B-1 ATPases contain signature motifs in M4 (CPC), M5 (YN[X]$_7$P) and M6 (MXXSS). P1B-2 ATPases show a CPC motif in M4, a conserved lysine in M5 and a DXG motif in M6. P1B-4 ATPases have an SPC motif in M4 and a HEXGT motif in M6. Transmembrane domains are shown in grey, the N-terminal metal binding domain (MBD) is shown in dark green, the A domain in blue, the P domain in purple, the N domain in light green and the C-terminal MBD in light blue (only present in P1B-2 ATPases). (B) Schematic overview of residues in M4 and M6 that are putatively involved in metal coordination. In P1B-1 ATPases, these are the two Cys residues of the CPC motif in M4, and the conserved Met of M6, which together coordinate Cu^{+} in a trigonal coordination geometry. In P1B-2 ATPases, these are the Cys residues in the CPC motif in M4 and the conserved Asp in M6, which together coordinate Zn^{2+} in a tetragonal coordination. In P1B-4, a possible metal binding mechanism involves the Cys residue of the SPC motif in M4, and possibly the backbone carbonyl oxygen of the Ser, and the conserved His of M6.

coordinated by two cysteines in M4 (in a CPC motif), but, in contrast to the copper pump, by an aspartate (D) in M6 (in a DxG motif) (Auld, 2001; Sitsel et al., 2015). The interaction with the aspartate involves both an oxygen and a nitrogen atom. Coordination of zinc by two sulphur atoms in M4 and two atoms in the aspartyl group in M6 leads to a characteristic tetrahedral coordination geometry for Zn^{2+} binding (Wang et al., 2014) (Fig. 3B). The fact that cysteines are required for heavy metal transport by P1B ATPases makes their biochemical studies difficult. Cysteines oxidize readily when not in a reducing environment, and for this reason many P1B ATPases are inactivated by oxidation when cells are disrupted (Leitenmaier et al., 2011). This easily becomes a problem when P1B ATPases are purified and assayed in vitro.

6. FUNCTION OF THE TERMINAL METAL BINDING DOMAINS

In addition to the core structure, most P1B ATPases contain N- or C-terminal soluble metal binding domains (MBDs). These are not included in the solved crystal structures as these domains were not ordered to allow for sufficient density to be visible. Common for the MBDs is that they are rich in histidine or cysteine residues. Copper transporting P1B ATPases have one or more N-terminal MBDs that each carry a CXXC motif in a predicted ferredoxin-like folding pattern (Banci et al., 2002; Jones, Daly, Cobine, Craik, & Dameron, 2003). The N-terminal MBDs of plant zinc transporting P1B ATPases have a similar ferredoxin fold. However, the exposed metal binding motif is CCXXE, rather than CXXC, thereby giving high affinity for zinc. These N-terminal zinc MBDs bind a single zinc atom (Eren, Gonzalez-Guerrero, Kaufman, & Arguello, 2007; Zimmermann et al., 2009).

Metallochaperones are involved in delivering copper to copper pumps. Metallochaperone proteins bind strongly to copper and secure that the free cytosolic concentrations of copper are kept at an extremely low level. One example is the copper metallochaperone Atx1 that interacts with P1B copper ATPases in fungi, animals and plants (Andres-Colas et al., 2006; Hamza, Schaefer, Klomp, & Gitlin, 1999; Huffman & O'Halloran, 2000; Li, Lacey, et al., 2017; Pufahl et al., 1997). Transfer of copper from metallochaperones to copper pumps has been observed both at the N terminal MBD (Singleton & Le Brun, 2007; Walker, Tsivkovskii, & Lutsenko, 2002) and the M domain (Gonzalez-Guerrero & Arguello, 2008). A potential docking site for ferredoxin-like copper chaperones is formed by TMB and TM1 at the putative entrance to the copper binding site in the M domain (Gourdon et al., 2011; Padilla-Benavides, McCann, & Arguello, 2013).

A major intracellular chelator of zinc in plants is nicotianamine (Clemens, Deinlein, Ahmadi, Horeth, & Uraguchi, 2013; Sinclair & Kramer, 2012) but it is not known whether nicotianamine interacts with zinc P1B ATPases.

Mutational deletions of the N-terminal MBD of AtHMA4 (Laurent et al., 2016; Verret et al., 2005), AtHMA2 (Wong, Jarvis, Sherson, & Cobbett, 2009) and HvHMA2 (Mills, Peaston, Runions, & Williams, 2012) show that transport activity is decreased when the N-terminal MBDs are removed, but that metal binding affinities of the pumps are retained, indicating that the MBD is not required for transport as such but rather plays a regulatory role. In zinc transporting plant P1B ATPases, extensive cytosolic C-terminal domains are present. These domains are highly variable both in length and in sequence but commonly enriched in either cysteine or histidine or sometimes both. HMA4 of *A. thaliana* has an extended C-terminal heavy metal binding domain containing 13 cysteine pairs and a terminal stretch of 11 histidines that serves as a high affinity chelator with capacity to bind 10 zinc ions per C terminus (Baekgaard et al., 2010). Sequential truncation of the C-terminal domain led to an increased ability to complement zinc sensitive yeast cells, indicating an autoregulatory role of the C-terminal domain (Baekgaard et al., 2010). Similarly, C-terminal truncation of the cadmium transporting P1B ATPase OsHMA3 from rice was demonstrated to increase cadmium tolerance (Kumagai et al., 2014). It is therefore possible that in zinc pumping P1B-2 ATPases with an extended C-terminal MBD, the MBD functions as a dual zinc sensor and autoinhibitor. One possibility is that upon zinc binding the sensor changes structure and in this way an inhibitory interaction with the core protein is neutralized (Baekgaard et al., 2010).

7. CLASSIFICATION OF P1B ATPases

Based on the phylogeny of P1B-ATPases, their substrate specificities and sequence signature motifs in their M domains, these pumps have been divided into seven subgroups (called P1B-1–P1B-7) (Arguello, 2003; Arguello, Eren, & Gonzalez-Guerrero, 2007; Arguello, Gonzalez-Guerrero, & Raimunda, 2011; Smith, Smith, & Rosenzweig, 2014). Specific signature sequences present in the last three TM domains reflect the substrate specificities of the pumps in each subgroup. Four of these subgroups (P1B-3, P1B-5, P1B-6 and P1B-7) are only found in prokaryotes. P1B-1 is the only subgroup found in animals, whereas the P1B-1, P1B-2 and P1B-4 subgroups are represented in plants. The overall secondary structure of P1B-1, P1B-2 and P1B-4 ATPases is shown in Fig. 3A. P1B-1 ATPases transport copper and are characterized by a CPC motif in M4, a $YN(X)_4P$ motif in M5 and a MXXSS motif in M6 (where 'X' can be any residue) (Fig. 3A). P1B-2 ATPases transport zinc and its toxic analogue cadmium and have a CPC motif in M4, a conserved K in M5 and a DXG motif in M6 (Fig. 3A). P1B-4 ATPases have a broad transport specificity (i.e. they can transport both Zinc and copper as well as other heavy metals) and are characterized by a SPC motif in M4 and a HEGT motif in M6 (Fig. 3A). Below we will restrict our discussion to these three subgroups, as they are the only ones found in plants.

8. THE ORIGIN OF P1B ATPases IN PLANTS

To present an overview of the repertoire and evolution of P1B ATPases present in plants, organisms that represent stages of plant evolution and have had their genomes fully sequenced were first selected. Plants will here be used as a synonym to Chloroplastida (Adl et al., 2012), i.e. organisms with primary chloroplasts with green chlorophyll. This supergroup is divided to Chlorophyta (green algae) and Charophyta (including all land plants). Then, sequences with significant similarity (BLAST E-score $<1e^{-20}$) to the *A. thaliana* HMA2 (a P1B-2 ATPase) were identified, and all sequences without a cysteine in M4 were discarded. However, variations in all other signature motifs were accepted. The sequences were aligned and subsequently subjected to phylogenetic analyses by two independent methods: Bayesian inference (where the maximal posterior probability is 1) and Maximum likelihood (where the maximal bootstrap value is 100). Posterior probabilities <0.95 and bootstrap values of $<60\%$ are generally considered non-support.

The tree, comprising a total of 203 sequences, is divided into two clades (marked by red and blue outer rings, respectively) (Fig. 4). At their base, each clade receives maximal statistical support (posterior probability = 1, bootstrap support = 100%) (Fig. 4). Strikingly, all sequences represented in the first clade (marked by red outer circle) have a CPC motif in M4 and a MXXSS motif in M6, and thus represents plant P1B-1 ATPases. Based on the signature motifs and from what we know from the structure, we would expect these pumps to have a ligand-binding site with trigonal coordination geometry that fits Cu^+ (Fig. 3B). In fact, proteins in the clade that have been characterized at the molecular level do indeed pump

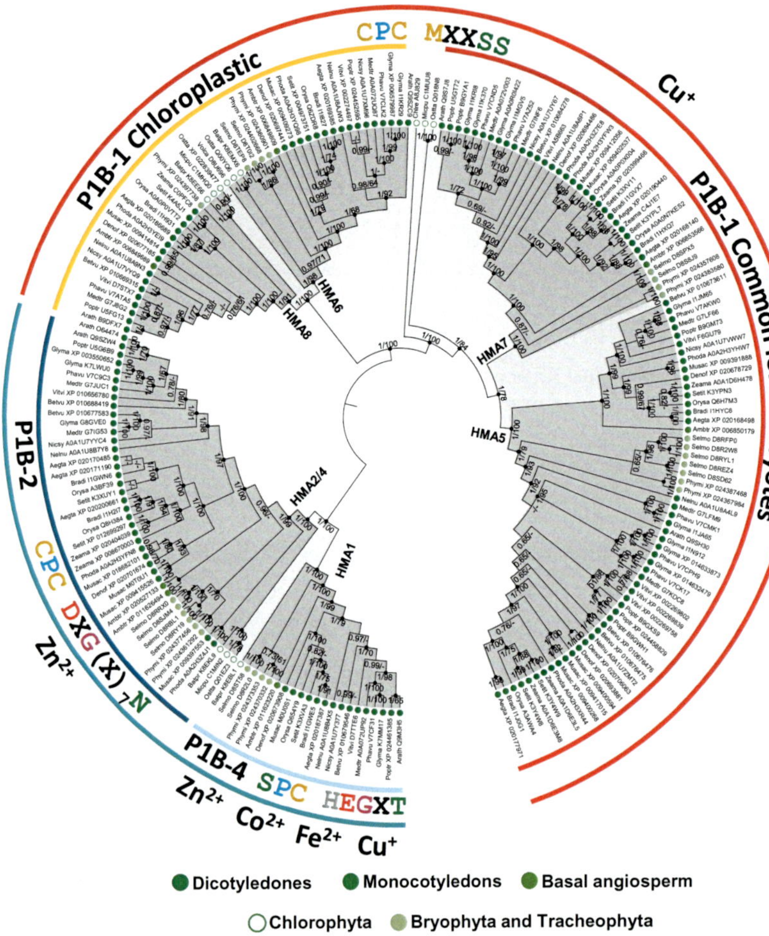

Fig. 4 Phylogenetic tree of heavy metal ATPases of P1B-1, P1B-2 and P1B-4 subgroups from chlorophyceae. The phylogenetic analysis of 203 P1B-1, P1B-2 and P1B-4 subgroups from chlorophyceae first shows the clear clades formed by copper and zinc pumps, indicated by red and blue outer rings, respectively. The analysis also shows the P1B-4 and P1B-2 subgroups within the Zinc transporting clade and the chloroplastic and original P1B-1 pumps within the copper transporting clade. The P1B-1 clade corresponds to all sequences in the analysis containing the CPC motif in M4 and MXXS motif in M6. The P1B-2 clade corresponds to all sequences containing CPC in M6 and DXG(X)$_7$N signature sequence, and the P1B-4 clade corresponds to all sequences containing the SPC motif in M4 and the HEGXT motif in M6. Sequences within the clusters divide into monocot, dicot, basal angiosperm, bryophyte, tracheophyte, and Chlorophyta clades. In the figure, sequences are marked with the following colours: Dicotyledons, dark green; Monocotyledons, medium green; Basal angiosperm, light green; Chlorophyta, light green circle; Bryophytae and Tracheophytae, bright green. The outermost half-rings mark

(Continued)

Cu^+ (Catty et al., 2011; Deng et al., 2013; Hirayama et al., 1999; Kobayashi et al., 2008; Sautron et al., 2015). The clade is divided into four major subclades (shaded in dark grey) that have been named at their base according to a representative HMA protein from *A. thaliana*. The HMA6 and HMA8 clades each receive maximal statistical support and both HMA6 and HMA8 localize to chloroplasts (labeled 'P1B-1 Chloroplastic' in Fig. 4). The HMA6 and HMA8 subclades both include sequences from green algae (Chlorophyta) and land plants (Fig. 4) suggesting that the split occurred before the time when lands plants diverged from green algae. It has been suggested that

Fig. 4—cont'd copper transporting sequences (in red), and zinc and other substrates (in blue). The inner ring marks chloroplastic copper P1B-1 ATPases (in orange), vertically transferred copper P1B-1 ATPases (marked 'P1B-1 Common for eukaryotes') (in *dark red*), P1B-4 ATPases (*light blue*) and P1B-2 ATPases (in dark blue). Branch points for copper and zinc transporting P1B ATPases under which all sequences of the subtype are contained are marked in light grey. The branch point after which homologues of a single *A. thaliana* pump is contained is marked in dark grey. For all branch points concurrent in Bayesian inference and maximum likelihood phylogenetic analysis, the score is noted as Bayesian score/maximum likelihood score, scores below Bayesian scores of 0.6 and maximum likelihood scores of 60 are omitted and replaced by a dash. The best score is 1 for Bayesian analysis and 100 for maximum likelihood analysis. Black circles designate nodes with a maximal Bayesian inference score of 1. Species abbreviations are: Arath, *Arabidopsis thaliana*; Ambtr, *Amborella trichopoda*; Aegta, *Aegilops tauschii*; Bradi, *Brachypodium distachyon*; Batpr, *Bathycoccus prasinos*; Betvu, *Beta vulgaris*; Chlre, *Chlamydomonas reinhardtii*; Denof, *Dendrobium officinale*; Glyma, *Glycine max*; Micpu, *Micromonas pusilla*; Medtr, *Medicago truncatula*; Musac, *Musa acuminata*; Nelnu, *Nelumbo nucifera*; Nicsy, *Nicotiana sylvestris*; Orysa, *Oryza sativa*; Ostta, *Ostreococcus tauri*; Phoda, *Phoenix dactylifera*; Poptr, *Populus trichocarpa*; Phymi, *Physcomitrella patens*; Phavu, *Phaseolus vulgaris*; Setit, *Setaria italica*; Selmo, *Selaginella moellendorffii*; Volca, *Volvox carteri*; Vitvi, *Vitis vinifera*; Zeama, *Zea mays*. Sequences were collected from the KEGG genome database (Kanehisa & Goto, 2000) from January to May 2018 using BLASTP with standard settings against the protein sequence of *A. thaliana* HMA2 (uniprot accession Q9SZW4), collecting all sequences with a return BLAST E-score below $1e^{-20}$. Collected sequences were aligned using the MUSCLE algorithm (Edgar, 2004) in MEGA6 (Tamura, Stecher, Peterson, Filipski, & Kumar, 2013). Sequences were selected for phylogenetic analysis based on the presence of a full DKTGT motif, preserved GE in the TGES motif, preserved transmembrane domains M1–M6 and the presence of metal coordinating residues in M4 and M6. Phylogenetic analysis was performed using Bayesian inference and maximum likelihood. Bayesian inference was performed using MrBayes 3.2.6 (Ronquist & Huelsenbeck, 2003) with the following settings: eight chains of Markov chain Monte Carlo iterations and a heated parameter of 0.05, trees being sampled every 100 generations. Average standard deviations of split frequencies at ended analysis after 1,000,000 generations were 0.012017. Maximum likelihood analysis was performed using RAxML 8.2.4 (Stamatakis, 2014) and running 1000 bootstraps using the LG+G substitution model.

HMA6 and HMA8 originated from a duplication of the cyanobacterial gene *CtaA*, which is expressed as a protein in thylakoids (Hanikenne & Baurain, 2013).

In contrast to this long history, the two other clades of copper pumps (named 'HMA5' and 'HMA7' in Fig. 4) seem to have separated in land plants, as they have no members from green algae but have a green algal clade at their common base. Based on an extensive phylogenetic analysis that included sequences from all domains of life, it was concluded that members of the chloroplastic clades evolved from cyanobacterial sequences, that were transferred to the nuclear genome by endosymbiotic gene transfer, whereas those of the other group (marked 'P1B-1 Common for eukaryotes') evolved by vertical gene transfer from eubacteria (Hanikenne & Baurain, 2013).

The second clade marked by blue outer circle is at first glance more heterogeneous, as it is not united by single signature motifs. However, what is common for all sequences represented in this clade is that they have a negatively charged residue in M6. The clade is divided into two subclades, each of which receives maximal statistical support. The first subclade (marked 'HMA2/4' at its base) represents plant P1B-2 ATPases, which are characterized by an aspartate (D; negatively charged) in M6. The second subclade (marked 'HMA1' at its base) represents plant P1B-4 ATPases, characterized by a glutamate (E; negatively charged) in M6 that is preceded by a histidine (H; not charged under physiological conditions). Importantly, when the sequences are aligned, the aspartate in P1B-2 ATPases aligns with the histidine in P1B-4 ATPases and not with the glutamate. In plant P1B-2 ATPases, the aspartate is at the same position as the M6 aspartate in the ZntA zinc pump, and is predicted to contribute to coordination of the bound metal (Wang et al., 2014). In analogy, the histidine in plant P1B-4 ATPase would be predicted to contribute to metal coordination and the glutamate side chain would not be part of the binding site but would only contribute with negative charge. The mutation of the M6 histidine to an alanine has indeed been demonstrated to severely diminish metal binding to the transport site (Smith, Ross, Hoffman, & Rosenzweig, 2017). In addition, all plant P1B-2 ATPases have a CPC motif in M4 whereas P1B-4 ATPases have a SPC motif at the same position, with the exception of a single microalgae sequence, *Bathycoccus prasinos* K8EBL1, which has an APC motif. These differences probably reflect the different metal ion specificities of the two subgroups: Those pumps from the P1B-2 group that have been characterized at the molecular level pump zinc (and the related toxic analogue cadmium) whereas those of the P1B-4 group have a broader metal ion specificity and have been implicated in the pumping of zinc, cadmium, cobalt, iron and copper.

Both the P1B-2 and P1B-4 subgroups include sequences from green algae and therefore must have an early origin (Fig. 4). The lack of division of the P1B-2 pumps into separate HMA2, HMA3 and HMA4 pumps is striking. These divide within or close to the furthest branch points within the P1B-2 cluster, indicating that duplications and subfunctionalizations of a single ancestral chlorophyceae P1B-2 pump must have occurred repeatedly through the evolution of land plants (Zorrig, Abdelly, & Berthomieu, 2011). Moreover, clustering of similar pumps from different species is evident within both the monocot and the dicot subclades indicating that duplication and subfunctionalization have occurred after the divergence of dicots from monocots. A similar pattern is notable for the subclustering of species in the HMA5 and HMA7 subclades of the P1B-1 sequences. These are likely to represent duplications and subfunctionalizations as well. An early gene duplication event has divided the HMA5 clade into the two distinct and highly supported subclades and most land plant species have retained both copies, an exception being Brassicaceae where one of the genes have been lost (Li, Iqbal, et al., 2017).

Taken together, the phylogenetic division of plant P1B ATPases in two major clades corresponds with their transport function and is associated with conserved amino acid residues in the transmembrane segments involved in substrate coordination (TMs 4 and 6). The analyses indicates that the different copper transporting P1B-1 ATPases (represented by HMA5, HMA6, HMA7 and HMA8) share a common origin whereas two subgroups that transport zinc (P1B-2 represented by HMA2/4 and P1B-4 represented by HMA1) share a common origin.

9. THE ORIGIN OF P1B-2 ATPases IN PLANTS

In order to look further into the origin of the plant zinc pumps, a phylogenetic analysis was generated that included P1B-2 and P1B-4 sequences from not only plants, but also from other eukaryotic supergroups and from prokaryotes. To be certain that all relevant eukaryotic sequences were considered, exhaustive searches were carried out in the Kyoto Encyclopedia of Genes and Genomes (KEGG) and National Center for Biotechnology Information (NCBI) databases and all statistically significant hits were inspected manually.

In the resulting tree (Fig. 5), two clades were evident and these corresponded to P1B-2 and P1B-4 ATPases, respectively. The P1B-2 clade further contains a number of eubacterial subclades, all of them basal to a

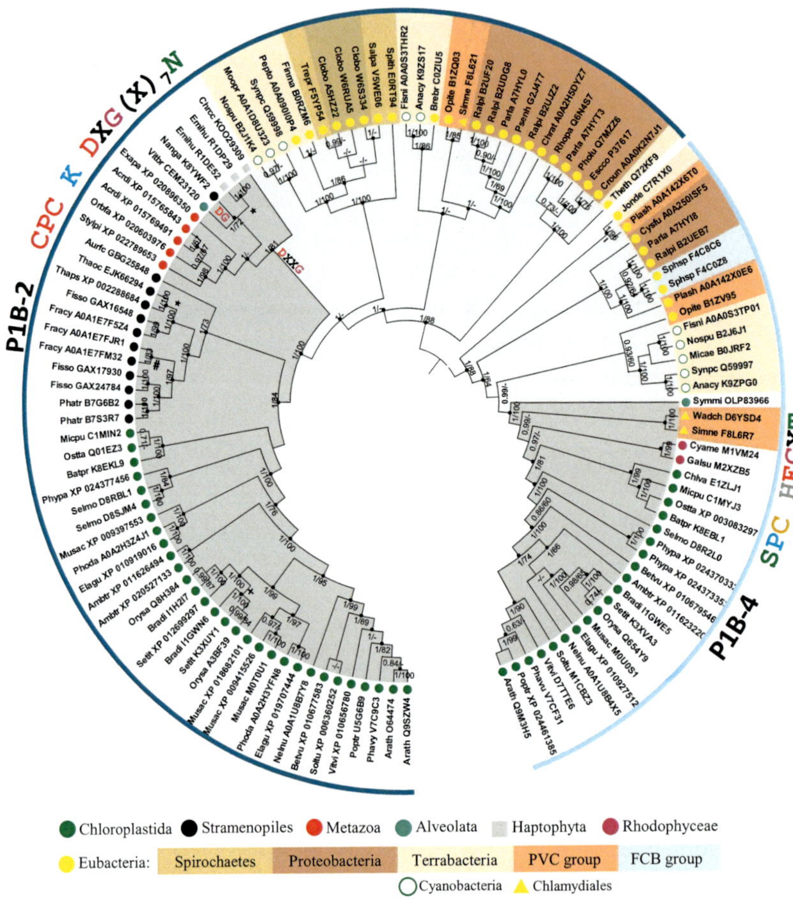

Fig. 5 Phylogenetic tree of heavy metal ATPases of the P1B-2 and P1B-4 subgroups from bacteria and eukaryotes. A phylogenetic analysis of 118 P1B ATPases from eubacteria and eukaryotes with either an Asp or His residue in M6 results in two clades each characterized its members having signature motifs of P1B-2 or P1B-4 ATPases, respectively. For P1B-2 ATPases these are the Cys-Pro-Cys (CPC) motif in transmembrane domain (TM) 4, a conserved Lys (K) in M5 and a conserved Asp, Gly and Asp (DXG(X)$_7$N in M6. For the P1B-4 ATPases these are the Ser-Pro-Cys (SPC) motif in M4 and the conserved His-Glu-Gly-X-Thr (HEGXT) motif in M6. Colored dots mark taxonomic origin: Chloroplastida, green; Stramenopiles, black; Metazoa, red; Alveolata, blue; Haptophyta, grey; Rhodophyceae, purple; Cyanobacteria, green open circle; Chlamydiales, yellow triangle; Other eubacteria, yellow circle. The outer ring marks P1B-2 ATPases (dark blue) and P1B-4 ATPases (light blue). Major clades are colored grey with eukaryotic subclades colored dark grey. Statistical support is marked at nodes as Bayesian inference score/maximum likelihood score. The best score is 1 for Bayesian analysis and 100 for maximum likelihood analysis. Black circles designate nodes with a maximal Bayesian inference score of 1. Deviations from conserved sequence motifs are marked as

(Continued)

single clade representing eukaryotic sequences. Likewise, the P1B-4 clade contains a number of bacterial subclades, which are basic to a single clade with eukaryotic sequences. This would suggest that P1B-2 and P1B-4 ATPases each have a eubacterial origin and the two subgroups separated very early in the evolution of eubacteria.

Fig. 5—cont'd follows: *, the M5 conserved K is replaced with an N; #, The M6 motif is DXXG; +, the M6 conserved N is not present; −, the M6 HEGXT is replaced by HEXXE. Abbreviations are Ambtr, *Amborella trichopoda*; Arath, *Arabidopsis thaliana*; Batpr, *Bathycoccus prasinos*; Betvu, *Beta vulgaris*; Bradi, *Brachypodium distachyon*; Chlva, *Chlorella variabilis*; Metazoa: Acrdi, *Acropora digitifera*; Cyanobacteria: Anacy, *Anabaena cylindrical*; Aurfc, *Aurantiochytrium* sp. *FCC1311*; Brebr, *Brevibacillus brevis*; Chrat, *Chromobacterium* sp. *ATCC53434*; Chrcc, *Chrysochromulina* sp. *CCMP291*; Clobo, *Clostridium botulinum*; Croun, *Cronobacter universalis*; Cyame, *Cyanidioschyzon merolae*; Cysfu, *Cystobacter fuscus*; Elagu, *Elaeis guineensis*; Emihu, *Emiliania huxleyi*; Escco, *Escherichia coli*; Exapa, *Exaiptasia pallida*; Finma, *Finegoldia magna*; Fisni, *Fischerella* sp. *NIES-3754*; Fisso, *Fistulifera solaris*; Fracy, *Fragilariopsis cylindrus*; Galsu, *Galdieria sulphuraria*; Jonde, *Jonesia denitrificans*; Micae, *Microcystis aeruginosa*; Micpu, *Micromonas pusilla*; Moopr, *Moorea producens*; Musac, *Musa acuminata*; Nanga, *Nannochloropsis gaditana*; Nelnu, *Nelumbo nucifera*; Nospu, *Nostoc punctiforme*; Opite, *Opitutus terrae*; Orbfa, *Orbicella faveolata*; Orysa, *Oryza sativa*; Ostta, *Ostreococcus tauri*; Parla, *Parvibaculum lavamentivorans*; Pepto, *Peptoniphilus* sp. *ING2-D1G*; Phatr, *Phaeodactylum tricornutum*; Phavu, *Phaseolus vulgaris*; Phoda, *Phoenix dactylifera*; Pholu, *Photorhabdus luminescens*; Phypa, *Physcomitrella patens*; Plash, *Planctomyces* sp. *SH-PL14*; Poptr, *Populus trichocarpa*; Psenh, *Pseudogulbenkiania* sp. *NH8B*; Ralpi, *Ralstonia pickettii*; Rhope, *Rhodopseudomonas palustris*; Salpa, *Salinispira pacifica*; Selmo, *Selaginella moellendorffii*; Setit, *Setaria italica*; Simne, *Simkania negevensis*; Soltu, *Solanum tuberosum; Sphsp, Sphingobacterium* sp. *21*; Spith, *Spirochaeta thermophile*; Stylpi, *Stylophora pistillata*; Symmi, *Symbiodinium microadriaticum*; Synpc, *Synechocystis* sp. *PCC6803*; Thaoc, *Thalassiosira oceanica*; Thaps, *Thalassiosira pseudonana*; Theth, *Thermus thermophilus*; Trepr, *Treponema primitia*; Vitbr, *Vitrella brassicaformis*; Vitvi, *Vitis vinifera*; Wadch, *Waddlia chondrophila*. The phylogenetic analysis was carried out essentially as described in the legend to Fig. 4. In addition to retrieving predicted protein sequences from the genomes of selected green plants, BLAST searches were performed against all eukaryotic organisms (but excluding Chloroplastida) in the KEGG and the NCBI nonredundant protein sequence databases. All hits with a BLAST E-score below $1e^{-25}$ were collected and selected for phylogenetic analysis. To discern genome contamination in metazoan sequences, all sequences returning >995 bacterial sequences of a total of 1000 in a standard BLASTP search were discarded from further analysis. Phylogenetic analysis was performed using Bayesian inference and maximum likelihood. Bayesian inference was performed using MrBayes 3.2.6 (Ronquist & Huelsenbeck, 2003) with the following settings: eight chains of Markov chain Monte Carlo iterations and a heated parameter of 0.05, trees being sampled every 100 generations. Average standard deviations of split frequencies at ended analysis after 1,000,000 generations were 0.003813. Maximum likelihood analysis was performed using RAxML 8.2.4 (Stamatakis, 2014) and running 1000 bootstraps using the LG+G substitution model.

In eukaryotes, P1B-4 sequences were identified in all investigated green plants (i.e. Chloroplastida; land plants and green algae) but also in red algae and in the genome of a photosynthetic diatom belonging to the supergroup Alveolata.

A striking feature of the otherwise eukaryotic P1B-4 clade is that two eubacterial sequences, both belonging to the group of Chlamydiales, branch off within the otherwise eukaryotic clade (Fig. 5). This has been observed before and it was proposed that eukaryotic P1B-4 ATPases originate from a horizontal gene transfer from Chlamydiales (Baum, 2013; Hanikenne & Baurain, 2013; Huang & Gogarten, 2007; Moustafa, Reyes-Prieto, & Bhattacharya, 2008). Supporting this notion is a close phylogenetic relation between other plastidic proteins in plants to Chlamydiales. If this is true, this event must have happened very early in eukaryotic evolution and most likely in a photosynthetic organism, as all members of the clade are photosynthetic. The chlamydial branch is situated close to the base of the eukaryotic clade but the statistical support for this position is low. If a correct placement is before the appearance of the eukaryotic clade, vertical 'ordinary' gene transfer is a possibility, but signs of a direct relationship between Chlamydiae and photosynthetic eukaryotes have never been observed. Another explanation to be considered is that horizontal gene transfer was indeed involved but rather the opposite way, i.e. from Clamydiae to eukaryotes. Horizontal gene transfers from eukaryotes to prokaryotes is well documented but rare (Alsmark et al., 2013; Guljamow et al., 2007; Jenkins et al., 2002). All characterized eukaryotic P1B-4 ATPases localize to chloroplasts. Thus, an alternative origin of P1B-4 ATPases could be via endosymbiotic gene transfer from Cyanobacteria. Indeed, sequences from Cyanobacteria occur in the eubacterial subclade that is closest to the base of the eukaryotic subclade (Fig. 4), which is in accordance with the model but still far from proving it.

Eukaryotic P1B-2 ATPases include sequences from all investigated members of Chloroplastida but also from photosynthetic microalgae belonging to the supergroup Stramenopiles, to photosynthetic phytoplankton belonging to the supergroup Haptophyta and to a photosynthetic alga belonging to the supergroup Alveolata.

However, P1B-2 ATPase like sequences in eukaryotes are not restricted to photosynthetic organisms. Thus, within a subclade of otherwise photosynthetic algae from Stramenopiles, Alveolata and Haptophyta, a cluster of animal sequences is present (Fig. 5). All of these sequences originate from a single class, the cnidarians Anthozoa, which are marine invertebrates many of which form corals. Many Anthozoa possess polyps that are filled with photosynthetic zooxanthellae, which are single-celled dinoflagellates that live in symbiosis with the coral. Up to half of Cnidarian species exist in symbiotic relationships with algae, e.g. dinoflagellates belonging to Alveolata (Douglas, Mcauley, & Davies, 1993). Cnidaria have been demonstrated to have acquired a relatively large proportion of their genetic material from horizontal gene transfer (Bhattacharya et al., 2016) and have been reported previously to have acquired genes from protists (Andersson, 2005). That the anthozoan P1B-2 sequences originate from protist algae, and not from any other group, is supported by the observation that sequences in the cluster including anthozoan and algal sequences contain a DXXG motif in M6, where all other P1B-2 ATPases have a DXG motif at this position (Fig. 4). Thus, it seems likely that animal P1B-2 sequences did evolve by horizontal gene transfer from algae.

The prokaryotic source of P1B-2 ATPases in photosynthetic eukaryotes is more uncertain. In plants, P1B-2 ATPases are never found in chloroplasts but rather in plasma membranes and vacuolar membranes, where they serve to extrude heavy metals from cells. Thus, a cyanobacterial origin is not evident. In the tree, Cyanobacteria are present in the eubacterial subclade basal to the eukaryotic clade, but they show no closer relation than does members of, e.g. Spirochaetes that are also present in this subclade. Thus, it is neither contradicted nor proven that eukaryotic P1B-2 ATPases originate from Cyanobacteria. It is striking however, that, with the exception of the Anthozoa, which likely represents a horizontal gene transfer event, only bikont, photosynthetic organisms possess P1B-2 ATPases within the eukarya. It is likely that these were either obtained at a point after bikonts and dikonts diverged, or that an ancient eukaryotic P1B-2 ATPase was lost in an early ancestor to the unikont clade.

Taken together, zinc pumps are widespread in prokaryotes and photosynthetic eukaryotes where they contribute to maintaining ion homeostasis. The origin of zinc pumps in diverse phylae appears very complex and may have involved lateral transfer of genes from prokaryotes to eukaryotes, from eukaryotes to prokaryotes and from eukaryotes to eukaryotes. In photosynthetic eukaryotes, it is possible that zinc pumps originated as the result of endosymbiotic gene transfer from cyanobacteria but at present, this model has neither been proven nor discarded. Chlamydiales, a group of eubacteria, may have obtained their zinc pumps by horizontal gene transfer from eukaryotic cells, but this possibility is also uncertain. Among animals, only marine invertebrates have zinc pumps, and these may originate from a photosynthesizing eukaryote. Other animals might not have zinc pumps because they had no need for them, but rather because they never had the chance to obtain them, and therefore had to evolve other strategies for maintaining cellular zinc homeostasis.

10. FUTURE PERSPECTIVES

The publication of structures of a copper and a zinc ATPase, both from eubacteria, has greatly stimulated the research on P1B ATPases. However, as the published structures are of proteins in the low-affinity E2 state with no ions bound, substrate binding has been inferred by modeling. Structures of P1B ATPases in the E1 state would shed further light on the binding site coordination of the metal substrates. In addition, structures of eukaryotic pumps could provide insights into the level of structure conservation across the tree of life, and would further the understanding of how plant P1B type ATPases function and are regulated. Terminal MBDs of P1B ATPases are not sufficiently ordered to be observed in crystal structures, but a promising alternative approach to obtain such structural information comes with the emergence of high-resolution cryo-electron microscopy of membrane proteins. Measuring the in vitro activity of heavy metal ATPases, especially that of P1B-2 ATPases, has proven challenging and robust assays capable of coping with both the low turnover, the high affinity and the oxidation sensitivity of essential cysteines should be optimized for each type of pump. Insights into the origin of the plant P1B ATPases are accumulating as the library of sequenced genomes increases. However, the picture is still not clear regarding the origin of plant P1B-2 and P1B-4 ATPases.

REFERENCES

Abdel-Ghany, S. E., Muller-Moule, P., Niyogi, K. K., Pilon, M., & Shikanai, T. (2005). Two P-type ATPases are required for copper delivery in Arabidopsis thaliana chloroplasts. *Plant Cell*, *17*(4), 1233–1251. https://doi.org/10.1105/tpc.104.030452.

Adl, S. M., Simpson, A. G., Lane, C. E., Lukes, J., Bass, D., Bowser, S. S., et al. (2012). The revised classification of eukaryotes. *The Journal of Eukaryotic Microbiology*, *59*(5), 429–493. https://doi.org/10.1111/j.1550-7408.2012.00644.x.

Aguirre, G., & Pilon, M. (2015). Copper delivery to chloroplast proteins and its regulation. *Frontiers in Plant Science*, *6*, 1250. https://doi.org/10.3389/fpls.2015.01250.

Alsmark, C., Foster, P. G., Sicheritz-Ponten, T., Nakjang, S., Martin Embley, T., & Hirt, R. P. (2013). Patterns of prokaryotic lateral gene transfers affecting parasitic microbial eukaryotes. *Genome Biology*, *14*(2), R19. https://doi.org/10.1186/gb-2013-14-2-r19.

Andersson, J. O. (2005). Lateral gene transfer in eukaryotes. *Cellular and Molecular Life Sciences*, *62*(11), 1182–1197. https://doi.org/10.1007/s00018-005-4539-z.

Andres-Colas, N., Sancenon, V., Rodriguez-Navarro, S., Mayo, S., Thiele, D. J., Ecker, J. R., et al. (2006). The Arabidopsis heavy metal P-type ATPase HMA5 interacts with metallochaperones and functions in copper detoxification of roots. *The Plant Journal, 45*(2), 225–236. https://doi.org/10.1111/j.1365-313X.2005.02601.x.
Andresen, E., Peiter, E., & Kupper, H. (2018). Trace metal metabolism in plants. *Journal of Experimental Botany, 69*(5), 909–954. https://doi.org/10.1093/jxb/erx465.
Appelhagen, I., Nordholt, N., Seidel, T., Spelt, K., Koes, R., Quattrochio, F., et al. (2015). TRANSPARENT TESTA 13 is a tonoplast P3A-ATPase required for vacuolar deposition of proanthocyanidins in Arabidopsis thaliana seeds. *The Plant Journal, 82*(5), 840–849. https://doi.org/10.1111/tpj.12854.
Arguello, J. M. (2003). Identification of ion-selectivity determinants in heavy-metal transport P1B-type ATPases. *The Journal of Membrane Biology, 195*(2), 93–108.
Arguello, J. M., Eren, E., & Gonzalez-Guerrero, M. (2007). The structure and function of heavy metal transport P1B-ATPases. *Biometals, 20*(3–4), 233–248. https://doi.org/10.1007/s10534-006-9055-6.
Arguello, J. M., Gonzalez-Guerrero, M., & Raimunda, D. (2011). Bacterial transition metal P(1B)-ATPases: Transport mechanism and roles in virulence. *Biochemistry, 50*(46), 9940–9949. https://doi.org/10.1021/bi201418k.
Auld, D. S. (2001). Zinc coordination sphere in biochemical zinc sites. *Biometals, 14*(3–4), 271–313. https://doi.org/10.1023/A:1012976615056.
Axelsen, K. B., & Palmgren, M. G. (1998). Evolution of substrate specificities in the P-type ATPase superfamily. *Journal of Molecular Evolution, 46*(1), 84–101.
Axelsen, K. B., & Palmgren, M. G. (2001). Inventory of the superfamily of P-type ion pumps in Arabidopsis. *Plant Physiology, 126*(2), 696–706.
Baekgaard, L., Mikkelsen, M. D., Sorensen, D. M., Hegelund, J. N., Persson, D. P., Mills, R. F., et al. (2010). A combined zinc/cadmium sensor and zinc/cadmium export regulator in a heavy metal pump. *The Journal of Biological Chemistry, 285*(41), 31243–31252. https://doi.org/10.1074/jbc.M110.111260.
Banci, L., Bertini, I., Ciofi-Baffoni, S., Finney, L. A., Outten, C. E., & O'Halloran, T. V. (2002). A new zinc-protein coordination site in intracellular metal trafficking: Solution structure of the Apo and Zn(II) forms of ZntA(46-118). *Journal of Molecular Biology, 323*(5), 883–897.
Baum, D. (2013). The origin of primary plastids: A pas de deux or a menage a trois? *Plant Cell, 25*(1), 4–6. https://doi.org/10.1105/tpc.113.109496.
Baxter, I., Tchieu, J., Sussman, M. R., Boutry, M., Palmgren, M. G., Gribskov, M., et al. (2003). Genomic comparison of P-type ATPase ion pumps in Arabidopsis and rice. *Plant Physiology, 132*(2), 618–628. https://doi.org/10.1104/pp.103.021923.
Bertini, I., Luchinat, C., & Monnanni, R. (1985). Zinc enzymes. *Journal of Chemical Education, 62*(11), 924. https://doi.org/10.1021/ed062p924.
Bhattacharya, D., Agrawal, S., Aranda, M., Baumgarten, S., Belcaid, M., Drake, J. L., et al. (2016). Comparative genomics explains the evolutionary success of reef-forming corals. *eLife, 5*, e13288. https://doi.org/10.7554/eLife.13288.
Bonza, M. C., Morandini, P., Luoni, L., Geisler, M., Palmgren, M. G., & De Michelis, M. I. (2000). At-ACA8 encodes a plasma membrane-localized calcium-ATPase of Arabidopsis with a calmodulin-binding domain at the N terminus. *Plant Physiology, 123*(4), 1495–1506.
Boutigny, S., Sautron, E., Finazzi, G., Rivasseau, C., Frelet-Barrand, A., Pilon, M., et al. (2014). HMA1 and PAA1, two chloroplast-envelope PIB-ATPases, play distinct roles in chloroplast copper homeostasis. *Journal of Experimental Botany, 65*(6), 1529–1540. https://doi.org/10.1093/jxb/eru020.
Burkhead, J. L., Reynolds, K. A., Abdel-Ghany, S. E., Cohu, C. M., & Pilon, M. (2009). Copper homeostasis. *The New Phytologist, 182*(4), 799–816. https://doi.org/10.1111/j.1469-8137.2009.02846.x.

Catty, P., Boutigny, S., Miras, R., Joyard, J., Rolland, N., & Seigneurin-Berny, D. (2011). Biochemical characterization of AtHMA6/PAA1, a chloroplast envelope Cu(I)-ATPase. *The Journal of Biological Chemistry*, *286*(42), 36188–36197. https://doi.org/10.1074/jbc.M111.241034.

Clemens, S., Deinlein, U., Ahmadi, H., Horeth, S., & Uraguchi, S. (2013). Nicotianamine is a major player in plant Zn homeostasis. *Biometals*, *26*(4), 623–632. https://doi.org/10.1007/s10534-013-9643-1.

Clemens, S., Palmgren, M. G., & Kramer, U. (2002). A long way ahead: Understanding and engineering plant metal accumulation. *Trends in Plant Science*, 7(7), 309–315.

Demmig, B., & Gimmler, H. (1983). Properties of the isolated intact chloroplast at cytoplasmic K concentrations: I. Light-induced cation uptake into intact chloroplasts is driven by an electrical potential difference. *Plant Physiology*, *73*(1), 169–174.

Deng, F., Yamaji, N., Xia, J., & Ma, J. F. (2013). A member of the heavy metal P-type ATPase OsHMA5 is involved in xylem loading of copper in rice. *Plant Physiology*, *163*(3), 1353–1362. https://doi.org/10.1104/pp.113.226225.

DeWitt, N. D., Hong, B., Sussman, M. R., & Harper, J. F. (1996). Targeting of two Arabidopsis H^+-ATPase isoforms to the plasma membrane. *Plant Physiology*, *112*(2), 833–844. https://doi.org/10.1104/pp.112.2.833.

Douglas, A. E., Mcauley, P. J., & Davies, P. S. (1993). Algal symbiosis in cnidaria. *Journal of Zoology*, *231*(1), 175–178. https://doi.org/10.1111/j.1469-7998.1993.tb05362.x.

Dunkley, T. P., Hester, S., Shadforth, I. P., Runions, J., Weimar, T., Hanton, S. L., et al. (2006). Mapping the Arabidopsis organelle proteome. *Proceedings of the National Academy of Sciences of the United States of America*, *103*(17), 6518–6523. https://doi.org/10.1073/pnas.0506958103.

Edgar, R. C. (2004). MUSCLE: A multiple sequence alignment method with reduced time and space complexity. *BMC Bioinformatics*, *5*, 113. https://doi.org/10.1186/1471-2105-5-113.

Eren, E., Gonzalez-Guerrero, M., Kaufman, B. M., & Arguello, J. M. (2007). Novel Zn^{2+} coordination by the regulatory N-terminus metal binding domain of Arabidopsis thaliana Zn^{2+}-ATPase HMA2. *Biochemistry*, *46*(26), 7754–7764. https://doi.org/10.1021/bi7001345.

Evans, D. E. (1994). Calmodulin-stimulated calcium pumping ATPases located at higher plant intracellular membranes: A significant divergence from other eukaryotes? *Physiologia Plantarum*, *90*(2), 420–426. https://doi.org/10.1111/j.1399-3054.1994.tb00408.x.

Garcia, L., Welchen, E., & Gonzalez, D. H. (2014). Mitochondria and copper homeostasis in plants. *Mitochondrion*, *19*(Pt. B), 269–274. https://doi.org/10.1016/j.mito.2014.02.011.

Geisler, M., Frangne, N., Gomes, E., Martinoia, E., & Palmgren, M. G. (2000). The ACA4 gene of Arabidopsis encodes a vacuolar membrane calcium pump that improves salt tolerance in yeast. *Plant Physiology*, *124*(4), 1814–1827.

Gonzalez-Guerrero, M., & Arguello, J. M. (2008). Mechanism of Cu^+-transporting ATPases: Soluble Cu^+ chaperones directly transfer Cu^+ to transmembrane transport sites. *Proceedings of the National Academy of Sciences of the United States of America*, *105*(16), 5992–5997. https://doi.org/10.1073/pnas.0711446105.

Gourdon, P., Liu, X. Y., Skjorringe, T., Morth, J. P., Moller, L. B., Pedersen, B. P., et al. (2011). Crystal structure of a copper-transporting PIB-type ATPase. *Nature*, *475*(7354), 59–64. https://doi.org/10.1038/nature10191.

Gravot, A., Lieutaud, A., Verret, F., Auroy, P., Vavasseur, A., & Richaud, P. (2004). AtHMA3, a plant P-1B-ATPase, functions as a Cd/Pb transporter in yeast. *FEBS Letters*, *561*(1–3), 22–28. https://doi.org/10.1016/S0014-5793(04)00072-9.

Gronberg, C., Sitsel, O., Lindahl, E., Gourdon, P., & Andersson, M. (2016). Membrane anchoring and ion-entry dynamics in P-type ATPase copper transport. *Biophysical Journal*, *111*(11), 2417–2429. https://doi.org/10.1016/j.bpj.2016.10.020.

Grotz, N., & Guerinot, M. L. (2006). Molecular aspects of Cu, Fe and Zn homeostasis in plants. *Biochimica et Biophysica Acta, 1763*(7), 595–608. https://doi.org/10.1016/j.bbamcr.2006.05.014.

Guljamow, A., Jenke-Kodama, H., Saumweber, H., Quillardet, P., Frangeul, L., Castets, A. M., et al. (2007). Horizontal gene transfer of two cytoskeletal elements from a eukaryote to a cyanobacterium. *Current Biology, 17*(17), R757–R759. https://doi.org/10.1016/j.cub.2007.06.063.

Hahn, A., Vonck, J., Mills, D. J., Meier, T., & Kuhlbrandt, W. (2018). Structure, mechanism, and regulation of the chloroplast ATP synthase. *Science, 360*(6389), eaat4318. https://doi.org/10.1126/science.aat4318.

Hamza, I., Schaefer, M., Klomp, L. W., & Gitlin, J. D. (1999). Interaction of the copper chaperone HAH1 with the Wilson disease protein is essential for copper homeostasis. *Proceedings of the National Academy of Sciences of the United States of America, 96*(23), 13363–13368.

Hanikenne, M., & Baurain, D. (2013). Origin and evolution of metal P-type ATPases in Plantae (Archaeplastida). *Frontiers in Plant Science, 4,* 544. https://doi.org/10.3389/fpls.2013.00544.

Hashimoto-Sugimoto, M., Higaki, T., Yaeno, T., Nagami, A., Irie, M., Fujimi, M., et al. (2013). A Munc13-like protein in Arabidopsis mediates H^+-ATPase translocation that is essential for stomatal responses. *Nature Communications, 4,* 2215. https://doi.org/10.1038/ncomms3215.

Hirayama, T., Kieber, J. J., Hirayama, N., Kogan, M., Guzman, P., Nourizadeh, S., et al. (1999). RESPONSIVE-TO-ANTAGONIST1, a Menkes/Wilson disease-related copper transporter, is required for ethylene signaling in Arabidopsis. *Cell, 97*(3), 383–393.

Huang, L., Berkelman, T., Franklin, A. E., & Hoffman, N. E. (1993). Characterization of a gene encoding a Ca^{2+}-ATPase-like protein in the plastid envelope. *Proceedings of the National Academy of Sciences of the United States of America, 90*(21), 10066–10070. https://doi.org/10.1073/pnas.90.21.10066.

Huang, J., & Gogarten, J. P. (2007). Did an ancient chlamydial endosymbiosis facilitate the establishment of primary plastids? *Genome Biology, 8*(6), R99. https://doi.org/10.1186/gb-2007-8-6-r99.

Huffman, D. L., & O'Halloran, T. V. (2000). Energetics of copper trafficking between the Atx1 metallochaperone and the intracellular copper transporter, Ccc2. *The Journal of Biological Chemistry, 275*(25), 18611–18614. https://doi.org/10.1074/jbc.C000172200.

Hung, Y. H., Layton, M. J., Voskoboinik, I., Mercer, J. F., & Camakaris, J. (2007). Purification and membrane reconstitution of catalytically active Menkes copper-transporting P-type ATPase (MNK; ATP7A). *The Biochemical Journal, 401*(2), 569–579. https://doi.org/10.1042/BJ20060924.

Hussain, D., Haydon, M. J., Wang, Y., Wong, E., Sherson, S. M., Young, J., et al. (2004). P-type ATPase heavy metal transporters with roles in essential zinc homeostasis in Arabidopsis. *Plant Cell, 16*(5), 1327–1339. https://doi.org/10.1105/tpc.020487.

Iwano, M., Igarashi, M., Tarutani, Y., Kaothien-Nakayama, P., Nakayama, H., Moriyama, H., et al. (2014). A pollen coat-inducible autoinhibited Ca^{2+}-ATPase expressed in stigmatic papilla cells is required for compatible pollination in the Brassicaceae. *Plant Cell, 26*(2), 636–649. https://doi.org/10.1105/tpc.113.121350.

Jakobsen, M. K., Poulsen, L. R., Schulz, A., Fleurat-Lessard, P., Moller, A., Husted, S., et al. (2005). Pollen development and fertilization in Arabidopsis is dependent on the MALE GAMETOGENESIS IMPAIRED ANTHERS gene encoding a type V P-type ATPase. *Genes & Development, 19*(22), 2757–2769. https://doi.org/10.1101/gad.357305.

Jenkins, C., Samudrala, R., Anderson, I., Hedlund, B. P., Petroni, G., Michailova, N., et al. (2002). Genes for the cytoskeletal protein tubulin in the bacterial genus Prosthecobacter. *Proceedings of the National Academy of Sciences of the United States of America, 99*(26), 17049–17054. https://doi.org/10.1073/pnas.012516899.

Jones, C. E., Daly, N. L., Cobine, P. A., Craik, D. J., & Dameron, C. T. (2003). Structure and metal binding studies of the second copper binding domain of the Menkes ATPase. *Journal of Structural Biology, 143*(3), 209–218.

Kanehisa, M., & Goto, S. (2000). KEGG: Kyoto encyclopedia of genes and genomes. *Nucleic Acids Research, 28*(1), 27–30.

Kim, Y. Y., Choi, H., Segami, S., Cho, H. T., Martinoia, E., Maeshima, M., et al. (2009). AtHMA1 contributes to the detoxification of excess Zn(II) in Arabidopsis. *The Plant Journal, 58*(5), 737–753. https://doi.org/10.1111/j.1365-313X.2009.03818.x.

Kobayashi, Y., Kuroda, K., Kimura, K., Southron-Francis, J. L., Furuzawa, A., Kimura, K., et al. (2008). Amino acid polymorphisms in strictly conserved domains of a P-type ATPase HMA5 are involved in the mechanism of copper tolerance variation in Arabidopsis. *Plant Physiology, 148*(2), 969–980. https://doi.org/10.1104/pp.108.119933.

Kramer, U. (2010). Metal hyperaccumulation in plants. *Annual Review of Plant Biology, 61*, 517–534. https://doi.org/10.1146/annurev-arplant-042809-112156.

Kramer, U., & Clemens, S. (2005). Functions and homeostasis of zinc, copper, and nickel in plants. In *Molecular biology of metal homeostasis and detoxification: From microbes to man* (pp. 215–271). Berlin, Heidelberg: Springer.

Kumagai, S., Suzuki, T., Tezuka, K., Satoh-Nagasawa, N., Takahashi, H., Sakurai, K., et al. (2014). Functional analysis of the C-terminal region of the vacuolar cadmium-transporting rice OsHMA3. *FEBS Letters, 588*(5), 789–794. https://doi.org/10.1016/j.febslet.2014.01.037.

Kupper, H., & Andresen, E. (2016). Mechanisms of metal toxicity in plants. *Metallomics, 8*(3), 269–285. https://doi.org/10.1039/c5mt00244c.

Laurent, C., Lekeux, G., Ukuwela, A. A., Xiao, Z., Charlier, J. B., Bosman, B., et al. (2016). Metal binding to the N-terminal cytoplasmic domain of the PIB ATPase HMA4 is required for metal transport in Arabidopsis. *Plant Molecular Biology, 90*(4–5), 453–466. https://doi.org/10.1007/s11103-016-0429-z.

Lee, S. M., Kim, H. S., Han, H. J., Moon, B. C., Kim, C. Y., Harper, J. F., et al. (2007). Identification of a calmodulin-regulated autoinhibited Ca^{2+}-ATPase (ACA11) that is localized to vacuole membranes in Arabidopsis. *FEBS Letters, 581*(21), 3943–3949. https://doi.org/10.1016/j.febslet.2007.07.023.

Leitenmaier, B., & Kupper, H. (2013). Compartmentation and complexation of metals in hyperaccumulator plants. *Frontiers in Plant Science, 4*, 374. https://doi.org/10.3389/fpls.2013.00374.

Leitenmaier, B., Witt, A., Witzke, A., Stemke, A., Meyer-Klaucke, W., Kroneck, P. M., et al. (2011). Biochemical and biophysical characterisation yields insights into the mechanism of a Cd/Zn transporting ATPase purified from the hyperaccumulator plant *Thlaspi caerulescens*. *Biochimica et Biophysica Acta, 1808*(10), 2591–2599. https://doi.org/10.1016/j.bbamem.2011.05.010.

Li, Y., Iqbal, M., Zhang, Q., Spelt, C., Bliek, M., Hakvoort, H. W. J., et al. (2017). Two Silene vulgaris copper transporters residing in different cellular compartments confer copper hypertolerance by distinct mechanisms when expressed in *Arabidopsis thaliana*. *The New Phytologist, 215*(3), 1102–1114. https://doi.org/10.1111/nph.14647.

Li, W., Lacey, R. F., Ye, Y., Lu, J., Yeh, K. C., Xiao, Y., et al. (2017). Triplin, a small molecule, reveals copper ion transport in ethylene signaling from ATX1 to RAN1. *PLoS Genetics, 13*(4), e1006703. https://doi.org/10.1371/journal.pgen.1006703.

Liang, F., Cunningham, K. W., Harper, J. F., & Sze, H. (1997). ECA1 complements yeast mutants defective in Ca^{2+} pumps and encodes an endoplasmic reticulum-type Ca^{2+}-ATPase in Arabidopsis thaliana. *Proceedings of the National Academy of Sciences of the United States of America, 94*(16), 8579–8584. https://doi.org/10.1073/pnas.94.16.8579.

Lichten, L. A., & Cousins, R. J. (2009). Mammalian zinc transporters: Nutritional and physiologic regulation. *Annual Review of Nutrition, 29,* 153–176. https://doi.org/10.1146/annurev-nutr-033009-083312.

Limonta, M., Romanowsky, S., Olivari, C., Bonza, M. C., Luoni, L., Rosenberg, A., et al. (2014). ACA12 is a deregulated isoform of plasma membrane Ca^{2+}-ATPase of *Arabidopsis thaliana. Plant Molecular Biology, 84*(4–5), 387–397. https://doi.org/10.1007/s11103-013-0138-9.

Linder, M. C., & Goode, C. A. (1991). *Biochemistry of copper.* Boston, MA: Springer.

Lippard, S. J., & Berg, J. M. (1994). *Principles of inorganic chemistry.* New York, USA: University Science Books.

Lopez-Marques, R. L., Poulsen, L. R., Hanisch, S., Meffert, K., Buch-Pedersen, M. J., Jakobsen, M. K., et al. (2010). Intracellular targeting signals and lipid specificity determinants of the ALA/ALIS P4-ATPase complex reside in the catalytic ALA alpha-subunit. *Molecular Biology of the Cell, 21*(5), 791–801. https://doi.org/10.1091/mbc.E09-08-0656.

Lopez-Marques, R. L., Poulsen, L. R., & Palmgren, M. G. (2012). A putative plant aminophospholipid flippase, the Arabidopsis P4 ATPase ALA1, localizes to the plasma membrane following association with a beta-subunit. *PLoS One,* 7(4), e33042. https://doi.org/10.1371/journal.pone.0033042.

Lucca, N., & Leon, G. (2012). Arabidopsis ACA7, encoding a putative auto-regulated Ca^{2+}-ATPase, is required for normal pollen development. *Plant Cell Reports, 31*(4), 651–659. https://doi.org/10.1007/s00299-011-1182-z.

Marschner, H. (1995). *Mineral nutrition of higher plants* (2nd ed.). London, UK: Academic Press.

Mattle, D., Zhang, L., Sitsel, O., Pedersen, L. T., Moncelli, M. R., Tadini-Buoninsegni, F., et al. (2015). A sulfur-based transport pathway in Cu^{+}-ATPases. *EMBO Reports, 16*(6), 728–740. https://doi.org/10.15252/embr.201439927.

McCall, K. A., Huang, C., & Fierke, C. A. (2000). Function and mechanism of zinc metalloenzymes. *The Journal of Nutrition, 130*(5S Suppl), 1437S–1446S. https://doi.org/10.1093/jn/130.5.1437S.

McDowell, S. C., Lopez-Marques, R. L., Cohen, T., Brown, E., Rosenberg, A., Palmgren, M. G., et al. (2015). Loss of the *Arabidopsis thaliana* P4-ATPases ALA6 and ALA7 impairs pollen fitness and alters the pollen tube plasma membrane. *Frontiers in Plant Science, 6,* 197. https://doi.org/10.3389/fpls.2015.00197.

Mikkelsen, M. D., Pedas, P., Schiller, M., Vincze, E., Mills, R. F., Borg, S., et al. (2012). Barley HvHMA1 is a heavy metal pump involved in mobilizing organellar Zn and Cu and plays a role in metal loading into grains. *PLoS One,* 7(11), e49027. https://doi.org/10.1371/journal.pone.0049027.

Mills, R. F., Doherty, M. L., Lopez-Marques, R. L., Weimar, T., Dupree, P., Palmgren, M. G., et al. (2008). ECA3, a Golgi-localized P2A-type ATPase, plays a crucial role in manganese nutrition in Arabidopsis. *Plant Physiology, 146*(1), 116–128. https://doi.org/10.1104/pp.107.110817.

Mills, R. F., Peaston, K. A., Runions, J., & Williams, L. E. (2012). HvHMA2, a P(1B)-ATPase from barley, is highly conserved among cereals and functions in Zn and Cd transport. *PLoS One,* 7(8), e42640. https://doi.org/10.1371/journal.pone.0042640.

Miyadate, H., Adachi, S., Hiraizumi, A., Tezuka, K., Nakazawa, N., Kawamoto, T., et al. (2011). OsHMA3, a P1B-type of ATPase affects root-to-shoot cadmium translocation in rice by mediating efflux into vacuoles. *The New Phytologist*, *189*(1), 190–199. https://doi.org/10.1111/j.1469-8137.2010.03459.x.

Moller, J. V., Olesen, C., Winther, A. M. L., & Nissen, P. (2010). What can be learned about the function of a single protein from its various X-ray structures: The example of the sarcoplasmic calcium pump. In J. J. Lacapere (Ed.), *Membrane protein structure determination: Methods and protocols: Vol. 654*. (pp. 119–140). Totowa, NJ: Humana Press.

Morel, M., Crouzet, J., Gravot, A., Auroy, P., Leonhardt, N., Vavasseur, A., et al. (2009). AtHMA3, a P1B-ATPase allowing Cd/Zn/Co/Pb vacuolar storage in Arabidopsis. *Plant Physiology*, *149*(2), 894–904. https://doi.org/10.1104/pp.108.130294.

Moreno, I., Norambuena, L., Maturana, D., Toro, M., Vergara, C., Orellana, A., et al. (2008). AtHMA1 is a thapsigargin-sensitive Ca^{2+}/heavy metal pump. *The Journal of Biological Chemistry*, *283*(15), 9633–9641. https://doi.org/10.1074/jbc.M800736200.

Moustafa, A., Reyes-Prieto, A., & Bhattacharya, D. (2008). Chlamydiae has contributed at least 55 genes to Plantae with predominantly plastid functions. *PLoS One*, *3*(5), e2205. https://doi.org/10.1371/journal.pone.0002205.

Navarro, J. A., & Schneuwly, S. (2017). Copper and zinc homeostasis: Lessons from *Drosophila melanogaster*. *Frontiers in Genetics*, *8*(223), 223. https://doi.org/10.3389/fgene.2017.00223.

Olsen, L. I., Hansen, T. H., Larue, C., Osterberg, J. T., Hoffmann, R. D., Liesche, J., et al. (2016). Mother-plant-mediated pumping of zinc into the developing seed. *Nature Plants*, *2*(5), 16036. https://doi.org/10.1038/nplants.2016.36.

Olsen, L. I., & Palmgren, M. G. (2014). Many rivers to cross: The journey of zinc from soil to seed. *Frontiers in Plant Science*, *5*, 30. https://doi.org/10.3389/fpls.2014.00030.

Omasits, U., Ahrens, C. H., Muller, S., & Wollscheid, B. (2014). Protter: Interactive protein feature visualization and integration with experimental proteomic data. *Bioinformatics*, *30*(6), 884–886. https://doi.org/10.1093/bioinformatics/btt607.

Padilla-Benavides, T., McCann, C. J., & Arguello, J. M. (2013). The mechanism of Cu^{+} transport ATPases: Interaction with CU^{+} chaperones and the role of transient metal-binding sites. *The Journal of Biological Chemistry*, *288*(1), 69–78. https://doi.org/10.1074/jbc.M112.420810.

Palmer, C. M., & Guerinot, M. L. (2009). Facing the challenges of Cu, Fe and Zn homeostasis in plants. *Nature Chemical Biology*, *5*(5), 333–340. https://doi.org/10.1038/nchembio.166.

Palmgren, M. G., & Nissen, P. (2011). P-type ATPases. *Annual Review of Biophysics*, *40*(1), 243–266. https://doi.org/10.1146/annurev.biophys.093008.131331.

Pedersen, J. T., & Palmgren, M. (2017). Why do plants lack sodium pumps and would they benefit from having one? *Functional Plant Biology*, *44*(5), 473. https://doi.org/10.1071/fp16422.

Petris, M. J., Mercer, J. F., Culvenor, J. G., Lockhart, P., Gleeson, P. A., & Camakaris, J. (1996). Ligand-regulated transport of the Menkes copper P-type ATPase efflux pump from the Golgi apparatus to the plasma membrane: A novel mechanism of regulated trafficking. *The EMBO Journal*, *15*(22), 6084–6095.

Poulsen, L. R., Lopez-Marques, R. L., McDowell, S. C., Okkeri, J., Licht, D., Schulz, A., et al. (2008). The Arabidopsis P4-ATPase ALA3 localizes to the Golgi and requires a beta-subunit to function in lipid translocation and secretory vesicle formation. *Plant Cell*, *20*(3), 658–676. https://doi.org/10.1105/tpc.107.054767.

Poulsen, L. R., Lopez-Marques, R. L., Pedas, P. R., McDowell, S. C., Brown, E., Kunze, R., et al. (2015). A phospholipid uptake system in the model plant *Arabidopsis thaliana*. *Nature Communications*, *6*, 7649. https://doi.org/10.1038/ncomms8649.

Pufahl, R. A., Singer, C. P., Peariso, K. L., Lin, S. J., Schmidt, P. J., Fahrni, C. J., et al. (1997). Metal ion chaperone function of the soluble Cu(I) receptor Atx1. *Science, 278*(5339), 853–856.

Rodrigues, R. B., Sabat, G., Minkoff, B. B., Burch, H. L., Nguyen, T. T., & Sussman, M. R. (2014). Expression of a translationally fused TAP-tagged plasma membrane proton pump in *Arabidopsis thaliana*. *Biochemistry, 53*(3), 566–578. https://doi.org/10.1021/bi401096m.

Rodriguez, F. I., Esch, J. J., Hall, A. E., Binder, B. M., Schaller, G. E., & Bleecker, A. B. (1999). A copper cofactor for the ethylene receptor ETR1 from Arabidopsis. *Science, 283*(5404), 996–998.

Roelofsen, H., Wolters, H., Van Luyn, M. J., Miura, N., Kuipers, F., & Vonk, R. J. (2000). Copper-induced apical trafficking of ATP7B in polarized hepatoma cells provides a mechanism for biliary copper excretion. *Gastroenterology, 119*(3), 782–793.

Ronquist, F., & Huelsenbeck, J. P. (2003). MrBayes 3: Bayesian phylogenetic inference under mixed models. *Bioinformatics, 19*(12), 1572–1574.

Sautron, E., Mayerhofer, H., Giustini, C., Pro, D., Crouzy, S., Ravaud, S., et al. (2015). HMA6 and HMA8 are two chloroplast Cu^{+}-ATPases with different enzymatic properties. *Bioscience Reports, 35*(3), e00201. https://doi.org/10.1042/BSR20150065.

Schiott, M., Romanowsky, S. M., Baekgaard, L., Jakobsen, M. K., Palmgren, M. G., & Harper, J. F. (2004). A plant plasma membrane Ca2+ pump is required for normal pollen tube growth and fertilization. *Proceedings of the National Academy of Sciences of the United States of America, 101*(25), 9502–9507. https://doi.org/10.1073/pnas.0401542101.

Seigneurin-Berny, D., Gravot, A., Auroy, P., Mazard, C., Kraut, A., Finazzi, G., et al. (2006). HMA1, a new Cu-ATPase of the chloroplast envelope, is essential for growth under adverse light conditions. *The Journal of Biological Chemistry, 281*(5), 2882–2892. https://doi.org/10.1074/jbc.M508333200.

Shikanai, T., Muller-Moule, P., Munekage, Y., Niyogi, K. K., & Pilon, M. (2003). PAA1, a P-type ATPase of Arabidopsis, functions in copper transport in chloroplasts. *Plant Cell, 15*(6), 1333–1346.

Sinclair, S. A., & Kramer, U. (2012). The zinc homeostasis network of land plants. *Biochimica et Biophysica Acta, 1823*(9), 1553–1567. https://doi.org/10.1016/j.bbamcr.2012. 05.016.

Singleton, C., & Le Brun, N. E. (2007). Atx1-like chaperones and their cognate P-type ATPases: Copper-binding and transfer. *Biometals, 20*(3–4), 275–289. https://doi.org/10.1007/s10534-006-9068-1.

Sitsel, O., Gronberg, C., Autzen, H. E., Wang, K., Meloni, G., Nissen, P., et al. (2015). Structure and function of Cu(I)- and Zn(II)-ATPases. *Biochemistry, 54*(37), 5673–5683. https://doi.org/10.1021/acs.biochem.5b00512.

Smith, A. T., Ross, M. O., Hoffman, B. M., & Rosenzweig, A. C. (2017). Metal selectivity of a Cd-, Co-, and Zn-transporting P1B-type ATPase. *Biochemistry, 56*(1), 85–95. https://doi.org/10.1021/acs.biochem.6b01022.

Smith, A. T., Smith, K. P., & Rosenzweig, A. C. (2014). Diversity of the metal-transporting P1B-type ATPases. *Journal of Biological Inorganic Chemistry, 19*(6), 947–960. https://doi.org/10.1007/s00775-014-1129-2.

Stamatakis, A. (2014). RAxML version 8: A tool for phylogenetic analysis and post-analysis of large phylogenies. *Bioinformatics, 30*(9), 1312–1313. https://doi.org/10.1093/bioinformatics/btu033.

Tamura, K., Stecher, G., Peterson, D., Filipski, A., & Kumar, S. (2013). MEGA6: Molecular evolutionary genetics analysis version 6.0. *Molecular Biology and Evolution, 30*(12), 2725–2729. https://doi.org/10.1093/molbev/mst197.

Toyoshima, C. (2009). How Ca^{2+}-ATPase pumps ions across the sarcoplasmic reticulum membrane. *Biochimica et Biophysica Acta, 1793*(6), 941–946. https://doi.org/10.1016/j.bbamcr.2008.10.008.

Ueno, D., Milner, M. J., Yamaji, N., Yokosho, K., Koyama, E., Clemencia Zambrano, M., et al. (2011). Elevated expression of TcHMA3 plays a key role in the extreme Cd tolerance in a Cd-hyperaccumulating ecotype of *Thlaspi caerulescens*. *The Plant Journal*, *66*(5), 852–862. https://doi.org/10.1111/j.1365-313X.2011.04548.x.

Verret, F., Gravot, A., Auroy, P., Leonhardt, N., David, P., Nussaume, L., et al. (2004). Overexpression of AtHMA4 enhances root-to-shoot translocation of zinc and cadmium and plant metal tolerance. *FEBS Letters*, *576*(3), 306–312. https://doi.org/10.1016/j.febslet.2004.09.023.

Verret, F., Gravot, A., Auroy, P., Preveral, S., Forestier, C., Vavasseur, A., et al. (2005). Heavy metal transport by AtHMA4 involves the N-terminal degenerated metal binding domain and the C-terminal His11 stretch. *FEBS Letters*, *579*(6), 1515–1522. https://doi.org/10.1016/j.febslet.2005.01.065.

Walker, J. M., Tsivkovskii, R., & Lutsenko, S. (2002). Metallochaperone Atox1 transfers copper to the NH2-terminal domain of the Wilson's disease protein and regulates its catalytic activity. *The Journal of Biological Chemistry*, *277*(31), 27953–27959. https://doi.org/10.1074/jbc.M203845200.

Wang, K., Sitsel, O., Meloni, G., Autzen, H. E., Andersson, M., Klymchuk, T., et al. (2014). Structure and mechanism of Zn^{2+}-transporting P-type ATPases. *Nature*, *514*(7523), 518–522. https://doi.org/10.1038/nature13618.

Williams, R. J. P. (1987). The biochemistry of zinc. *Polyhedron*, *6*(1), 61–69. https://doi.org/10.1016/S0277-5387(00)81239-5.

Woeste, K. E., & Kieber, J. J. (2000). A strong loss-of-function mutation in RAN1 results in constitutive activation of the ethylene response pathway as well as a rosette-lethal phenotype. *Plant Cell*, *12*(3), 443–455. https://doi.org/10.1105/Tpc.12.3.443.

Wong, C. K., Jarvis, R. S., Sherson, S. M., & Cobbett, C. S. (2009). Functional analysis of the heavy metal binding domains of the Zn/Cd-transporting ATPase, HMA2, in *Arabidopsis thaliana*. *The New Phytologist*, *181*(1), 79–88. https://doi.org/10.1111/j.1469-8137.2008.02637.x.

Yamaji, N., Xia, J., Mitani-Ueno, N., Yokosho, K., & Feng Ma, J. (2013). Preferential delivery of zinc to developing tissues in rice is mediated by P-type heavy metal ATPase OsHMA2. *Plant Physiology*, *162*(2), 927–939. https://doi.org/10.1104/pp.113.216564.

Zimmermann, M., Clarke, O., Gulbis, J. M., Keizer, D. W., Jarvis, R. S., Cobbett, C. S., et al. (2009). Metal binding affinities of Arabidopsis zinc and copper transporters: Selectivities match the relative, but not the absolute, affinities of their amino-terminal domains. *Biochemistry*, *48*(49), 11640–11654. https://doi.org/10.1021/bi901573b.

Zorrig, W., Abdelly, C., & Berthomieu, P. (2011). The phylogenetic tree gathering the plant Zn/Cd/Pb/Co P1B-ATPases appears to be structured according to the botanical families. *Comptes Rendus Biologies*, *334*(12), 863–871. https://doi.org/10.1016/j.crvi.2011.09.004.

FURTHER READING

Adle, D. J., Sinani, D., Kim, H., & Lee, J. (2007). A cadmium-transporting P1B-type ATPase in yeast Saccharomyces cerevisiae. *The Journal of Biological Chemistry*, *282*(2), 947–955. https://doi.org/10.1074/jbc.M609535200.

Fan, B., & Rosen, B. P. (2002). Biochemical characterization of CopA, the *Escherichia coli* Cu(I)-translocating P-type ATPase. *The Journal of Biological Chemistry*, *277*(49), 46987–46992. https://doi.org/10.1074/jbc.M208490200.

Forbes, J. R., Hsi, G., & Cox, D. W. (1999). Role of the copper-binding domain in the copper transport function of ATP7B, the P-type ATPase defective in Wilson disease. *The Journal of Biological Chemistry*, *274*(18), 12408–12413.

Sorensen, D. M., Moller, A. B., Jakobsen, M. K., Jensen, M. K., Vangheluwe, P., Buch-Pedersen, M. J., et al. (2012). Ca^{2+} induces spontaneous dephosphorylation of a novel P5A-type ATPase. *The Journal of Biological Chemistry*, *287*(34), 28336–28348. https://doi.org/10.1074/jbc.M112.387191.

Walker, J. M., Huster, D., Ralle, M., Morgan, C. T., Blackburn, N. J., & Lutsenko, S. (2004). The N-terminal metal-binding site 2 of the Wilson's disease protein plays a key role in the transfer of copper from Atox1. *The Journal of Biological Chemistry*, *279*(15), 15376–15384. https://doi.org/10.1074/jbc.M400053200.

CHAPTER FOUR

Metal Transport in the Developing Plant Seed

Seckin Eroglu[1]
Department of Genetics and Bioengineering, Izmir University of Economics, Izmir, Turkey
[1]Corresponding author: e-mail address: erogluseckin@gmail.com

Contents

Abstract

Healthy plant growth depends on a balanced metal homeostasis at the organ, tissue and sub-cellular levels, which is mediated principally by plasma and vacuolar membrane metal transporters. The genetic bases of metal acquisition in developing seeds has long remained poorly understood. Recent technical advances have helped circumvent the difficulties of conducting metal nutrient research on the extremely small seeds of *Arabidopsis thaliana*. The review presents recent advances in our understanding of seed metal homeostasis focussing on this model plant. Metals are loaded from phloem to the seed coat and must pass through the endosperm to reach the embryo. The embryo comprises several apoplastic and symplastic pathways that strictly depend on the changing physiology of the developing seed organs. Metals that reach the developing embryo fuel immediate cellular processes or accumulate in vacuoles to support forthcoming germination. In the mature embryo, metal distribution is homogeneous, with the exception of iron and manganese which localize to distinct cell layers. These metal localizations are strictly dependent on expression of specific tonoplast transporters, with putative functions that go beyond the storage of metals. Accumulating evidence indicates that they can control the timing of metal entry into the embryo.

Advances in Botanical Research, Volume 87
ISSN 0065-2296
https://doi.org/10.1016/bs.abr.2018.09.010

1. GENERAL PRINCIPLES IN PLANT METAL HOMEOSTASIS

A healthy plant growth requires a continuous supply of metal nutrients. In plant cells, metals activate enzymes, contribute to protein synthesis and function as signalling molecules. Particularly, transition metals mediate electron transport reactions, such as photosynthesis, respiration, and detoxification of reactive oxygen species, due to their ability to take several different charges (Andresen, Peiter, & Küpper, 2018). In hyperaccumulator species, metals can accumulate in leaf vacuoles at concentrations that are poisonous for insects, thereby contributing to the protection of the plant from herbivory (Boyd, 2007). However, meeting the demand for metals first requires the scavenging of positively charged metal cations from the negatively charged soil particles and subsequently their mobilization from roots to sink tissues. Whenever plants fail to provide sufficient amounts of metals to sink tissues, growth is restricted. In agriculture, plants often cannot meet the need for metals (e.g. iron and zinc), making metal deficiencies one of the most widespread problems of agricultural production.

Membrane metal transporters determine the majority of metal content in plant organs, tissues, and organelles. For example, whether ferroportin2 (FPN2) transporter is truncated or full length can account for differential cobalt levels in shoots of distinct *Arabidopsis thaliana* natural accessions (Morrissey et al., 2009). How plants gain metal hypersensitivity (Castaings, Caquot, Loubet, & Curie, 2016; Gao et al., 2017) or hypertolerance (Morrissey et al., 2009) can often be explained by assessing a change in activity of their membrane transporters. Comparison of hypertolerants with their non-hypertolerant relatives shows a striking difference in the expression of metal transporters upon metal stress. For example, in roots under cadmium stress, *cation exchanger1* (*CAX1*) of hypertolerant *Arabidopsis halleri* was upregulated 15 times more than in *A. thaliana* (Baliardini, Meyer, Salis, Saumitou-Laprade, & Verbruggen, 2015; Becher, Talke, Krall, & Krämer, 2003). Sufficient amounts of metals must be taken up from the soil and transported to the sink tissues in order to complete the plant's life cycle. Metal transport from the soil to the seeds involves (i) root acquisition, comprising uptake from the rhizosphere to the cytosol of epidermal and cortex cells of the root, (ii) long-distance transport, including loading from root pericycle into the xylem, root-to-shoot transport via the xylem and its subsequent unloading, (iii) remobilization from source (e.g. leaves) to sink organs (e.g. developing seeds) via the phloem and (iv) transport within the seed, including translocation from maternal to filial tissues and storage. At each of these steps, metals must pass through several membranes with the help of diverse families of metal transporters.

Metal transporters belong to several protein families such as the natural resistance-associated macrophage proteins (NRAMPs), the metal-tolerance proteins (MTPs), the vacuolar iron transporter-like family (VTLs), the ZRT/IRT-like proteins (ZIPs) and yellow stripe1-like family (YSLs). NRAMPs transport divalent cations, particularly Mn^{2+} and Fe^{2+} into the cytosol (Thomine, Wang, Ward, Crawford, & Schroeder, 2000). Their primary physiological function appears to be maintaining iron (Fe) and manganese (Mn) homeostasis. Particularly, NRAMPs are associated with the acquisition and radial transport of Fe and Mn from the outer to inner parts of the root (Castaings et al., 2016) and their remobilization from the vacuole (Mary et al., 2015). MTPs localize to the tonoplast to detoxify excess of metals from the cytosol by sequestering them into the vacuole (Delhaize, Kataoka, Hebb, White, & Ryan, 2003). According to their type, MTPs principally catalyze Mn, Mn/Fe or zinc (Zn) transport (Montanini, Blaudez, Jeandroz, Sanders, & Chalot, 2007). MTP expression profile often overlaps with the presence of large concentrations of metals in vacuoles of different tissues such as shoots of hyperaccumulators (Gustin et al., 2009; Persans, Nieman, & Salt, 2001), the metal-rich aleurone layer of wheat kernels (Vatansever, Filiz, & Eroglu, 2017) or the Mn-rich subepidermis of *A. thaliana* embryos (Eroglu, Meier, von Wirén, & Peiter, 2016). Thus, MTPs are good targets for developing biofortification strategies due to their metal sequestration capacity (Ricachenevsky, Menguer, Sperotto, Williams, & Fett, 2013). vacuolar iron transporter1 (VIT1) is an ortholog of yeast tonoplast Fe/Mn transporter CCC1 (Kim et al., 2006; Li, Chen, Ward, & Kaplan, 2001). Both VIT1 and a group of proteins that share homology with VIT1, the family of VTLs, transport Fe and Mn into the vacuole and participate in Fe homeostasis (Gollhofer, Timofeev, Lan, Schmidt, & Buckhout, 2014). The ZIP family comprises 15 members in Arabidopsis, which transport Zn, Fe, Mn and Cu (Milner, Seamon, Craft, & Kochian, 2013). Several ZIP homologs are induced by Zn or Fe deficiency, suggesting their involvement in homeostasis of these metals (Waters & Sankaran, 2011). Although a number of ZIP proteins have not been yet characterized, a protein belonging to this group, iron regulated transporter1 (IRT1), represents one of the most extensively characterized metal transporters and is reviewed elsewhere (Brumbarova, Bauer, & Ivanov, 2015; Jeong, Merkovich, Clyne, & Connolly, 2017). YSLs transport metals that are chelated with nicotianamine (NA), a nonpeptidyl chelation agent. NA is found in phloem sap and can form stable complexes with metals under neutral or slightly basic pH, suggesting that it is the principal metal chelator in the phloem (Stephan & Scholz, 1993; von Wirén et al., 1999). Thus, the crucial role of YSLs in long-distance transport and seed metal loading is emerging (Curie et al., 2009; Waters et al., 2006). Members of the oligopeptide transporter (OPT) family are characterized as peptide carriers and share a considerable homology with YSLs (Koh et al., 2002). In contrast to other members of the family which generally carry glutathione or glutathione complexes (Wongkaew et al., 2018), OPT3 transports Fe ions (Zhai et al., 2014).

Acquisition of Fe by the root is well understood and provides a good example of how transporters operate cooperatively to retrieve metals from the rhizosphere. First, Fe has to be mobilized, as most of the Fe in soil is immobile due to its precipitation in sparingly soluble oxides. Fe is mobilized through acidification by the activity of proton pumps, particularly AHA2 (Santi & Schmidt, 2009), and chelation by low molecular weight compounds, such as coumarins, which are secreted into the rhizosphere (Schmid et al., 2014; Tsai & Schmidt, 2017). The mobilized ferric Fe [Fe(III)] diffuses in the apoplast, followed by the reduction of Fe(III) to ferrous Fe [Fe(II)] via plasma membrane-bound ferric chelate reductase enzyme (Robinson, Procter, Connolly, & Guerinot, 1999) and coumarins (Rajniak et al., 2018; Sisó-Terraza et al., 2016). From the apoplast, Fe enters the cytosol of epidermal and cortex cells through divalent cation transporter IRT1. IRT1 is the only high-affinity Fe transporter involved in root Fe acquisition (Korshunova, Eide, Clark, Guerinot, & Pakrasi, 1999). In contrast to IRT1, NRAMP1 and perhaps other low-affinity plasma membrane transporters contribute to Fe transport provided that extracellular Fe concentration is sufficiently high (Castaings et al., 2016).

IRT1 shows a broad range of affinity for divalent cations, which has direct consequences on Fe acquisition and general plant health. While taking up Fe, IRT1 also non-specifically takes up other metals as indicated by the strong correlation of IRT1 expression with accumulation of Mn, Co, Zn, Cd, and Ni in the root (Korshunova et al., 1999; Nishida et al., 2011). This poses a heavy metal stress unless these metals are detoxified by sequestration into the vacuole (Thomine & Vert, 2013). Vacuolar sequestration is mediated by tonoplast-localized metal transporters that belong to diverse families (Arrivault, Senger, & Krämer, 2006; Eroglu et al., 2016; Morrissey et al., 2009; Schaaf et al., 2006). Interestingly, expression of these transporters is tightly connected to that of IRT1 by a common transcription factor, FER-like iron deficiency induced transcription factor (FIT), suggesting that heavy metal intake concomitant with Fe intake is coupled with subsequent sequestration of the former into the vacuole (Colangelo & Guerinot, 2004). In addition, similar to a safety fuse, Fe accompanying metals may shut down Fe acquisition machinery to curb further intake. This can be achieved either by inhibiting IRT1 (Dubeaux, Neveu, Zelazny, & Vert, 2018) or ferric chelate reductase activity (Eroglu et al., 2016). Therefore, strategies in the roots provide a perfect example of how several transporters and other helper proteins have to work in concert, to ensure that an adequate amount of Fe is taken up while the inhibitory or toxic effects of other metals are suppressed.

In the cytoplasm, free Fe triggers oxidative stress through Fenton reactions; therefore, most of the Fe should be kept chelated. Fe species differ at the tissue and subcellular levels, depending on the availability and concentration of their chelators and the stability of complexes in that particular pH. In plastids, Fe is stored in protein nanocages called ferritins which are suggested to release Fe for photosynthetic reactions (Briat, Duc, Ravet, & Gaymard, 2010; Harrison & Arosio, 1996).

Plant embryos are young plants that do need metals for development. Unlike mature plants, they do not possess functional roots; thus, the developing embryo completely depends on the metal flow that is derived from the maternal tissues. Knowledge of seed metal homeostasis at the molecular level has been very poor until recently, due to technical difficulties in quantifying metals in small seeds of Arabidopsis. However, recent technical advances now permit measurements in these organs with higher accuracy and increased resolution. This review will discuss recent advances in seed metal nutrient homeostasis focussing on Arabidopsis.

2. ARABIDOPSIS SEED METAL HOMEOSTASIS

Metal translocation from the mother plant to the seed and its subsequent distribution within the seed is critical for successful germination and for food safety. Traditionally, seed metal nutrient research has ignored Arabidopsis, since the tools and techniques that were available were simply too limiting to conduct analyses on such small seeds. However, recent advances seem to put Arabidopsis at the centre of seed nutrient research. These advances include imaging techniques with increased resolution (Punshon, Ricachenevsky, Hindt, Socha, & Zuber, 2013; Zhao, Moore, Lombi, & Zhu, 2014), the availability of transcriptomes of individual seed tissues (Belmonte et al., 2013; Dekkers et al., 2013; Le et al., 2010), and protocols that allow collecting higher biomass in a shorter time (Bates, Jewell, & Browse, 2013; Perry & Wang, 2003; Raissig, Gagliardini, Jaenisch, Grossniklaus, & Baroux, 2013).

Metal transporters must ensure that filial tissues (embryo and endosperm) receive sufficient amounts of metals so that the metal demand of the developing seed is fulfilled, but not exceeded. The developing seed requires metals for both immediate use and use after detachment from the mother plant until the root of the germinating seed sets in. Regarding its immediate use, the developing seed is not totally different from the mature plant, since it carries out photosynthesis (although the rate is much lower than in the mature leaves and its function may be rather to provide oxygen to heterotrophic embryos; Rolletschek, Weber, & Borisjuk, 2003), it respires and detoxifies radicals all of which require metal-bearing enzymes. Characterization of metal transport mutants which can be rescued by additional metal supply has indicated that embryos might be aborted when metal supply to the seed is not sufficient (Stacey, Koh, Becker, & Stacey, 2002; Waters et al., 2006; Wong, Jarvis, Sherson, & Cobbett, 2008). In addition, harvested seeds contain fairly constant levels of metals, independent of how the mother plant was fed (Rengel, Batten, & Crowley, 1999). As an example, in nature and in agronomic practice, limited, i.e., basic soils, or excess Fe, i.e., acidic or fertilized soils, can impact seed yield dramatically but not the final seed Fe concentration (Tyler, 1998; Wiersma, 2005, 2012). Alternatively, recent findings show that seed Fe concentrations are regulated genetically, by expression of certain transcription factors such as basic helix-loop-helix 104 (bHLH104), Brutus (BTS), and Brutus-like (BTSL). Increase in expression of bHLH104 (Li, Zhang, Ai, Liang, & Yu, 2016) and decrease in expression of BTS (Li et al., 2016; Long et al., 2010) as well as its two paralogs, BTSL1 and BTSL2, resulted in a striking increase in Fe concentration of the seed (Hindt et al., 2017; Li et al., 2016; Long et al., 2010) Future research will have to pinpoint the metal transporter genes that are regulated by these transcription factors.

3. POST-PHLOEM METAL TRANSPORT

Upon fertilization, the ovule transforms into a seed and becomes a sink for metal nutrients. The micropyler region, which represents the open space where the pollen tube enters into the ovule to fertilize the egg, subsequently fuses to close the seed. The seed coat contains two integuments of the ovule. Its chalazal region is connected to the silique via the funiculus. The funiculus itself contains vascular bundles, comprising xylem, and phloem. Due to the low level of transpiration in the seed, the xylem does not contribute significantly to the feeding of the seed. Thus, the phloem has been suggested as the main source of metals arriving at the seed coat, and ultimately in the embryo (Fig. 1).

In phloem, metal nutrients can be carried in chelated forms. Because the phloem sap is neutral to basic, metals such as Fe tend to precipitate unless they are chelated. A major chelator of metals in phloem is nicotianamine (NA) (Ishimaru et al., 2010; Schmidke & Stephan, 2006; von Wirén et al., 1999). Consistent with the chelation function of NA, plant lines that carry a mutation in NA-metal carrier proteins, YSLs, may show differential

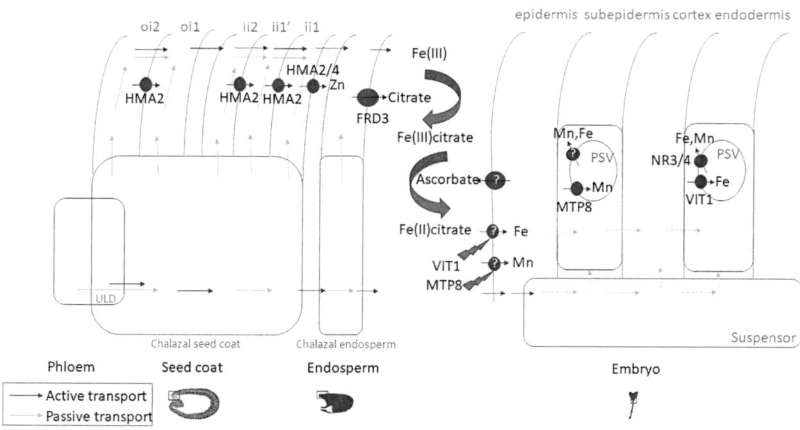

Fig. 1 Metal translocation from phloem to the embryo inside the Arabidopsis seed. Phloem ends in unloading domain of the chalazal seed coat. Metals that are unloaded here diffuse through the outer and inner integuments, and chalazal seed coat. HMA2 is expressed in all integuments, whereas HMA4 is mainly expressed in the innermost layer of the seed coat, the endothelium. Both transport Zn. FRD3 transports citrate to the space between the endosperm and embryo and chelates Fe. FRD3 expression localizes to the protodermis of the embryo (not shown in the figure) and in the endosperm. Chelated Fe(III) is reduced by ascorbate and taken up into the embryo cells. VIT1 and MTP8, which are tonoplast metal transporters, control Fe and Mn intake to the embryo. Metals can be either directly taken up by the embryo or via the suspensor. The embryo and suspensor form one single symplast at the globular stage, but as the embryo develops, several distinct symplastic domains appear. In mature seeds, the suspensor gets degraded. Fe is taken into vacuoles of endodermal cells of the embryo by VIT1 and in subepidermal cells of the abaxial side of the cotyledons by MTP8. Fe in endodermal cells is effluxed by NRAMP3/4. Red sketches at the bottom are seed organs and zoom outs of the regions above them. Green arrows represent symplastic (passive) pathways while red arrows represent apoplastic barriers, where active transport is necessary. Dashed green arrows represent symplastic pathways that only exist during the early stages of developing embryos. *Oi's*, outer integument cell layers; *ii's*, inner integument cell layers. Lightning-like symbol indicates signalling. *ULD*, unloading domain. *NR3/4*, NRAMP3 and 4. *PSV*, protein storage vacuole.

metal concentrations in the seed. For instance, *ysl1* single and *ysl1 ysl3* double knock out lines in Arabidopsis and a *ysl2* mutant in rice exhibit lower Fe accumulation in seeds (Jean, Schikora, Mari, Briat, & Curie, 2005; Koike et al., 2004; Waters et al., 2006). In order to address why *ysl* mutants show such seed metal phenotype, Fe distribution in the shoot of NA-less Arabidopsis mutants was investigated (Schuler, Rellán-Álvarez, Fink-Straube, Abadía, & Bauer, 2012). NA-less mutant leaves accumulated an excess of Fe in the vascular tissues although the leaf mesophyll was severely Fe-deficient, suggesting a role for NA in the unloading of Fe from the phloem to the nearby tissues (Schuler et al., 2012). It is possible that YSLs that are localized to the phloem are also needed to unload NA-metal complexes in the chalaza of the seed coat.

4. FROM SEED COAT TO THE ENDOSPERM, AND FURTHER TO THE EMBRYO

Metals need to move from the seed coat to the endosperm in order to reach the developing embryo. Since metal trafficking inside seed layers cannot be observed in real time, the specific regions that are responsible for efflux and subsequent uptake must be inferred from indirect evidence. These include (i) the presence and density of plasmodesmatal connections between cells, which indicate whether tissues are symplastically connected, (ii) the cell morphology, i.e., whether cells are specialized as transfer-cell-like and (iii) whether tissues are covered with a waxy substance, such as suberin, which blocks solute transport. Wherever tissues are not symplastically connected, metal transport will depend on plasma membrane transport proteins. Therefore, identifying new metal transporters and localizing the site of their expression will aid our understanding of metal trafficking between seed tissues.

Metals that are unloaded in chalazal seed coat are destined for translocation to the endosperm (Fig. 1). Metals can then diffuse from the chalaza throughout the three cell layers of the outer integument which are symplastically connected (Stadler, Lauterbach, & Sauer, 2005). From the innermost cell layer of the outer integument, metals are transferred to the outer cell layer of the inner integument by means of an active transport. They subsequently diffuse throughout the two cell layers of this integument (Stadler et al., 2005). It should be noted that, even in tissues that are symplastically connected, the additional presence of an apoplastic pathway cannot be ruled out (Fig. 1). Such a pathway has been proposed for sucrose and aminoacid transport, based on expression of membrane transport proteins between inner cell layers of each of the integuments (Chen et al., 2015; Karmann, Müller, & Hammes, 2018; Müller et al., 2015).

From the innermost cell layer to the endosperm, metals must be transported via active processes (Stadler et al., 2005). Two proteins involved in this process, heavy metal ATPase2 and 4 (HMA2/4), were first identified in the vasculature of mature plants and associated with xylem loading of Zn (Hussain et al., 2004; Verret et al., 2004). In seeds, expression of the two proteins localized to the seed coat: HMA2 is expressed in all cell layers, whereas HMA4 is mainly present in the innermost layer. In addition, in *hma2/4* double knock-out seeds, Zn translocation from the seed coat to the developing embryo was reduced (Olsen et al., 2016). However, HMA2/4 are not the only Zn exporters in the seed coat, as the embryo of double knock-out lines can still receive Zn. Thus, lower affinity transporters may come into play upon accumulation of Zn in the seed coat (Olsen et al., 2016).

The endosperm continuously transforms throughout seed development. From the coenocyte stage where it is formed of a single multinuclei cell harbouring a large central vacuole, the endosperm goes through uneven cellularization events which lead to gradual shrinking of the central vacuole. Depending on the location of cellularizations, the endosperm can be subdivided as micropyler (where the pollen tube enters the ovule), chalazal (above the end of vascular bundles) and peripheral (starting from the peripheries then extending to the centre) endosperms (Olsen, 2004). In contrast to the others, the chalazal endosperm is cellularized much later and contains structures resembling those of transfer cells (Nguyen, Brown, & Lemmon, 2000). Furthermore, the chalazal endosperm contains a high number of mitochondria and an abundant endoplasmic reticulum (Nguyen et al., 2000).

The above properties point to the chalazal endosperm as a promising candidate for nutrient transport between the seed coat and the developing embryo or the micropyler endosperm. A direct transport pathway from the chalazal endosperm to the embryo proper and/or the suspensor (a single cell layer attaching the embryo proper to the micropyler endosperm) has been suggested to operate during the very early phase of embryogenesis. This pathway would involve the central vacuole that is protruded into the chalazal endosperm at one end, and at the other end placed in close proximity of the embryo proper (Otegui, Capp, & Staehelin, 2002). However, such a direct transport may only be operational during the early stages of seed development; since, later on, the central vacuole shrinks, and the embryo becomes completely surrounded by the micropyler endosperm.

Seed tissues are covered with waxy substances known to restrict metal transport in certain organs of mature plants. In particular, the cuticle, which is a hydrophobic film over epidermis, is a major factor decreasing the effect of foliar fertilizers (Fernández & Eichert, 2009). Hydrophilic nutrients can still pass through the cuticle, albeit at much lower rates. Recent research showed that in the developing Arabidopsis seed, both the endosperm (Beeckman, De Rycke, Viane, & Inzé, 2000; De Giorgi et al., 2015) and the embryo proper (Delude, Moussu, Joubès, Ingram, & Domergue, 2016) are covered with a cuticle. Very recently, the cuticle surrounding the endosperm has been reported to have gaps in the micropyler and the chalazal regions (Loubéry, Giorgi, Utz-Pugin, Demonsais, & Lopez-Molina, 2018). Besides cuticle, suberin is another waxy substance found in seeds. In contrast to the cuticle which localizes to the outer surface of cell walls, suberin localizes to their inner side. In roots, suberization of the endodermis was recently suggested as nutrient-stress responsive, thereby regulating nutrient acquisition (Barberon et al., 2016). In seeds, suberization occurs in the outer integument (Molina, Ohlrogge, & Pollard, 2007) and chalazal plug (Franke et al., 2009; Lashbrooke et al., 2016), and is mostly developmentally regulated. The exact distribution of waxy substances and their implication in nutrient acquisition of seed organs should be addressed in future research.

Nutrient transfer from endosperm to embryo may involve the suspensor. This assumption is based on indirect evidence including the differentiation of suspensor cells with transfer cell-like features in some species, absence of the cuticle layer in suspensor in contrast to the cuticle-covered embryo proper and expression of nutrient transporter gene families in the suspensor (Ingram & Nawrath, 2017; Stacey et al., 2008; Yeung & Meinke, 1993). Furthermore, histochemical analyses by Roschzttardtz, Conéjéro, Curie, and Mari (2009) have revealed a Fe accumulation in the suspensor of developing Arabidopsis embryos. However, this Fe accumulation extends to the symplastically connected embryo proper (Stadler et al., 2005), making it impossible to interpret whether and to which extent the embryo proper and the suspensor are responsible for Fe intake. The suspensor can only feed the embryo proper for a limited time since its symplastic connections are weakened and then crushed by the developing embryo following heart stage (Stadler et al., 2005). Thus, it is believed that, later on, the embryo proper acquires nutrients from the surrounding endosperm tissues autonomously.

Metal acquisition by the embryo appears to be controlled in a nutrient-specific manner and by complex mechanisms, having similarities and distinctions to root metal acquisition. In the xylem sap of roots, the main chelator of Fe is citrate (Durrett, Gassmann, & Rogers, 2007; Rellán-Álvarez et al., 2009). It is secreted to the xylem sap by an efflux transporter called ferric reductase defective3 (FRD3) (Durrett et al., 2007). In addition to the root vasculature, *FRD3* is expressed in the chalazal endosperm throughout development (Belmonte et al., 2013) and in the aleurone (Roschzttardtz, Séguéla-Arnaud, Briat, Vert, & Curie, 2011), a single cell layer that is the remainder of consumed endosperm in mature seeds. A direct evidence that citrate chelates Fe in seeds came from the analysis of Fe ligands in the liquid endosperm of pea (Grillet et al., 2014). Therefore, it has been suggested that FRD3-mediated citrate efflux mobilizes Fe in the apoplast between the endosperm and embryo (Grillet et al., 2014; Roschzttardtz et al., 2011). In roots, Fe acquisition is dependent on a Fe(III) reduction step carried out by a membrane-bound ferric chelate reductase (Robinson et al., 1999) and reductants, i.e., coumarins, that are exuded to the rhizosphere (Mladěnka et al., 2010; Rajniak et al., 2018). Interestingly, the maturing embryo is also able to perform a Fe(III) reduction in vitro, albeit by a different mechanism involving an efflux of ascorbate. Fe(III) reduction capacity is impaired in the presence of ascorbate scavengers in in vitro essays and mutants that are defective in ascorbate synthesis could accumulate lower Fe levels (Grillet et al., 2014). Well-known members of the Fe acquisition machinery of roots (IRT1, FRO2, etc.) that are all regulated by FIT are not expressed in the embryo (Belmonte et al., 2013). Therefore, the proteins determining Fe uptake in the embryo seem to be distinct from those in roots and are yet to be elucidated.

Comparison of elemental concentrations of developing embryos revealed that embryo acquisition of Fe takes place prior to that of Mn (Otegui et al., 2002; Roschzttardtz et al., 2009; Socha, 2016). While Fe enters early into the developing embryo, i.e., at torpedo stage (Roschzttardtz et al., 2009; Socha, 2016), Mn is sequestered in the endoplasmic reticulum of chalazal and micropyler endosperms and not released until the early phase of the photosynthetic bent-cotyledon stage (Otegui et al., 2002). In mature plants, Mn is transported into the endoplasmic reticulum by a P-type Ca^{2+} pump named ECA1 (Liang, Cunningham, Harper, & Sze, 1997). Since the corresponding gene is also transcribed in the chalazal and micropyler endosperms (Belmonte et al., 2013); ECA1 might mediate the transitory accumulation of Mn in these tissues. Because Fe and Mn have similar atomic radii and charges, membrane transporters often cannot differentiate between these two ions. Thus, transient storage of Mn in the endosperm could be a strategy to prioritize Fe transport to the embryo since essentiality of Fe for an organism is more widespread.

5. METAL TRANSPORT WITHIN THE EMBRYO

Metals are essential for metabolic processes occurring in rapidly developing embryos especially during early embryogenesis. Later on, metals are directed to storage compartments, as a preparation for desiccation stage. This is well illustrated in the case of Fe and photosynthesis. In green leaves of mature plants, 80% of Fe is located in the chloroplast (Terry & Abadía, 1986). In seeds, chloroplasts develop and the embryo becomes photosynthetically active at late globule stage (Tejos, Mercado, & Meisel, 2010). Therefore, similar to what occurs in leaves, a large part of embryonic Fe is stored in chloroplasts to support photosynthesis. In agreement with this view, Perls/DAB staining revealed an enrichment of Fe under the shoot apex of torpedo stage embryo (Fig. 2) showing an overlap with the region

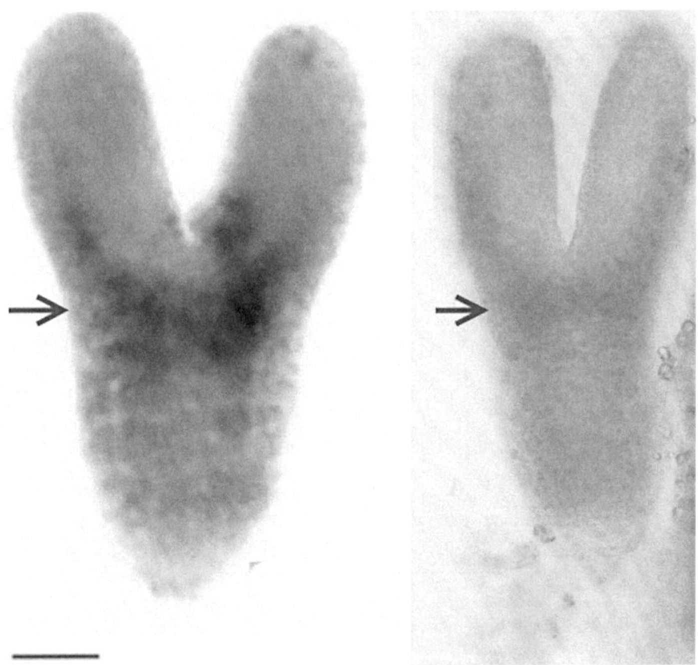

Fig. 2 Overlap of Fe and chlorophyll distribution in torpedo stage embryos. On the left, an isolated embryo was stained by Perls'/DAB and observed under a microscope. Fe is revealed as a black band under the shoot apex. Note that chlorophyll was removed during the staining protocol. On the right, distribution of chlorophyll in a similar embryo was observed under the microscope without any treatment. Bar is 50 μm. Red arrows indicate preferential Fe (left) and chlorophyll (right) accumulation under the shoot apex.

where strong chlorophyll accumulation is observed (Fig. 2 and see the chlorophyll fluorescence in fig. 2R in Tejos et al., 2010). However, Fe pool in the chloroplasts is obscured as the embryo develops, because the massive amount of Fe localizes around the provascular strands and also possibly chloroplast distribution no longer shows an obvious pattern (Fras, Smolen, & Maluszynska, 2008; Roschzttardtz et al., 2009; Tejos et al., 2010). During the subsequent maturation stages, Fe is effluxed from chloroplasts (Divol et al., 2013). Therefore, prior to desiccation stage, the largest Fe pool is localized around provascular strands which accounts for half of total seed Fe (Ramos, Khodja, Mary, & Thomine, 2013). This means that the use of metals that enter the embryo progressively shifts from maintaining metabolic needs to building up reserves.

Several genetic factors that mediate metal homeostasis within the embryo have recently been identified. Fe pools in dedifferentiating chloroplasts in maturing embryos are mobilized by two YSL proteins, YSL4/6 (Divol et al., 2013). In support of this, mutant plants that showed inactivation of these two genes were susceptible to Fe toxicity and contained an excess of Fe in the chloroplasts. Furthermore, at least one of these YSLs, YSL6 accumulated in seeds during late maturation and localized to the chloroplast membrane in embryos (Divol et al., 2013). However, it should be noted that this chloroplast localization could not be confirmed by a different study in which YSL6-GFP localized instead to the tonoplast or endoplasmic reticulum (Conte et al., 2013). The release of Fe from the chloroplast may contribute to Fe accumulation around the provascular strands. Transport of Fe into this hotspot has been shown to be mediated by VIT1, an ortholog of the vacuolar metal transporter CCC1 in yeast, based on complete loss of Fe enrichment around provascular strands in *vit1* knock out mutants (Kim et al., 2006). Vacuolar Fe stores are mobilized during germination in order to support metabolism, such as photosynthesis, before the root metal uptake machinery steps in (Bastow et al., 2018; Lanquar et al., 2005). Mobilization of Fe stores from the endodermal vacuoles depends on two tonoplast-localized proteins from the NRAMP family, NRAMP3/4 (Lanquar et al., 2005). In germinating seeds of wild-type plants, the staining of Fe around provascular strands disappears during the first 3 days (Roschzttardtz et al., 2009) and thereafter, root iron acquisition takes over, according to promoter activity of IRT1 (Lanquar et al., 2005). By contrast, in a *nramp3/4* double knock-out, Fe release from the endodermis was delayed (Roschzttardtz et al., 2009) and germinating seeds could not survive under Fe deficiency. Interestingly, in a *vit1* background, where Fe stores are mislocalized, the susceptibility of *nramp3/4* to Fe deficiency during early germination disappears. Thus, it is proposed that NRAMP3/4 specifically mediate an efflux of Fe from endodermal cells, and together with VIT1 constitute an influx-efflux unit in these cells (Mary et al., 2015).

Among the micronutrients investigated so far, in addition to Fe, only Mn appears as clearly enriched in certain cell types of the embryo. This enrichment is observed in the subepidermal cell layer of the abaxial sides of the cotyledons and in hypocotyl cortex. It is mediated by the tonoplast metal transporter MTP8 (Eroglu, 2015; Eroglu et al., 2017, 2016). MTP8-mediated enrichment accounts for one-third of total Mn in the seed (Ramos et al., 2013). Characterization of seeds that show genetic inactivation of *VIT1*, *MTP8* and other metal transporter genes resulted in seeds that mislocalized Fe and Mn in several different ways (Table 1). In contrast to the single large lytic vacuole in cells of mature plants, embryo cells contain several smaller vacuoles with one or few globules in each, containing phytic acid salts that chelate metals (Bolte et al., 2011; Feeney, Kittelmann, Menassa, Hawes, & Frigerio, 2018). Although Fe and Mn were differentially localized at cell-layer level; at the subcellular levels, both were observed in these globules (Eroglu et al., 2017; Lanquar et al., 2005; Roschzttardtz et al., 2009), probably associated with phytate (Bruch, Thomine, Tabares, & Un, 2015). Cell layer-specific distribution of Fe and Mn immediately raised the question of whether these highly specialized localizations have any kind of adaptive significance.

Table 1 Iron and Manganese Enriched Tissues in Mature Embryos of Mutant Arabidopsis Lines

Genotype	Fe Hotspots	Mn Hotspots	References
Col-0	e	s	Kim et al. (2006)
vit1	s	s	Kim et al. (2006)
mtp8	e	e	Eroglu (2015)
vit1 mtp8	h	h	Eroglu et al. (2017)
cax1/3	e	s (l.p)	Punshon et al. (2012)
nramp3/4	e	s + e	Socha (2016)
mtp8 nramp3/4	e	h + e	Socha (2016)
Imbibed Col-0	e + s	s	Eroglu et al. (2017)
Imbibed *mtp8*	e	s	Eroglu et al. (2017)

The table mostly refers to dry seeds. Imbibed seeds were incubated in water for 2 days. The metal distribution is described as follows: h: homogeneously distributed, e: enriched in endodermal cell layer, s: enriched in sub-epidermal cell layer, (l.p): enriched but less pronounced than in wild type (Col-0).

6. DO TONOPLAST TRANSPORTERS CONTROL METAL ACQUISITION IN THE EMBRYO?

The functional significance of cell-layer specific accumulation of Fe and Mn in the embryo is yet unclear. None of the cell layers, including the Mn-enriched subepidermal or the Fe-enriched endodermal ones, differ from each other under the microscope, regarding the number, distribution, and size of storage organelles (Busse & Evert, 1999), thereby failing to indicate any kind of specialization for these cell layers. Although a few studies have suggested that Fe enrichment in the endodermis is critical for germination under Fe deficient conditions (Gollhofer et al., 2014; Kim et al., 2006; Mary et al., 2015), mechanistic explanations for the phenotypes are lacking. In addition, the specific enrichment of Mn in subepidermal cells could not be associated with any relevant physiological function.

Alternatively, the two transporters responsible for the cell layer specific accumulations, VIT1 and MTP8, may be involved in embryo metal loading. Three lines of evidence suggest that VIT1 and MTP8 affect metal translocation from the endosperm to the embryo. Firstly, the activities of the *VIT1* and *MTP8* promoters coincide with the entry into the embryo of Fe and Mn, respectively (Belmonte et al., 2013; Eroglu et al., 2017; Otegui et al., 2002). *VIT1* is expressed as early as the torpedo stage, in contrast to *MTP8*, which is not expressed until the bent-cotyledon stage. Secondly, qualitative X-ray fluorescence imaging has shown that the absence of VIT1 delays Fe loading to the embryo (Punshon et al., 2013; Socha, 2016). Whereas Fe can be observed in endodermal cells of wild-type embryos, at the same developmental stage of *vit1* seeds, it is localized in the coat. Thirdly, the absence of MTP8 somehow triggered embryo to switch on Mn uptake in the early phase of germination (Eroglu et al., 2017). In wild-type plants, seed embryos take up external Mn at millimolar concentrations only, suggesting that during germination Mn transport into the embryo is mediated by a low-affinity transport system. In contrast, *mtp8* embryos efficiently accumulate Mn at micromolar concentrations, suggesting that an alternative, high-affinity acquisition system is activated. Taken together, these data suggest that tonoplast transporters can control metal acquisition in the embryo possibly by signalling the embryo metal status to plasma membrane transporters expressed at its surface (Fig. 1). Further exploration of this model offers a promising research direction.

Seed is the final target for nutrients in the life cycle of annual plants, including staple crops and model plants such as *A. thaliana*. Up to date, many metal transporter genes have been characterized in mature plants, but it has been neglected most often, to take one step further to investigate whether these metal transporter genes also contribute to seed homeostasis. Therefore, despite mature plants have been extensively studied, metal transportation to the seed and within the seed is still poorly understood. Yet, metals in seeds must be optimized for their concentration and localization to improve stress tolerance in agriculture and to combat micronutrient deficiencies in humans. The present review compiled the relevant literature and drew a picture of our current understanding of seed metal nutrition. By interpreting existing data and discussing knowledge gaps, we tried to generate new hypotheses and pointed out some new research directions, which will hopefully be fruitful.

ACKNOWLEDGEMENTS

The author would like to thank Dr. Elizabeth Grabau, Dr. Alexandra Leskova and Dr. Katarina Vogel-Mikus for constructive criticism of the manuscript.

REFERENCES

Andresen, E., Peiter, E., & Küpper, H. (2018). Trace metal metabolism in plants. *Journal of Experimental Botany*, *69*, 909–954.

Arrivault, S., Senger, T., & Krämer, U. (2006). The Arabidopsis metal tolerance protein AtMTP3 maintains metal homeostasis by mediating Zn exclusion from the shoot under Fe deficiency and Zn oversupply. *The Plant Journal*, *46*, 861–879. https://doi.org/10.1111/j.1365-313X.2006.02746.x.

Baliardini, C., Meyer, C. L., Salis, P., Saumitou-Laprade, P., & Verbruggen, N. (2015). CATION EXCHANGER1 cosegregates with cadmium tolerance in the metal hyperaccumulator Arabidopsis halleri and plays a role in limiting oxidative stress in Arabidopsis Spp. *Plant Physiology*, *169*, 549.

Barberon, M., Vermeer, J. E. M., De Bellis, D., Wang, P., Naseer, S., Andersen, T. G., et al. (2016). Adaptation of root function by nutrient-induced plasticity of endodermal differentiation. *Cell*, *164*, 447–459. https://doi.org/10.1016/j.cell.2015.12.021.

Bastow, E. L., de la Torre, V. S. G., Maclean, A. E., Green, R. T., Merlot, S., Thomine, S., et al. (2018). Vacuolar iron stores gated by NRAMP3 and NRAMP4 are the primary source of iron in germinating seeds. *Plant Physiology*, *177*, 1267–1276. https://doi.org/10.1104/pp.18.00478.

Bates, P. D., Jewell, J. B., & Browse, J. (2013). Rapid separation of developing Arabidopsis seeds from siliques for RNA or metabolite analysis. *Plant Methods*, *9*, 9. https://doi.org/10.1186/1746-4811-9-9.

Becher, M., Talke, I. N., Krall, L., & Krämer, U. (2003). Cross-species microarray transcript profiling reveals high constitutive expression of metal homeostasis genes in shoots of the zinc hyperaccumulator Arabidopsis halleri. *The Plant Journal*, *37*, 251–268. https://doi.org/10.1046/j.1365-313X.2003.01959.x.

Beeckman, T., De Rycke, R., Viane, R., & Inzé, D. (2000). Histological study of seed coat development in Arabidopsis thaliana. *Journal of Plant Research*, *113*, 139–148. https://doi.org/10.1007/PL00013924.

Belmonte, M. F., Kirkbride, R. C., Stone, S. L., Pelletier, J. M., Bui, A. Q., Yeung, E. C., et al. (2013). Comprehensive developmental profiles of gene activity in regions and subregions of the Arabidopsis seed. *Proceedings of the National Academy of Sciences of the United States of America*, *110*, E435–E444.

Bolte, S., Lanquar, V., Soler, M.-N., Beebo, A., Satiat-Jeunemaître, B., Bouhidel, K., et al. (2011). Distinct lytic vacuolar compartments are embedded inside the protein storage vacuole of dry and germinating Arabidopsis thaliana seeds. *Plant & Cell Physiology, 52*, 1142–1152. https://doi.org/10.1093/pcp/pcr065.

Boyd, R. S. (2007). The defense hypothesis of elemental hyperaccumulation: Status, challenges and new directions. *Plant and Soil, 293*, 153–176.

Briat, J.-F., Duc, C., Ravet, K., & Gaymard, F. (2010). Ferritins and iron storage in plants. *Biochimica et Biophysica Acta (BBA)—General Subjects, 1800*, 806–814. https://doi.org/10.1016/j.bbagen.2009.12.003.

Bruch, E. M., Thomine, S., Tabares, L. C., & Un, S. (2015). Variations in Mn (II) speciation among organisms: What makes D. radiodurans different. *Metallomics, 7*, 136–144.

Brumbarova, T., Bauer, P., & Ivanov, R. (2015). Molecular mechanisms governing Arabidopsis iron uptake. *Trends in Plant Science, 20*, 124–133. https://doi.org/10.1016/j.tplants.2014.11.004.

Busse, J. S., & Evert, R. F. (1999). Pattern of differentiation of the first vascular elements in the embryo and seedling of *Arabidopsis thaliana*. *International Journal of Plant Sciences, 160*, 1–13. https://doi.org/10.1086/314098.

Castaings, L., Caquot, A., Loubet, S., & Curie, C. (2016). The high-affinity metal transporters NRAMP1 and IRT1 team up to take up iron under sufficient metal provision. *Scientific Reports, 6*, 37222.

Chen, L.-Q., Lin, I. W., Qu, X.-Q., Sosso, D., McFarlane, H. E., Londoño, A., et al. (2015). A cascade of sequentially expressed sucrose transporters in the seed coat and endosperm provides nutrition for the Arabidopsis embryo. *The Plant Cell, 27*, 607–619. https://doi.org/10.1105/tpc.114.134585.

Colangelo, E. P., & Guerinot, M. L. (2004). The essential basic helix-loop-helix protein FIT1 is required for the iron deficiency response. *The Plant Cell, 16*, 3400–3412. Online.

Conte, S. S., Chu, H. H., Chan-Rodriguez, D., Punshon, T., Vasques, K. A., Salt, D. E., et al. (2013). Arabidopsis thaliana Yellow Stripe1-Like4 and Yellow Stripe1-Like6 localize to internal cellular membranes and are involved in metal ion homeostasis. *Frontiers in Plant Science, 4*, 283. https://doi.org/10.3389/fpls.2013.00283.

Curie, C., Cassin, G., Couch, D., Divol, F., Higuchi, K., Le Jean, M., et al. (2009). Metal movement within the plant: Contribution of nicotianamine and yellow stripe 1-like transporters. *Annals of Botany, 103*, 1–11.

De Giorgi, J., Piskurewicz, U., Loubery, S., Utz-Pugin, A., Bailly, C., Mène-Saffrané, L., et al. (2015). An endosperm-associated cuticle is required for Arabidopsis seed viability, dormancy and early control of germination. *PLoS Genetics, 11*, e1005708. https://doi.org/10.1371/journal.pgen.1005708.

Dekkers, B. J., Pearce, S., van Bolderen-Veldkamp, R. P., Marshall, A., Widera, P., Gilbert, J., et al. (2013). Transcriptional dynamics of two seed compartments with opposing roles in Arabidopsis seed germination. *Plant Physiology, 163*, 205–215.

Delhaize, E., Kataoka, T., Hebb, D. M., White, R. G., & Ryan, P. R. (2003). Genes encoding proteins of the cation diffusion facilitator family that confer manganese tolerance. *The Plant Cell, 15*, 1131–1142. Online. https://doi.org/10.1105/tpc.009134.

Delude, C., Moussu, S., Joubès, J., Ingram, G., & Domergue, F. (2016). Plant surface lipids and epidermis development. In Y. Nakamura & Y. Li-Beisson (Eds.), *Subcellular biochemistry: Vol. 86. Lipids in plant and algae development* (pp. 287–313). Cham: Springer. https://doi.org/10.1007/978-3-319-25979-6_12.

Divol, F., Couch, D., Conéjéro, G., Roschzttardtz, H., Mari, S., & Curie, C. (2013). The Arabidopsis YELLOW STRIPE LIKE4 and 6 transporters control iron release from the chloroplast. *The Plant Cell, 25*, 1040–1055.

Dubeaux, G., Neveu, J., Zelazny, E., & Vert, G. (2018). Metal sensing by the IRT1 transporter-receptor orchestrates its own degradation and plant metal nutrition. *Molecular Cell, 69*, 953–964.e5. https://doi.org/10.1016/j.molcel.2018.02.009.

Durrett, T. P., Gassmann, W., & Rogers, E. E. (2007). The FRD3-mediated efflux of citrate into the root vasculature is necessary for efficient iron translocation. *Plant Physiology, 144*, 197–205.

Eroglu, S. (2015). *Characterization of MTP8 as a tonoplast Fe/Mn transporter essential for Fe efficiency and for Fe and Mn localization in the subepidermis of Arabidopsis embryos*. Halle (Saale): Universitäts-und Landesbibliothek Sachsen-Anhalt, Diss.

Eroglu, S., Giehl, R. F. H., Meier, B., Takahashi, M., Terada, Y., Ignatiev, K., et al. (2017). Metal tolerance protein 8 mediates manganese homeostasis and iron re-allocation during seed development and germination. *Plant Physiology, 174*, 1633–1647. https://doi.org/10.1104/pp.16.01646.

Eroglu, S., Meier, B., von Wirén, N., & Peiter, E. (2016). The vacuolar manganese transporter MTP8 determines tolerance to iron deficiency-induced chlorosis in Arabidopsis. *Plant Physiology, 170*, 1030–1045.

Feeney, M., Kittelmann, M., Menassa, R., Hawes, C., & Frigerio, L. (2018). Protein storage vacuoles originate from remodelled pre-existing vacuoles in Arabidopsis thaliana. *Plant Physiology, 177*, 241–254.

Fernández, V., & Eichert, T. (2009). Uptake of hydrophilic solutes through plant leaves: Current state of knowledge and perspectives of foliar fertilization. *Critical Reviews in Plant Sciences, 28*, 36–68. https://doi.org/10.1080/07352680902743069.

Franke, R., Höfer, R., Briesen, I., Emsermann, M., Efremova, N., Yephremov, A., et al. (2009). The DAISY gene from Arabidopsis encodes a fatty acid elongase condensing enzyme involved in the biosynthesis of aliphatic suberin in roots and the chalaza-micropyle region of seeds. *The Plant Journal, 57*, 80–95. https://doi.org/10.1111/j.1365-313X.2008.03674.x.

Fras, A., Smolen, B., & Maluszynska, J. (2008). Vascularization of zygotic and somatic embryos of Arabidopsis thaliana. *Acta Biologica Cracoviensia Series Botanica, 50*, 43–48.

Gao, H., Xie, W., Yang, C., Xu, J., Li, J., Wang, H., et al. (2017). NRAMP2, a trans-Golgi network-localized manganese transporter, is required for Arabidopsis root growth under manganese deficiency. *The New Phytologist, 217*, 179–193. https://doi.org/10.1111/nph.14783.

Gollhofer, J., Timofeev, R., Lan, P., Schmidt, W., & Buckhout, T. J. (2014). Vacuolar-iron-transporter1-like proteins mediate iron homeostasis in Arabidopsis. *PLoS ONE, 9*, e110468.

Grillet, L., Ouerdane, L., Flis, P., Hoang, M. T. T., Isaure, M.-P., Lobinski, R., et al. (2014). Ascorbate efflux as a new strategy for Iron reduction and transport in plants. *The Journal of Biological Chemistry, 289*, 2515–2525. https://doi.org/10.1074/jbc.M113.514828.

Gustin, J. L., Loureiro, M. E., Kim, D., Na, G., Tikhonova, M., & Salt, D. E. (2009). MTP1-dependent Zn sequestration into shoot vacuoles suggests dual roles in Zn tolerance and accumulation in Zn-hyperaccumulating plants. *The Plant Journal, 57*, 1116–1127. https://doi.org/10.1111/j.1365-313X.2008.03754.x.

Harrison, P. M., & Arosio, P. (1996). The ferritins: Molecular properties, iron storage function and cellular regulation. *Biochimica et Biophysica Acta, 1275*, 161–203.

Hindt, M. N., Akmakjian, G. Z., Pivarski, K. L., Punshon, T., Baxter, I., Salt, D. E., et al. (2017). BRUTUS and its paralogs, BTS LIKE1 and BTS LIKE2, encode important negative regulators of the iron deficiency response in Arabidopsis thaliana. *Metallomics, 9*, 876–890. https://doi.org/10.1039/C7MT00152E.

Hussain, D., Haydon, M. J., Wang, Y., Wong, E., Sherson, S. M., Young, J., et al. (2004). P-type ATPase heavy metal transporters with roles in essential zinc homeostasis in Arabidopsis. *The Plant Cell, 16*, 1327–1339. https://doi.org/10.1105/tpc.020487.

Ingram, G., & Nawrath, C. (2017). The roles of the cuticle in plant development: Organ adhesions and beyond. *Journal of Experimental Botany, 68*, 5307–5321.

Ishimaru, Y., Masuda, H., Bashir, K., Inoue, H., Tsukamoto, T., Takahashi, M., et al. (2010). Rice metal-nicotianamine transporter, OsYSL2, is required for the long-distance transport of iron and manganese. *The Plant Journal, 62*, 379–390. https://doi.org/10.1111/j.1365-313X.2010.04158.x.

Jean, M. L., Schikora, A., Mari, S., Briat, J.-F., & Curie, C. (2005). A loss-of-function mutation in AtYSL1 reveals its role in iron and nicotianamine seed loading. *The Plant Journal, 44*, 769–782.

Jeong, J., Merkovich, A., Clyne, M., & Connolly, E. L. (2017). Directing iron transport in dicots: Regulation of iron acquisition and translocation. *Current Opinion in Plant Biology, 39*, 106–113. https://doi.org/10.1016/j.pbi.2017.06.014.

Karmann, J., Müller, B., & Hammes, U. Z. (2018). The long and winding road: Transport pathways for amino acids in *Arabidopsis* seeds. *Plant Reproduction, 31*, 253–261. https://doi.org/10.1007/s00497-018-0334-5.

Kim, S. A., Punshon, T., Lanzirotti, A., Li, L., Alonso, J. M., Ecker, J. R., et al. (2006). Localization of iron in Arabidopsis seed requires the vacuolar membrane transporter VIT1. *Science, 314*, 1295.

Koh, S., Wiles, A. M., Sharp, J. S., Naider, F. R., Becker, J. M., & Stacey, G. (2002). An oligopeptide transporter gene family in Arabidopsis. *Plant Physiology, 128*, 21–29.

Koike, S., Inoue, H., Mizuno, D., Takahashi, M., Nakanishi, H., Mori, S., et al. (2004). OsYSL2 is a rice metal-nicotianamine transporter that is regulated by iron and expressed in the phloem. *The Plant Journal, 39*, 415–424. https://doi.org/10.1111/j.1365-313X.2004.02146.x.

Korshunova, Y. O., Eide, D., Clark, W. G., Guerinot, M. L., & Pakrasi, H. B. (1999). The IRT1 protein from Arabidopsis thaliana is a metal transporter with a broad substrate range. *Plant Molecular Biology, 40*, 37–44. https://doi.org/10.1023/A:1026438615520.

Lanquar, V., Lelièvre, F., Bolte, S., Hamès, C., Alcon, C., Neumann, D., et al. (2005). Mobilization of vacuolar iron by AtNRAMP3 and AtNRAMP4 is essential for seed germination on low iron. *The EMBO Journal, 24*, 4041–4051. https://doi.org/10.1038/sj.emboj.7600864.

Lashbrooke, J., Cohen, H., Levy-Samocha, D., Tzfadia, O., Panizel, I., Zeisler, V., et al. (2016). MYB107 and MYB9 homologs regulate suberin deposition in angiosperms. *The Plant Cell, 28*, 2097–2116. https://doi.org/10.1105/tpc.16.00490.

Le, B. H., Cheng, C., Bui, A. Q., Wagmaister, J. A., Henry, K. F., Pelletier, J., et al. (2010). Global analysis of gene activity during Arabidopsis seed development and identification of seed-specific transcription factors. *Proceedings of the National Academy of Sciences of the United States of America, 107*, 8063–8070. https://doi.org/10.1073/pnas.1003530107.

Li, L., Chen, O. S., Ward, D. M., & Kaplan, J. (2001). CCC1 is a transporter that mediates vacuolar iron storage in yeast. *The Journal of Biological Chemistry, 276*, 29515–29519.

Li, X., Zhang, H., Ai, Q., Liang, G., & Yu, D. (2016). Two bHLH transcription factors, bHLH34 and bHLH104, regulate iron homeostasis in Arabidopsis thaliana. *Plant Physiology, 170*, 2478–2493.

Liang, F., Cunningham, K. W., Harper, J. F., & Sze, H. (1997). ECA1 complements yeast mutants defective in Ca2+ pumps and encodes an endoplasmic reticulum-type Ca2+-ATPase in Arabidopsis thaliana. *Proceedings of the National Academy of Sciences of the United States of America, 94*, 8579–8584.

Long, T. A., Tsukagoshi, H., Busch, W., Lahner, B., Salt, D. E., & Benfey, P. N. (2010). The bHLH transcription factor POPEYE regulates response to iron deficiency in Arabidopsis roots. *The Plant Cell, 22*, 2219–2236. Online. https://doi.org/10.1105/tpc.110.074096.

Loubéry, S., Giorgi, J. D., Utz-Pugin, A., Demonsais, L., & Lopez-Molina, L. (2018). A maternally-deposited endosperm cuticle contributes to the physiological defects of transparent testa seeds. *Plant Physiology, 177*, 1218–1233. https://doi.org/10.1104/pp.18.00416.

Mary, V., Ramos, M. S., Gillet, C., Socha, A. L., Giraudat, J., Agorio, A., et al. (2015). Bypassing iron storage in endodermal vacuoles rescues the iron mobilization defect in the natural resistance associated-macrophage protein3natural resistance associated-macrophage protein4 double mutant. *Plant Physiology*, *169*, 748–759.

Milner, M. J., Seamon, J., Craft, E., & Kochian, L. V. (2013). Transport properties of members of the ZIP family in plants and their role in Zn and Mn homeostasis. *Journal of Experimental Botany*, *64*, 369–381. https://doi.org/10.1093/jxb/ers315.

Mladěnka, P., Macáková, K., Zatloukalová, L., Řeháková, Z., Singh, B. K., Prasad, A. K., et al. (2010). In vitro interactions of coumarins with iron. *Biochimie*, *92*, 1108–1114.

Molina, I., Ohlrogge, J. B., & Pollard, M. (2007). Deposition and localization of lipid polyester in developing seeds of Brassica napus and Arabidopsis thaliana: Lipid polyesters in seed coats. *The Plant Journal*, *53*, 437–449. https://doi.org/10.1111/j.1365-313X.2007.03348.x.

Montanini, B., Blaudez, D., Jeandroz, S., Sanders, D., & Chalot, M. (2007). Phylogenetic and functional analysis of the cation diffusion facilitator (CDF) family: Improved signature and prediction of substrate specificity. *BMC Genomics*, *8*, 107.

Morrissey, J., Baxter, I. R., Lee, J., Li, L., Lahner, B., Grotz, N., et al. (2009). The ferroportin metal efflux proteins function in Iron and cobalt homeostasis in Arabidopsis. *The Plant Cell*, *21*, 3326–3338. https://doi.org/10.1105/tpc.109.069401.

Müller, B., Fastner, A., Karmann, J., Mansch, V., Hoffmann, T., Schwab, W., et al. (2015). Amino acid export in developing Arabidopsis seeds depends on UmamiT facilitators. *Current Biology*, *25*, 3126–3131.

Nguyen, H., Brown, R. C., & Lemmon, B. E. (2000). The specialized chalazal endosperm in Arabidopsis thaliana and Lepidium virginicum (Brassicaceae). *Protoplasma*, *212*, 99–110. https://doi.org/10.1007/BF01279351.

Nishida, S., Tsuzuki, C., Kato, A., Aisu, A., Yoshida, J., & Mizuno, T. (2011). AtIRT1, the primary iron uptake transporter in the root, mediates excess nickel accumulation in Arabidopsis thaliana. *Plant & Cell Physiology*, *52*, 1433–1442.

Olsen, O.-A. (2004). Nuclear endosperm development in cereals and Arabidopsis thaliana. *The Plant Cell*, *16*, S214–S227. Online.

Olsen, L. I., Hansen, T. H., Larue, C., Østerberg, J. T., Hoffmann, R. D., Liesche, J., et al. (2016). Mother-plant-mediated pumping of zinc into the developing seed. *Nature Plants*, *2*, 16036. https://doi.org/10.1038/nplants.2016.36.

Otegui, M. S., Capp, R., & Staehelin, L. A. (2002). Developing seeds of Arabidopsis store different minerals in two types of vacuoles and in the endoplasmic reticulum. *The Plant Cell*, *14*, 1311–1327. Online. https://doi.org/10.1105/tpc.010486.

Perry, S. E., & Wang, H. (2003). Rapid isolation of Arabidopsis thaliana developing embryos. *Biotechniques*, *35*, 278–280. 282.

Persans, M. W., Nieman, K., & Salt, D. E. (2001). Functional activity and role of cation-efflux family members in Ni hyperaccumulation in Thlaspi goesingense. *Proceedings of the National Academy of Sciences of the United States of America*, *98*, 9995–10000. https://doi.org/10.1073/pnas.171039798.

Punshon, T., Hirschi, K., Yang, J., Lanzirotti, A., Lai, B., & Guerinot, M. L. (2012). The role of CAX1 and CAX3 in elemental distribution and abundance in Arabidopsis seed. *Plant Physiology*, *158*, 352–362. http://www.plantphysiology.org/content/158/1/352.short.

Punshon, T., Ricachenevsky, F. K., Hindt, M. N., Socha, A. L., & Zuber, H. (2013). Methodological approaches for using synchrotron X-ray fluorescence (SXRF) imaging as a tool in ionomics: Examples from Arabidopsis thaliana. *Metallomics*, *5*, 1133–1145. https://doi.org/10.1039/C3MT00120B.

Raissig, M. T., Gagliardini, V., Jaenisch, J., Grossniklaus, U., & Baroux, C. (2013). Efficient and rapid isolation of early-stage embryos from Arabidopsis thaliana seeds. *Journal of Visualized Experiments: JoVE*, *76*, e50371. https://doi.org/10.3791/50371.

Rajniak, J., Giehl, R. F., Chang, E., Murgia, I., Wirén, N., & Sattely, E. S. (2018). Biosynthesis of redox-active metabolites in response to iron deficiency in plants. *Nature Chemical Biology*, *14*, 442–450.

Ramos, M. S., Khodja, H., Mary, V., & Thomine, S. (2013). Using μPIXE for quantitative mapping of metal concentration in Arabidopsis thaliana seeds. *Frontiers in Plant Science*, *4*, 168. https://doi.org/10.3389/fpls.2013.00168.

Rellán-Álvarez, R., Giner-Martínez-Sierra, J., Orduna, J., Orera, I., Rodríguez-Castrillón, J. Á., García-Alonso, J. I., et al. (2009). Identification of a tri-iron (III), tri-citrate complex in the xylem sap of iron-deficient tomato resupplied with iron: New insights into plant iron long-distance transport. *Plant & Cell Physiology*, *51*, 91–102.

Rengel, Z., Batten, G. D., & Crowley, D. E. (1999). Agronomic approaches for improving the micronutrient density in edible portions of field crops. *Field Crops Research*, *60*, 27–40. https://doi.org/10.1016/S0378-4290(98)00131-2.

Ricachenevsky, F. K., Menguer, P. K., Sperotto, R. A., Williams, L. E., & Fett, J. P. (2013). Roles of plant metal tolerance proteins (MTP) in metal storage and potential use in biofortification strategies. *Frontiers in Plant Science*, *4*, 144. https://doi.org/10.3389/fpls.2013.00144.

Robinson, N. J., Procter, C. M., Connolly, E. L., & Guerinot, M. L. (1999). A ferric-chelate reductase for iron uptake from soils. *Nature*, *397*, 694–697. https://doi.org/10.1038/17800.

Rolletschek, H., Weber, H., & Borisjuk, L. (2003). Energy status and its control on embryogenesis of legumes. Embryo photosynthesis contributes to oxygen supply and is coupled to biosynthetic fluxes. *Plant Physiology*, *132*, 1196–1206. https://doi.org/10.1104/pp.102.017376.

Roschzttardtz, H., Conéjéro, G., Curie, C., & Mari, S. (2009). Identification of the endodermal vacuole as the iron storage compartment in the Arabidopsis embryo. *Plant Physiology*, *151*, 1329–1338. https://doi.org/10.1104/pp.109.144444.

Roschzttardtz, H., Séguéla-Arnaud, M., Briat, J.-F., Vert, G., & Curie, C. (2011). The FRD3 citrate effluxer promotes Iron nutrition between symplastically disconnected tissues throughout Arabidopsis development. *The Plant Cell*, *23*, 2725–2737. Online. https://doi.org/10.1105/tpc.111.088088.

Santi, S., & Schmidt, W. (2009). Dissecting iron deficiency-induced proton extrusion in Arabidopsis roots. *The New Phytologist*, *183*, 1072–1084. https://doi.org/10.1111/j.1469-8137.2009.02908.x.

Schaaf, G., Honsbein, A., Meda, A. R., Kirchner, S., Wipf, D., & von Wirén, N. (2006). AtIREG2 encodes a tonoplast transport protein involved in iron-dependent nickel detoxification in Arabidopsis thaliana roots. *The Journal of Biological Chemistry*, *281*, 25532–25540. https://doi.org/10.1074/jbc.M601062200.

Schmid, N. B., Giehl, R. F., Döll, S., Mock, H.-P., Strehmel, N., Scheel, D., et al. (2014). Feruloyl-CoA 6′-Hydroxylase1-dependent coumarins mediate iron acquisition from alkaline substrates in Arabidopsis. *Plant Physiology*, *164*, 160–172.

Schmidke, I., & Stephan, U. W. (2006). Transport of metal micronutrients in the phloem of castor bean (Ricinus communis) seedlings. *Physiologia Plantarum*, *95*, 147–153. https://doi.org/10.1111/j.1399-3054.1995.tb00821.x.

Schuler, M., Rellán-Álvarez, R., Fink-Straube, C., Abadía, J., & Bauer, P. (2012). Nicotianamine functions in the phloem-based transport of iron to sink organs, in pollen development and pollen tube growth in Arabidopsis. *The Plant Cell*, *24*, 2380–2400. Online. https://doi.org/10.1105/tpc.112.099077.

Sisó-Terraza, P., Luis-Villarroya, A., Fourcroy, P., Briat, J.-F., Abadía, A., Gaymard, F., et al. (2016). Accumulation and secretion of coumarinolignans and other coumarins in Arabidopsis thaliana roots in response to Iron deficiency at high pH. *Frontiers in Plant Science*, 7, 1711. https://doi.org/10.3389/fpls.2016.01711.

Socha, A. L. (2016). *Uncovering the mechanisms controlling metal micronutrient homeostasis in plants*. Dissertation Dartmouth College.
Stacey, M. G., Koh, S., Becker, J., & Stacey, G. (2002). AtOPT3, a member of the oligopeptide transporter family, is essential for embryo development in Arabidopsis. *The Plant Cell, 14*, 2799–2811. https://doi.org/10.1105/tpc.005629.
Stacey, M. G., Patel, A., McClain, W. E., Mathieu, M., Remley, M., Rogers, E. E., et al. (2008). The Arabidopsis AtOPT3 protein functions in metal homeostasis and movement of iron to developing seeds. *Plant Physiology, 146*, 589.
Stadler, R., Lauterbach, C., & Sauer, N. (2005). Cell-to-cell movement of green fluorescent protein reveals post-phloem transport in the outer integument and identifies symplastic domains in Arabidopsis seeds and embryos. *Plant Physiology, 139*, 701–712. https://doi.org/10.1104/pp.105.065607.
Stephan, U. W., & Scholz, G. (1993). Nicotianamine: Mediator of transport of iron and heavy metals in the phloem? *Physiologia Plantarum, 88*, 522–529.
Tejos, R. I., Mercado, A. V., & Meisel, L. A. (2010). Analysis of chlorophyll fluorescence reveals stage specific patterns of chloroplast-containing cells during Arabidopsis embryogenesis. *Biological Research, 43*, 99–111.
Terry, N., & Abadía, J. (1986). Function of iron in chloroplasts. *Journal of Plant Nutrition, 9*, 609–646.
Thomine, S., & Vert, G. (2013). Iron transport in plants: Better be safe than sorry. *Current Opinion in Plant Biology, 16*, 322–327. Physiology and metabolism. https://doi.org/10.1016/j.pbi.2013.01.003.
Thomine, S., Wang, R., Ward, J. M., Crawford, N. M., & Schroeder, J. I. (2000). Cadmium and iron transport by members of a plant metal transporter family in Arabidopsis with homology to Nramp genes. *Proceedings of the National Academy of Sciences of the United States of America, 97*, 4991–4996. https://doi.org/10.1073/pnas.97.9.4991.
Tsai, H. H., & Schmidt, W. (2017). Mobilization of Iron by plant-borne coumarins. *Trends in Plant Science, 22*, 538–548. https://doi.org/10.1016/j.tplants.2017.03.008.
Tyler, G. (1998). Plant seeds as mineral nutrient resource for seedlings—A comparison of plants from calcareous and silicate soils. *Annals of Botany, 81*, 455–459. https://doi.org/10.1006/anbo.1997.0581.
Vatansever, R., Filiz, E., & Eroglu, S. (2017). Genome-wide exploration of metal tolerance protein (MTP) genes in common wheat (Triticum aestivum): Insights into metal homeostasis and biofortification. *Biometals, 30*, 217–235. https://doi.org/10.1007/s10534-017-9997-x.
Verret, F., Gravot, A., Auroy, P., Leonhardt, N., David, P., Nussaume, L., et al. (2004). Overexpression of AtHMA4 enhances root-to-shoot translocation of zinc and cadmium and plant metal tolerance. *FEBS Letters, 576*, 306–312. https://doi.org/10.1016/j.febslet.2004.09.023.
von Wirén, N., Klair, S., Bansal, S., Briat, J.-F., Khodr, H., Shioiri, T., et al. (1999). Nicotianamine chelates both FeIII and FeII. Implications for metal transport in plants. *Plant Physiology, 119*, 1107–1114.
Waters, B. M., Chu, H.-H., DiDonato, R. J., Roberts, L. A., Eisley, R. B., Lahner, B., et al. (2006). Mutations in Arabidopsis yellow stripe-Like1 and yellow stripe-Like3 reveal their roles in metal ion homeostasis and loading of metal ions in seeds. *Plant Physiology, 141*, 1446–1458. https://doi.org/10.1104/pp.106.082586.
Waters, B. M., & Sankaran, R. P. (2011). Moving micronutrients from the soil to the seeds: Genes and physiological processes from a biofortification perspective. *Plant Science, 180*, 562–574.
Wiersma, J. V. (2005). High rates of Fe-EDDHA and seed iron concentration suggest partial solutions to iron deficiency in soybean. *Agronomy Journal, 97*, 924–934. https://doi.org/10.2134/agronj2004.0309.

Wiersma, J. V. (2012). Is seed iron concentration predictive of resistance to iron deficiency in soybean? *Crop Science*, *52*, 471. https://doi.org/10.2135/cropsci2011.10.0549.

Wong, C. K. E., Jarvis, R. S., Sherson, S. M., & Cobbett, C. S. (2008). Functional analysis of the heavy metal binding domains of the Zn/Cd-transporting ATPase, HMA2, in Arabidopsis thaliana. *The New Phytologist*, *181*, 79–88. https://doi.org/10.1111/j.1469-8137.2008.02637.x.

Wongkaew, A., Asayama, K., Kitaiwa, T., Nakamura, S., Kojima, K., Stacey, G., et al. (2018). AtOPT6 protein functions in long distance transport of glutathione in Arabidopsis thaliana. *Plant & Cell Physiology*, *59*, 1443–1451. https://doi.org/10.1093/pcp/pcy074.

Yeung, E. C., & Meinke, D. W. (1993). Embryogenesis in angiosperms: Development of the suspensor. *The Plant Cell*, *5*, 1371–1381. Onlinehttps://doi.org/10.1105/tpc.5.10.1371.

Zhai, Z., Gayomba, S. R., Jung, H., Vimalakumari, N. K., Piñeros, M., Craft, E., et al. (2014). OPT3 is a phloem-specific iron transporter that is essential for systemic iron signaling and redistribution of iron and cadmium in Arabidopsis. *The Plant Cell*, *26*, 2249–2264.

Zhao, F.-J., Moore, K. L., Lombi, E., & Zhu, Y.-G. (2014). Imaging element distribution and speciation in plant cells. *Trends in Plant Science*, *19*, 183–192. https://doi.org/10.1016/j.tplants.2013.12.001.

CHAPTER FIVE

Transporters and Mechanisms of Hormone Transport in Arabidopsis

Rashed Abualia*, Eva Benkova*[,1], Benoît Lacombe[†,1]
*IST Austria, Klosterneuburg, Austria
[†]Biochimie et Physiologie Moléculaire des Plantes (BPMP), Université de Montpellier, CNRS, INRA, SupAgro, Montpellier, France
[1]Corresponding authors: e-mail address: eva.benkova@ist.ac.at; benoit.lacombe@supagro.fr

Contents

Abstract

Plant hormones as signalling molecules play an essential role in the control of plant growth and development. Typically, sites of hormonal action are usually distant from the site of biosynthesis thus relying on efficient transport mechanisms. Over the last decades, molecular identification of proteins and protein complexes involved in hormonal transport has started. Advanced screens for genes involved in hormonal transport in combination with transport assays using heterologous systems such as yeast, insect, or tobacco BY2 cells or *Xenopus* oocytes provided important insights into mechanisms underlying distribution of hormones in plant body and led to identification of principal transporters for each hormone. This review gives a short overview of the mechanisms of hormonal transport and transporters identified in *Arabidopsis thaliana*.

Advances in Botanical Research, Volume 87
ISSN 0065-2296
https://doi.org/10.1016/bs.abr.2018.09.007

1. INTRODUCTION

Plants differ from most animals in having a sessile life style that impairs their 'fight' or 'flight' responses to adverse environmental conditions. To compensate for such restrictions plants have evolved a wide range of mechanisms to adapt their growth and development to highly heterogeneous and often challenging environment. Plant hormones as essential endogenous regulators of growth and development play an irreplaceable role in mediating adaptive responses to a whole spectrum of external stimuli, including abiotic and biotic stresses. Overall, the actions of plant hormones are determined by the combined output of specific pathways, which control their biosynthesis and metabolism, transport, perception and signal transduction. Darwin's initial observations of wheat coleoptile phototropic bending (Darwin, 1880); which led to the postulation of hormones in plants, had already indicated that controlled distribution of these substances from source to site of action might be crucial for their biological function. Since then, many plant hormones have been identified and similarly to the first hypothesized phytohormone, auxin, their sites of synthesis and action were found to be distant. This inevitably triggered a question on mechanisms that mediate and control movement of hormones within the plant body.

When compared to animals, the lack of blood and lymphatic system streams in plants considerably limits effective distribution of substances. In order to overcome such constraints, plants evolved other means of transport to secure effective delivery of nutrients and other low molecular weight components, including hormones, within their body. Vascular systems encompassing phloem and xylem are considered the major routes for long-distance transport in plants. Most plant hormones, including auxin, abscisic acid (ABA), cytokinins and jasmonates, have been identified as substrates for a long-distance transport (reviewed in Lacombe and Achard, 2016). Long distance delivery of hormones relies on mechanisms that mediate their loading from the source into the xylem and/or phloem and unloading at the target tissue. Hence, over the last period identification of proteins that mediate uptake of hormones and their unloading became a challenging task for researchers.

On the other hand, a slower, short-distance movement of hormones takes place either via apoplast, symplast or transcellularly (Tester & Leigh, 2001). Unlike apoplastic transport occurring exclusively via extracellular space and symplastic transport via intracellular space, transcellular (cell to cell) transport requires precisely located transporters that mediate uptake of a compound into and its export out of the cell (Amtmann & Blatt, 2009). This cell to cell transport must be mediated via importers and exporters located at specific polar membrane domains and provide means for efficient vectorial flows of molecules.

Since the first observations that have outlined the biological significance of tightly regulated transport of hormones within the plant body, transporters for most hormones have been recognized and their transport properties characterized in plants as well as in heterologous systems. The main focus of our review is on major families of transporters that mediate transport of hormones including auxin, abscisic acid, cytokinins, gibberellins, jasmonates, ethylene, brassinosteroids and strigolactones (Fig. 1). Evidences for their transport activity, selectivity and directionality in heterologous systems (*Xenopus* oocyte, yeast and other systems) and *in planta* will be discussed.

Hormone	Name	Directionality	Subcellular Localization	Structure
Auxin	PIN1, PIN2, PIN3, PIN4, PIN7	Efflux	Plasma Membrane	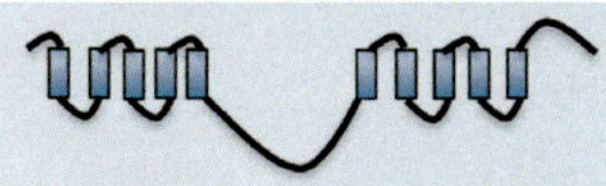
Auxin	PIN5, PIN6, PIN8, PIN9	Efflux	Endo-membranes	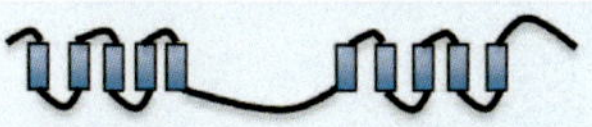
Auxin	ABCB1, ABCB19	Efflux	Plasma Membrane	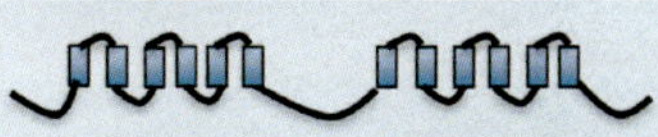
Auxin	ABCB4, ABCB21	Influx/Efflux	Plasma membrane	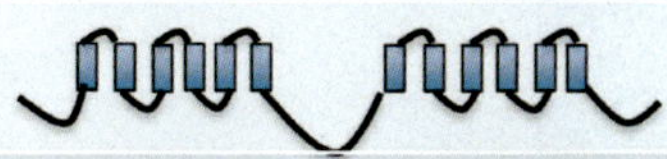
Auxin	AUX1, LAX1, LAX2, LAX3	Influx	Plasma membrane	
Auxin	NRT1.1/NPF6.3	Influx	Plasma membrane	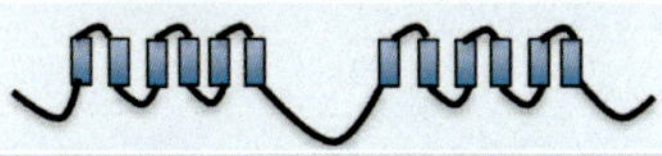
Auxin	WAT1		vacuole	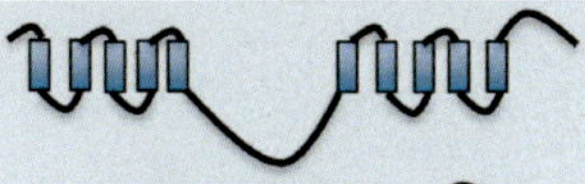
Cytokinin	AGCG14	Efflux	Plasma membrane	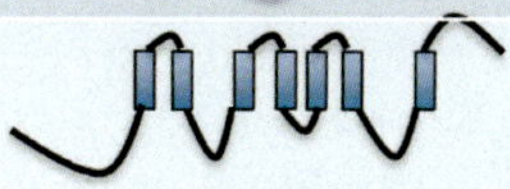

Fig. 1 Hormone transporters identified in *Arabidopsis thaliana*. Hormone substrates, names, directionality, subcellular localization and secondary structure are presented, see text for detail.

Hormone	Transporter	Influx/Efflux	Localization	Structure
Cytokinin	ENT3, ENT6, ENT7, ENT8	Influx	Plasma membrane	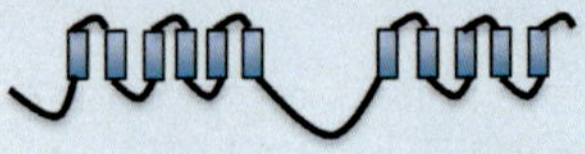
Cytokinin	PUP14	Influx	Plasma membrane	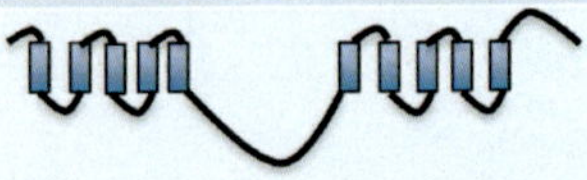
ABA	ABCG31	Efflux	Plasma membrane	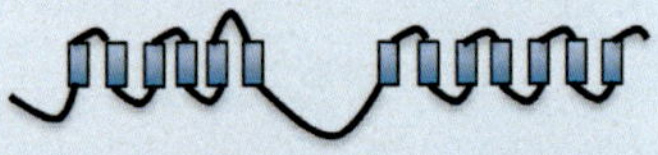
ABA	ABCG40	Influx	Plasma membrane	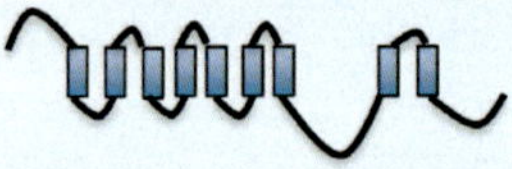
ABA	DTX50	Efflux	Plasma membrane	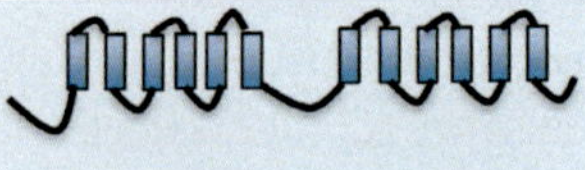
ABA	NPF4.1, 4.2, 4.5, 4.6	Influx	Plasma Membrane	
GA	NPF3.1	Influx	Plasma membrane	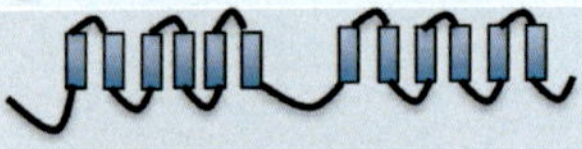
GA	SWEET13, SWEET14	Influx	Plasma membrane	
GA	NPF4.1 and 18 other NPFs	Influx	Plasma membrane	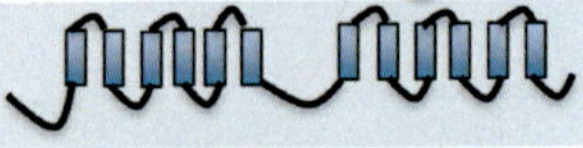
Jasmonate	ABCG16	Influx	Nuclear Envelop/ Plasma membrane	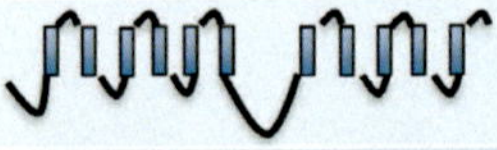
Jasmonate	NPF2.10 and 12 other NPF	Influx	Plasma membrane	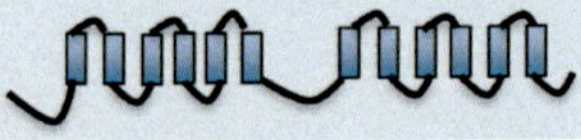

Fig. 1—Cont'd

2. AUXIN

Auxin is a group of naturally occurring molecules derived from tryptophan with the indole-3-acetic acid (IAA) as the major form of auxin. Auxin exhibits a broad range of regulatory functions in plant development including embryogenesis, postembryonic initiation and formation of organs, and tropic responses (Benková et al., 2003; Friml et al., 2003; Reinhardt, 2003). In higher plants, the long-distance transport through the phloem vasculature enables fast movement of auxin from the sources, typically young shoots, to sink tissues such as roots (Cambridge & Morris, 1996). On the other hand, auxin cell to cell movement, mediated via a set of membrane located transporters, the so-called polar auxin transport (PAT) machinery, represents a slower way of auxin delivery. Noteworthy, this active influx–efflux-dependent transfer of auxin between cells is of great biological significance. It enables directional movement of the hormone as well as its graded distribution across tissues, thus increasing the regulatory power of auxin. Indeed local accumulation of auxin and its graded distribution directs initiation of new organs, patterning and morphogenesis of tissues and organs, as well as differential cell growth, a process essential for tropic bending of plant organs (Benková et al., 2003; Friml et al., 2003; Friml, Wiśniewska, Benková, Mendgen, & Palme, 2002; Reinhardt, 2003).

This essential role of auxin in plant growth and development might justify why during plant evolution a complex transport system encompassing different transporter families has evolved. Over the last years auxin transport capacity of several protein families including PIN, ABCB, PILS, AUX/LAX and NPF has been recognized and auxin transport activity for key members of each family has been demonstrated biochemically (Barbez et al., 2012; Bennett et al., 1996; Krouk et al., 2010; Okada, Ueda, Komaki, Bell, & Shimura, 1991; Ye, Liu, Xing, & Kang, 2013). However, how these different transporters form dynamic systems that direct auxin transport and flexibly respond to various developmental, physiological and external cues remains to be elucidated.

2.1 PINs as Polar Auxin Efflux Transporters

The molecular basis of auxin cell to cell movement is largely determined by biochemical properties of cellular compartments and the weak acid character of the IAA molecule. Due to an acidic environment in the apoplast, IAA as a weak acid remains highly protonated and can diffuse passively across the plasma membrane into the cells. Inside cells, at near-neutral pH, IAA dissociates into an ionic form (IAA^-) and H^+ and as a consequence cannot exit cells passively. This biochemical basis of auxin cell to cell movement has been conceptualized in the chemiosmotic hypothesis (Rubery & Sheldrake, 1974, Raven, 1975), which also postulated the existence of active transporters that would facilitate export of auxin from cells (reviewed by Goldsmith, 1977). Only decades later, implementation of genetics and molecular biology tools have enabled identification of the transporters predicted by the chemiosmotic theory. Auxin efflux carrier PIN1 has been among the first to be recognized, following the isolation of the *pin-formed1 (pin1)* mutant in *Arabidopsis thaliana* (Okada et al., 1991). Striking resemblance of the *pin1* phenotype, characterized by a loss of lateral organs resulting in pin-like stems, to plants treated with chemical inhibitors of auxin transport hinted at the role of *PIN1* in auxin transport (Gälweiler et al., 1998; Okada et al., 1991). Later studies provided essential biochemical and biological evidence including transport assays in heterologous system that PIN1 functions as an auxin efflux transporter. Importantly, in perfect agreement with postulations of the chemiosmotic hypothesis, the membrane localization of PIN1 in compliance with directionality of auxin stream has been demonstrated (Benková et al., 2003; Petrasek, 2006; Wiśniewska et al., 2006).

In parallel, mutant screens aiming at identification of molecular factors controlling root gravitropism led to identification of *PIN2*, a gene exhibiting a high homology to *PIN1* (Chen et al., 1998; Luschnig, Gaxiola, Grisafi, & Fink, 1998; Müller et al., 1998; Utsuno, Shikanai, Yamada, & Hashimoto, 1998). PIN2 shares membrane topology similar to PIN1 and its polar membrane localization in root tissues is in line with its role in basipetal (shoot-ward) transport of auxin that coordinates root gravitropic bending (Müller et al., 1998).

Nowadays we know that the PIN family encompasses eight members that can be classified into 'long PINs' and 'short PINs' based on the presence of a hydrophilic loop between transmembrane domains (Křeček et al., 2009). Membrane localization of long PINs (PIN1, PIN2, PIN3, PIN4 and PIN7) is consistent with their activity in exporting auxin from cells. Rather surprising was the identification of short PINs (PIN5, PIN6 and PIN8) that localize to the endoplasmic reticulum (ER). Functional characterization of this subgroup of PINs pointed to a role in maintaining cellular haemostasis of auxin by balancing translocation of the hormone between ER and cytosol (Mravec et al., 2009). Recently, PIN6 was also found to localize at the plasma membrane suggesting that this protein might possess both types of transport capacity (Simon et al., 2016).

The auxin transport activity of PINs has been demonstrated in various heterologous systems (Barbez et al., 2013; Petrasek, 2006; Yang & Murphy, 2009; Zourelidou et al., 2014). Whereas characterization of PIN transport in yeast, HeLa, insect or BY2 cells corroborated their efflux activities, measurements in *Xenopus* oocytes have revealed that to transport auxin PIN proteins require activation via specific AGC family kinases (Barbosa et al., 2016; Habets & Offringa, 2014; Marhava et al., 2018; Zourelidou et al., 2014).

2.2 ABCBs as Non-Polar Auxin Efflux Transporters

Concomitantly to the study of PINs, physiological characterization of other auxin-related mutants led to the identification of the ABCBs, a subgroup of ABC transporters (see chapter "The ABC of ABC transporters" by Pierman et al. for complementary information). ABCBs act as primary active auxin pumps, which are activated by the binding of ATP (Geisler et al., 2005) and are able to transport auxin against its concentration gradient (Hwang et al., 2016; Park, Lee, Martinoia, & Geisler, 2017).

Several ABCBs isoforms have been associated with PAT but an auxin transport activity has only been demonstrated for ABCB1, 4, 19 and 21 (reviewed by Do, Martinoia, & Lee, 2018). ABCB1 (previously named PGP1) and ABCB19 (previously named PGP19/MDR1) were shown to function as specific auxin exporters (Bouchard et al., 2006; Geisler et al., 2005), while ABCB4 and ABCB21 were suggested to have a dual auxin transport directionality, for influx and efflux (Cho, Lee, & Cho, 2007; Kamimoto et al., 2012; Santelia et al., 2005). ABCBs are highly expressed in meristematic tissues including shoots and roots. Although ABCBs typically do not exhibit any cell polarity, lack of their function, as shown for *abcb1*, and *abcb19* mutant, results in severely impaired polar auxin transport and phenotypes typical for excess of apical auxin such as epinastic leaves and dwarfism (Geisler et al., 2003). Plant ABCBs seem to be very selective transporters having a few auxin-related molecules including indole-3-acetic acid (IAA), 1-naphthaleneacetic acid (1-NAA) and 2,4-dichlorophenoxy-acetic acid (2,4-D) as substrates (Bouchard et al., 2006). Membrane localization of ABCB1 and ABCB19 is dependent on interaction with ER located immunophilin TWD1 (Twisted Dwarf1) (Wang et al., 2013; Wu, Otegui, & Spalding, 2010). TWD1, besides playing a role in targeting ABCBs to the plasma membrane, it might also impact on their transport activity (Geisler, Aryal, di Donato, & Hao, 2017). Co-expression of ABCB1 or ABCB19 with TWD1 in yeast induced an increase in IAA efflux, although 1-NAA transport remained unaffected. Moreover, interaction of the ABCB19 with the photoreceptor kinase PHOTOTROPIN 1 (PHOT1) might provide an additional level of control. Heterologous expression in HeLa cells and physiological analyses indicate that phosphorylation of ABCB19 by PHOT1 inhibits its efflux activity (Christie et al., 2011).

Recent studies have indicated a tight interplay between ABCBs and PINs mediated auxin transport machineries. Besides physical interaction detected for subsets of PINs and ABCBs (Blakeslee et al., 2007; Kim et al., 2010; Rojas-Pierce et al., 2007), highly synergistic effects on auxin transport activity in heterologous system has been reported (Blakeslee et al., 2007). These observations suggest that PINs and ABCBs contribute to tightly coordinated auxin transport mechanisms, which function interactively to coordinate auxin distribution in a tissue-specific manner.

Finally, two ABCB transporters have been identified for their role in transport of another auxinic molecule, indole 3-butyric acid (IBA), which is a precursor of IAA (Strader & Bartel, 2011). Though with a low selectivity, ABCG36 and ABCG37 were found to mediate the export of IBA (Ruzicka et al., 2010; Strader & Bartel, 2011) and transport a wide range of secondary metabolites (Hwang et al., 2016).

2.3 AUX/LAX as Auxin Influx Transporters

Although the chemiosmotic hypothesis suggested that auxin diffuses passively into the cells across the plasma membrane, studies performed at the beginning of the 1980s suggested that cellular auxin uptake can also be facilitated by specific transporters (Hertel, 1983). Genetic screens aiming at identification of molecular factors controlling root sensitivity to auxin uncovered *AUX1*, which encodes an amino acid permease (Bennett et al., 1996). Developmental studies, revealed an altered sensitivity of *aux1* mutants to 2,4D and IAA, that represent typical substrates of the auxin influx carrier (Delbarre, Muller, Imhoff, & Guern, 1996), whereas a wild-type level of sensitivity to the lipophilic auxin, 1-NAA, was observed. These observations support a role of AUX1 in auxin transport (Bennett et al., 1996; Swarup et al., 2001). The AUX1 capacity to transport auxin has been confirmed by biochemical approaches using heterologous systems including yeast, insect cells and *Xenopus* oocytes (Carrier et al., 2008; Yang, Hammes, Taylor, Schachtman, & Nielsen, 2006). Later on, three AUX1 homologues named LAX1, LAX2 and LAX3 have been functionally characterized revealing developmentally specific, and partially redundant roles of AUX/LAX family members in regulation of various plant developmental processes (Péret et al., 2012).

2.4 Transporters Controlling Intracellular Auxin Homeostasis

A in silico screen for novel PIN-like putative carrier proteins with a predicted topology similar to that of PINs led to identification of PILS, a novel protein family in *Arabidopsis thaliana* encompassing seven members (Barbez et al., 2012). In contrast to a high similarity in their predicted protein topology, PIN and PILS proteins do not show pronounced protein sequence identity. However, the PILS proteins, similarly to the PINs, contain a putative auxin carrier domain supporting their role in auxin transport. Accordingly, analysis of plants with modulated activity of PILS proteins corroborated their role in auxin-dependent regulation of plant growth. Transport assays along with profiling of auxin metabolites suggested that PILS proteins acting at the endoplasmatic reticulum control intracellular distribution of auxin thereby fine-tuning its availability for nuclear auxin signalling. Hence, PILS proteins seem to perform a function complementary to that of 'short PINs', as discussed above, and they jointly contribute to maintenance of the intracellular auxin homeostasis.

Besides the PINS and PILS subgroups, which balance fluxes of auxin between the apoplasm, cytosol and ER, the tonoplast importer WALLS ARE THIN 1 (WAT1) was demonstrated to retrieve auxin from vacuoles suggesting that the vacuolar apparatus also contributes to auxin homeostasis by subcellular compartmentalization (Ranocha et al., 2013).

2.5 Auxin Transporters Mediating Plant Adaptive Responses

Auxin as an essential endogenous factor steering a large set of developmental processes is at the centre of mechanisms that mediate plant adaptations to environmental signals. Indeed, pathways controlling auxin metabolism, transport and signalling are modulated in response to various stresses and fluctuations in external conditions. Intriguingly, functional characterization of NRT1.1/NPF6.3/CHL1 hinted at a novel mechanism that integrates an environmental signal with an endogenously acting hormonal pathway. NRT1.1/NPF6.3/CHL1 originally identified as a transporter of nitrate has been found to exhibit auxin transport activity (Krouk et al., 2010). In various experimental systems including *Xenopus* oocytes, yeast and tobacco BY2 cells NRT1.1 auxin import activity was detected (Bouguyon et al., 2015; Krouk et al., 2010). Noteworthy, NRT1.1-dependent auxin uptake is negatively controlled by nitrate, suggesting that transduction of nitrate signal by NRT1.1 is directly associated with a modification of auxin transport. It has been proposed that this nitrate regulated auxin transport by

NRT1.1 might be a mechanism connecting nutrient and hormone signalling to coordinate root branching (see chapter "Root nitrate uptake" by Lejay and Gojon for complementary information).

3. CYTOKININS

Cytokinins belong to a heterogeneous group of N6-substituted adenine derivatives that regulate root and shoot meristem activity, vascular and organ development as well as plant responses to various environmental cues (Hwang, Sheen, & Müller, 2012). Like auxin, cytokinins are highly mobile molecules. However, in contrast to the well-characterized transport machinery of auxin, the nature of cytokinin transport is less clear. Two major groups of cytokinins encompassing trans-zeatin (tZ) and N6-(D2-isopentyl)adenine (iP) derivatives were found to be a substrate for a long-distance transport. Interestingly, whereas iP-type cytokinins, a major group of shoot-derived molecules are transported through the phloem, basipetally towards the root, the tZ type cytokinins, prominent in roots, are translocated through the xylem, acropetally towards shoots (Hirose et al., 2007; Kudo, Kiba, & Sakakibara, 2010). Recently two independent research groups have identified the first molecular components mediating this long-distance transport of cytokinins in *Arabidopsis* (Ko et al., 2014; Zhang, Novak, et al., 2014). Reverse genetics approaches applied to systematically characterizing the ATP-binding cassette transporter proteins recovered a G-type ABC transporter, ABCG14 as a transporter involved in the long-distance acropetal (root to shoot) translocation of the root-synthesized cytokinin. In accordance with a prevailing *ABCG14* expression in the root vasculature, transport assays using radiolabelled substrates supported the role of ABCG14 in acropetal movement of tZ. Loss of ABCG14 activity resulting in retarded shoot development was recovered by exogenous treatment with tZ further corroborating ABCG14 function in acropetal cytokinin transport (Ko et al., 2014).

The molecular components responsible for loading of cytokinins at the synthesis site and unloading them from phloem/xylem at the sink organ are largely unknown (Hirose et al., 2007; Kudo et al., 2010). So far, the equilibrative nucleoside transporter (ENT) family and the purine permease (PUP) family have been found to facilitate cytokinin transport (Bürkle et al., 2003; Hirose, Makita, Yamaya, & Sakakibara, 2005; Li, Liu, Baldwin, & Wang, 2003). The influx transport function of *Arabidopsis* ENT3, ENT6, ENT7 and ENT8 has been demonstrated in yeast (Hirose et al., 2007; Sun et al., 2005) being able to transport both iP-riboside and tZ-riboside cytokinin derivatives. Yet, *ent3* and *ent8* mutant plants showed negligible reduction in sensitivity to cytokinins, and the roles of ENT3 and ENT8 in cytokinin transport in vivo require further examination (Girke, Daumann, Niopek-Witz, & Möhlmann, 2014).

Several members of the purine permease (PUP) family have also been suggested to be cytokinin transporters. PUP1 and PUP2 cytokinin uptake transport activity in yeast has been reported, yet neither their genetic nor biochemical function *in planta* has been verified (Bürkle et al., 2003; Gillissen et al., 2000).

Recently, the PUP14 homologue has been recognized as a cytokinin uptake transporter (Zürcher, Liu, di Donato, Geisler, & Müller, 2016). The PUP14 protein localizes to the plasma membrane, and as revealed by transport assays in mesophyll protoplasts using radiolabelled tZ, it imports bioactive cytokinins, which might lead to depletion of apoplastic cytokinin pools. The silencing of *PUP14* using an artificial microRNA led to ectopic cytokinin signalling in the embryo and shoot apical meristem, whereas *PUP14* overexpression attenuated cytokinin response. Based on these findings, PUP14 has been proposed to mediate uptake of bioactive cytokinins from apoplast thereby inhibiting their binding to the plasma membrane cytokinin receptors (Zürcher et al., 2016). Hence, PUP14 might contribute to establishment of the spatiotemporal cytokinin sink patterns that coordinate tissue morphogenesis and organogenesis.

4. ABSCISIC ACID (ABA)

ABA is an important plant hormone involved in many physiological and developmental processes including seed maturation, germination and senescence. Moreover, ABA plays a pivotal role in plant responses to biotic and abiotic stresses such as infection, salinity, drought and wounding (Lee & Luan, 2012; Zeevaart & Creelman, 1988).

Although ABA biosynthesis and metabolism occur predominantly in vascular tissues and guard cells, ABA regulatory function is not limited to these cells and tissues, suggesting that it is transported throughout the plant. Stomatal closure occurs in leaves even when roots only experience drought stress, indicating that signals generated in the roots are transmitted to trigger a response in the leaves (reviewed by Boursiac et al., 2013; Dodd, 2013). On the other hand, stomatal closure can occur even in the absence of root-derived ABA. Reciprocal grafting between ABA-deficient mutants and wild-type plants in tomato and *Arabidopsis* demonstrated that stomatal closure is affected by the leaf (shoot) genotype, not the root genotype. However, increased ABA levels are not observed if roots are exposed to water stress without changing the water status in leaves, indicating that leaves are the main sites of ABA biosynthesis during water stress (Hu, Cao, Ge, & Li, 2016). While there has been accumulating evidence for transport of ABA over the years, this is only recently that the molecular bases of the ABA transport have started to be uncovered.

Similar to auxin, ABA is a weak acid, which exists in either protonated, uncharged forms (e.g. ABA-H) or in anionic forms (e.g. ABA^{-}) depending on the prevailing pH relative to its pK_a (Dodd, Tan, & He, 2003; Kaiser & Hartung, 1981). Because the uncharged forms of these molecules can permeate the cell membrane, it was assumed that this diffusive process would obviate a requirement for specific uptake transporters. However, under stress conditions, due to the alkalization of xylem sap and apoplastic fluid (Hartung, Sauter, & Hose, 2002; Sauter, Dietz, & Hartung, 2002; Wilkinson & Davies, 1997), the proton is dissociated from ABA-H and the membrane permeability of ABA is dramatically reduced. Under these conditions, ABA relies on transporters to move into cells.

Recently, genetic screens for potential ABA transporters identified two members of the G-subfamily of ATP-binding cassette (ABC) transporters, ABCG25, ABCG40/PDR12 (Kang et al., 2010; Kuromori et al., 2010). ABCG25/WBC25 was reported to transport ABA out of the vasculature cells to the xylem. Its identity as ABA exporter was confirmed by transport assays conducted in membrane vesicles from insect cells (*Sf9* lines) expressing ABCG25. Additionally, *Xenopus* oocytes expressing *ABCG25* accumulated lower levels of ABA than control oocytes when incubated in an ABA-containing bath solution (Kuromori et al., 2010). ABCG40/PDR12 was reported to mediate transport of ABA into guard cells, and its uptake activity was verified in yeast cells, or in a tobacco BY2 cell line expressing ABCG40, and in mesophyll protoplasts of *abcg40* mutant plants. Accordingly, *abcg40* plants exhibited an impaired stomatal response to ABA further supporting a role of ABCG40 in ABA uptake into guard cells (Kang et al., 2010, reviewed by Merilo et al., 2015).

Besides ABCG25 and ABCG40, two other *Arabidopsis* G-type ABC transporters were found to be involved in ABA transport. ABCG30 and ABCG31 were shown to transport ABA from the endosperm of the seed coats to the embryo, thereby maintaining seed dormancy (Kang et al., 2015).

Zhang, Novak, et al. (2014) and Zhang, Zhu, et al. (2014) have reported identification of a DTX/Multidrug and Toxic Compound Extrusion

(MATE) family member in *Arabidopsis thaliana*, AtDTX50, which functions as an ABA efflux transporter. When expressed heterologously in either an *Escherichia coli* strain or *Xenopus* oocytes, *AtDTX50* was found to facilitate ABA efflux. Accordingly, mesophyll cells of the *dtx50* mutant which were preloaded with ABA released less ABA compared with the wild-type controls. *AtDTX50* expression in the vascular tissues and guard cells was strongly up-regulated by exogenous ABA. The *dtx50* mutant plants exhibited increased sensitivity to ABA, but they were more tolerant to drought due to a lower stomatal conductance, consistent with DTX50 function as an ABA efflux carrier in guard cells.

Interestingly, several members of the NRT1/PTR Family (NPF) have been found to display ABA transport activity, among which NPF4.6/AIT1/NRT1.2, NPF4.5/AIT2, NPF4.1/AIT3 and NPF4.2/AIT4. Their ABA transporting activities were tested and confirmed by ABA accumulation studies in either yeast cells or Sf9 cells (Kanno et al., 2012). Although multiple NPFs have been detected as potential ABA transporters, their *in planta* function remains to be further elucidated.

In rice, OsPM1 (*Oryza sativa* plasma membrane protein 1), belonging to a protein family named AWPM-19 (ABA-induced wheat plasma membrane polypeptide-19) was found to mediate ABA import through the plasma membrane (Yao et al., 2018). OsPM1 is predominantly expressed in vascular tissues, guard cells and mature embryos. Phenotypic analysis of over-expression, RNA interference (RNAi) and knockout (KO) lines showed that OsPM1 is involved in drought responses and seed germination (Yao et al., 2018). Whether the ABA transport function of this family is conserved in *Arabidopsis* and other plant species remains to be further clarified.

5. GIBBERELLINS (GA)

The gibberellins (GAs) are a large family of tetracyclic diterpenoid carboxylic acid substances associated with various growth and developmental processes. Although more than 100 GAs have been discovered only a few such as GA_1 and GA_4 were demonstrated to be biologically active (Gupta & Chakrabarty, 2013).

The use of labelled GAs provided substantial evidence for GA local and long distance movement. However, it has remained unclear in which form and how endogenously made GAs are transported. Based on various strategies including grafting and transport assays, GA_{12}, the common precursor of all GAs, was identified as the major form of the graft transmissible signal in *Arabidopsis* (Binenbaum, Weinstain, & Shani, 2018; Regnault et al., 2015). Similarly to other acidic hormones, the chemiosmotic mechanism of cell to cell movement depending on activity of influx and efflux transporters applies to GAs. Whereas GA efflux carriers have not yet been discovered, a number of GA influx carriers have been identified over the past (Binenbaum et al., 2018).

A screen for mutant plants defective in root accumulation of fluorescein-tagged GA resulted in identification of NPF3.1 as a putative GA transporter. Examining its transport activity in *Xenopus* oocytes revealed that NPF3.1 imports GA_4 and to a lesser extent GA_1 and GA_3 in a pH-dependent manner. Lack of any striking GA-related phenotypes in *npf3.1* mutant suggests a functional redundancy with other related homologues (Tal et al., 2016).

Using a modified yeast two hybrid system that utilized the interaction of GA receptor, GID1, with GA insensitive (GAI) DELLA proteins as a mean to identify putative GA transporters, NPF4.1/AIT3 and 18 NPF family members, among a total of 45 investigated, were shown to possess a GA transport activity (Corratgé-Faillie & Lacombe, 2017; Kanno et al., 2012, 2016). Interestingly, individual NPF transporters differ in their substrate specificity. Twelve NPF members transport GA_1, GA_3 and GA_4, three members transport only GA_1 and GA_3 and three NPF transport only GA_1 and GA_4 (Corratgé-Faillie & Lacombe, 2017). Among them NPF2.1/GTR1 was found to specifically mediate import of GA_3, but no other GA derivatives. Stamen defects observed in the *npf2.1* loss-of-function mutant could be rescued by exogenous GA_3 application, suggesting that *npf2.1* mutant reduced fertility results from an aberrant transport of GA (Nour-Eldin et al., 2012).

The experimental design used for revealing the GA transporting activity of NPF family members, also led to the identification of several SWEET family members as potential GA transporters. Transport assays in yeast and *Xenopus* oocytes clearly demonstrated GA intake transport mediated by SWEET13 and 14, thus confirming their dual transport activity for both sugar and GAs. A double *sweet13 sweet14* mutant exhibited altered long-distance transport of exogenously applied GA and altered responses to GA during germination and seedling development, suggesting that SWEET13 and SWEET14 may be involved in modulating GA response in *Arabidopsis* (Kanno et al., 2016).

GAs are beautiful example of how establishment of novel screening strategies might enable identification of unknown molecular components of hormonal transport machineries. Such new and innovative methods are expected to reveals transporters whose function *in planta* might be have been masked by their ability to transport different hormones/substances.

6. JASMONATES

Jasmonates (JAs) are a class of plant hormones that plays a major role in development of reproductive organs, wound responses and immune signalling. Jasmonic acid (JA) and methyl jasmonate (MeJA) are highly mobile molecules that can be transported from cell to cell as well as from shoot to root (Gasperini et al., 2015). For example, when exogenously applied, they efficiently modulate gene expression in tissues distal to the treatments. Similarly, jasmonates elevated after shoot wounding modulate growth of roots to balance development of above- and underground plant organs. The molecular bases underlying transport of these regulatory molecules are among the least understood so far.

Studies implicating a yeast two hybrid method, using the JA receptor COI1 and its interaction with JAZ3, identified several members of NPF family, including NPF1.1, 1.2, 4.1, 5.7, 8.1, 8.2, and to a lesser extent NPF2.4, 2.6, 2.7, 2.10, 2.13, 3.1, 5.1, as potential JA-Ile transporters (Chiba et al., 2015; Nguyen, Martinoia, & Farmer, 2017). Although, JA transport by most of these candidates has been demonstrated in heterologous systems such as yeast and *Xenopus* oocytes, the function of these proteins in JA/JA-Ile transport remains to be evaluated in the plant (Corratgé-Faillie & Lacombe, 2017). NPF2.10/GTR1 is among the best characterized JA transporters of the NPF subfamily. In *Xenopus* oocyte NPF2.10 was found to exhibit a low substrate specificity, having the capacity to transport different substrate such as JA, GA and glucosinolates (Ishimaru et al., 2017; Saito et al., 2015). Loss of the NPF2.10 activity in a *npf2.10/gtr1* mutant severely affected JA transport from wounded to undamaged leaves providing support for in vivo function of this gene in JA transport (Ishimaru et al., 2017).

A candidate approach to screen the ABC transporter family for homologues able to transport JAs yielded ABCG16/jasmonate transporter 1 (JAT1). Interestingly, unlike other members of this family, ABCG16/JAT1 was detected at both plasma membrane and nuclear envelope. In depth transport studies indicated that plasma membrane located JAT1 facilitates efflux of JA to the apoplast, whereas at nuclear envelope it promotes influx to nucleus (Li et al., 2017). Hence, JAT1-mediated transport might be part of a homeostatic mechanism to balance levels of JA among cellular compartments. Role of ABCG16/JAT1 in planta was characterized using *Arabidopsis* plants with modulated activity of this gene. Loss of the ABCG16 function resulted in reduced sensitivity of root growth to exogenous JA; decreased expression of jasmonate marker genes in response to exogenously applied JA; enhanced susceptibility to biotic stress, thus further supporting role of this gene in JA regulated plant development.

7. ETHYLENE

Ethylene is a gaseous hormone involved in physiological processes such as fruit ripening, senescence, abscission and immune signalling. Although ethylene molecules can freely diffuse through plasma membranes and move from cell to cell (Yang & Hoffman, 1984), its immediate precursor 1-aminocyclopropane-1-carboxylic acid (ACC) is a non-gaseous substance whose mobility is dependent on specialized transporters (Van de Poel & Van Der Straeten, 2014). ACC resistant 2 (ARE2) is among the first potential candidates to mediate ACC transport. Loss of ARE2 function significantly attenuates sensitivity to ACC, while not interfering with sensitivity to ethylene. *ARE2* encodes for a plasma membrane located lysine histidine transporter 1 (LHT1) a member of amino acid transporter family. Transport assays in protoplasts using radiolabelled substrate corroborated LHT1 function in ACC uptake (Shin et al., 2015).

8. BRASSINOSTEROIDS

Brassinosteroids (BRs) are a group of plant steroidal hormones, which regulate processes as diverse as cell elongation, xylem differentiation and fruit ripening (Bajguz & Tretyn, 2003). Over the last years, there have been significant progresses in our understanding of BR biology, including their distribution in the plant, biosynthesis, perception and signal transduction, and the interactions between BRs and other plant hormones (Tang, Han, & Chai, 2016). However, there had been little focus until recently on the mechanism of BR transport in the plant body over short and/or long distances. Although current evidence does not support a long-distance BR transport within the plant, it is possible, that endogenous BRs do move over short distances both within and between cells. Generally, BR synthesis occurs inside the cells and their receptors are localized at the plasma membrane (Symons, Ross, Jager, & Reid, 2007). Thus BR transport from the cell interior to the extracellular space, actively or passively, is essential to initiate BR signal transduction (Vukašinović & Russinova, 2018). Even though this efflux transport has been hypothesized for many years, BR specific transporters are yet to be identified.

9. STRIGOLACTONES

Strigolactones were originally discovered as root exudate signals that trigger parasitic weed seed germination (Yoneyama et al., 2008). This is only lately that the function of these carotenoid-derived compounds as endogenous plant signalling substances controlling shoot and root branching has been recognized. Although strigolactones are produced both in roots and shoots, grafting experiments have shown that they might subjected to long-distance transport presumably through xylem (Kohlen et al., 2011). Intriguingly, first hints at strigolactone transport came from the *Petunia* model system, in which the G-type ABC transporter, PDR1, has been identified as a strigolactone transporter. PDR1 loss-of-function plants excreted less strigolactone, were impaired in mycorrhization, and produced more branches in the aerial parts of the plant. Further studies showed that the transfer of exogenously applied GR24, a synthetic strigolactone, from the root to the shoot was impaired in *pdr1* mutant plants, suggesting a role of PDR1 in root to shoot transport (Kretzschmar et al., 2012). Whereas the functional characterization of a close PDR1 homologue in *Nicotiana tabacum* supports findings in *Petunia*, no functional *PDR1* homologue has been identified in Arabidopsis. Hence, whether the mechanisms of strigolactone transport are conserved among different plant species awaits to be resolved.

10. CONCLUSION

Over the last 15 years, significant progresses have been made in the identification and functional characterization of transporters that mediate the distribution of plant hormones. Although to different extent, these progresses concern all major groups of plant hormones. We described here a striking complexity of networks controlling distribution of plant hormones within a plant body encompassing long- and short-distance transport as well as translocation of hormones between individual cellular compartments. Functional studies of mutants defective in either long distance, cell to cell or subcellular transport clearly show that all of these transport systems play crucial, largely complementary functions in fine-tuning hormonal activity. Intriguingly, the majority of so far identified hormone transporters belong to either NPF or ABC gene families (Fig. 1) (for complementary information on ABC transporters, see chapter "The ABC of ABC transporters" by Pierman et al.). Future research will show whether this comes from a bias of approaches applied in screens for hormone transporters so far, or whether natural selection led to a true dominance of these two families in regulation of hormone distribution. Number of hormonal transporters show capacity to mediate transport of more than one hormone, or other molecules such as macro- or micronutrients apart from hormones. This suggests that transporters might act as important convergence points of hormonal cross-talk or as integrators of environmental and hormonal signals to drive plant adaptive responses.

ACKNOWLEDGEMENTS

A.B., E.V. and B.L. are supported by a binational grant of the Austrian Science Fund (FWF01_I1774S) and Agence Nationale de la Recherche (ANR-14-CE34-0007-01-HONIT).

REFERENCES

Amtmann, A., & Blatt, M. R. (2009). Regulation of macronutrient transport. *The New Phytologist*, *181*, 35–52.

Bajguz, A., & Tretyn, A. (2003). The chemical characteristic and distribution of brassinosteroids in plants. *Phytochemistry*, *62*, 1027–1046.

Barbez, E., Kubeš, M., Rolčík, J., Béziat, C., Pěnčík, A., Wang, B., et al. (2012). A novel putative auxin carrier family regulates intracellular auxin homeostasis in plants. *Nature*, *485*, 119–122.

Barbez, E., Laňková, M., Pařezová, M., Maizel, A., Zažímalová, E., Petrášek, J., et al. (2013). Single-cell-based system to monitor carrier driven cellular auxin homeostasis. *BMC Plant Biology*, *13*, 20.

Barbosa, I. C. R., Shikata, H., Zourelidou, M., Heilmann, M., Heilmann, I., & Schwechheimer, C. (2016). Phospholipid composition and a polybasic motif determine D6 PROTEIN KINASE polar association with the plasma membrane and tropic responses. *Development*, *143*, 4687–4700.

Benková, E., Michniewicz, M., Sauer, M., Teichmann, T., Seifertová, D., Jürgens, G., et al. (2003). Local, efflux-dependent auxin gradients as a common module for plant organ formation. *Cell*, *115*, 591–602.

Bennett, M. J., Marchant, A., Green, H. G., May, S. T., Ward, S. P., Millner, P. A., et al. (1996). Arabidopsis AUX1 gene: A permease-like regulator of root gravitropism. *Science*, *273*, 948–950.

Binenbaum, J., Weinstain, R., & Shani, E. (2018). Gibberellin localization and transport in plants. *Trends in Plant Science*, *23*, 410–421.

Blakeslee, J. J., Bandyopadhyay, A., Lee, O. R., Mravec, J., Titapiwatanakun, B., Sauer, M., et al. (2007). Interactions among PIN-FORMED and P-glycoprotein auxin transporters in Arabidopsis. *Plant Cell*, *19*, 131–147.

Bouchard, R., Bailly, A., Blakeslee, J. J., Oehring, S. C., Vincenzetti, V., Lee, O. R., et al. (2006). Immunophilin-like TWISTED DWARF1 modulates auxin efflux activities of *Arabidopsis* P-glycoproteins. *The Journal of Biological Chemistry*, *281*, 30603–30612.

Bouguyon, E., Brun, F., Meynard, D., Kubeš, M., Pervent, M., Leran, S., et al. (2015). Multiple mechanisms of nitrate sensing by Arabidopsis nitrate transceptor NRT1.1. *Nature Plants*, *1*, 15015.

Boursiac, Y., Léran, S., Corratgé-Faillie, C., Gojon, A., Krouk, G., & Lacombe, B. (2013). ABA transport and transporters. *Trends in Plant Science*, *18*, 325–333.

Bürkle, L., Cedzich, A., Döpke, C., Stransky, H., Okumoto, S., Gillissen, B., et al. (2003). Transport of cytokinins mediated by purine transporters of the PUP family expressed in phloem, hydathodes, and pollen of Arabidopsis. *The Plant Journal*, *34*, 13–26.

Cambridge, A. P., & Morris, D. A. (1996). Transfer of exogenous auxin from the phloem to the polar auxin transport pathway in pea (*Pisum sativum* L.). *Planta*, *199*, 583–588.

Carrier, D. J., Bakar, N. T. A., Swarup, R., Callaghan, R., Napier, R. M., Bennett, M. J., et al. (2008). The binding of auxin to the Arabidopsis auxin influx transporter AUX1. *Plant Physiology*, *148*, 529–535.

Chen, R., Hilson, P., Sedbrook, J., Rosen, E., Caspar, T., & Masson, P. H. (1998). The Arabidopsis thaliana AGRAVITROPIC 1 gene encodes a component of the polar-auxin-transport efflux carrier. *Proceedings of the National Academy of Sciences*, *95*, 15112–15117.

Chiba, Y., Shimizu, T., Miyakawa, S., Kanno, Y., Koshiba, T., Kamiya, Y., et al. (2015). Identification of Arabidopsis thaliana NRT1/PTR FAMILY (NPF) proteins capable of transporting plant hormones. *Journal of Plant Research*, *128*, 679–686.

Cho, M., Lee, S. H., & Cho, H.-T. (2007). P-glycoprotein 4 displays auxin efflux transporter–like action in Arabidopsis root hair cells and tobacco cells. *Plant Cell*, *19*, 3930–3943.

Christie, J. M., Yang, H., Richter, G. L., Sullivan, S., Thomson, C. E., Lin, J., et al. (2011). phot1 inhibition of ABCB19 primes lateral auxin fluxes in the shoot apex required for phototropism. *PLoS Biology*, *9*, e1001076.

Corratgé-Faillie, C., & Lacombe, B. (2017). Substrate (un)specificity of Arabidopsis NRT1/PTR FAMILY (NPF) proteins. *Journal of Experimental Botany*, *68*, 3107–3113.

Darwin, C. R. (1880). *The power of movement in plants*. London: John Murray.

Delbarre, A., Muller, P., Imhoff, V., & Guern, J. (1996). Comparison of mechanisms controlling uptake and accumulation of 2,4-dichlorophenoxy acetic acid, naphthalene-1-acetic acid, and indole-3-acetic acid in suspension-cultured tobacco cells. *Planta*, *198*, 532–541.

Do, T. H. T., Martinoia, E., & Lee, Y. (2018). Functions of ABC transporters in plant growth and development. *Current Opinion in Plant Biology*, *41*, 32–38.

Dodd, I. C. (2013). Abscisic acid and stomatal closure: A hydraulic conductance conundrum? *New Phytolologist*, *197*, 6–8.

Dodd, I. C., Tan, L. P., & He, J. (2003). Do increases in xylem sap pH and/or ABA concentration mediate stomatal closure following nitrate deprivation? *Journal of Experimental Botany*, *54*, 1281–1288.

Friml, J., Vieten, A., Sauer, M., Weijers, D., Schwarz, H., Hamann, T., et al. (2003). Efflux-dependent auxin gradients establish the apical–basal axis of *Arabidopsis*. *Nature*, *426*, 147–153.

Friml, J., Wiśniewska, J., Benková, E., Mendgen, K., & Palme, K. (2002). Lateral relocation of auxin efflux regulator PIN3 mediates tropism in *Arabidopsis*. *Nature*, *415*, 806–809.

Gälweiler, L., Guan, C., Müller, A., Wisman, E., Mendgen, K., Yephremov, A., et al. (1998). Regulation of polar auxin transport by AtPIN1 in Arabidopsis vascular tissue. *Science*, *282*, 2226–2230.

Gasperini, D., Chételat, A., Acosta, I. F., Goossens, J., Pauwels, L., Goossens, A., et al. (2015). Multilayered organization of jasmonate signalling in the regulation of root growth. *PLoS Genetics*, *11*, e1005300.

Geisler, M., Aryal, B., di Donato, M., & Hao, P. (2017). A critical view on ABC transporters and their interacting partners in auxin transport. *Plant & Cell Physiology*, *58*, 1601–1614.

Geisler, M., Blakeslee, J. J., Bouchard, R., Lee, O. R., Vincenzetti, V., Bandyopadhyay, A., et al. (2005). Cellular efflux of auxin catalyzed by the Arabidopsis MDR/PGP transporter AtPGP1. *The Plant Journal*, *44*, 179–194.

Geisler, M., Kolukisaoglu, H. Ü., Bouchard, R., Billion, K., Berger, J., Saal, B., et al. (2003). TWISTED DWARF1, a unique plasma membrane-anchored immunophilin-like protein, interacts with Arabidopsis multidrug resistance-like transporters AtPGP1 and AtPGP19. *Molecular Biology of the Cell, 14*, 4238–4249.

Gillissen, B., Bürkle, L., André, B., Kühn, C., Rentsch, D., Brandl, B., et al. (2000). A new family of high-affinity transporters for adenine, cytosine, and purine derivatives in Arabidopsis. *Plant Cell, 12*, 291–300.

Girke, C., Daumann, M., Niopek-Witz, S., & Möhlmann, T. (2014). Nucleobase and nucleoside transport and integration into plant metabolism. *Frontiers in Plant Science, 5*.

Goldsmith, M. H. M. (1977). The polar transport of auxin. *Annual Review of Plant Physiology, 28*, 439–478.

Gupta, R., & Chakrabarty, S. K. (2013). Gibberellic acid in plant. *Plant Signaling & Behavior, 8*.

Habets, M. E. J., & Offringa, R. (2014). PIN-driven polar auxin transport in plant developmental plasticity: A key target for environmental and endogenous signals. *The New Phytologist, 203*, 362–377.

Hartung, W., Sauter, A., & Hose, E. (2002). Abscisic acid in the xylem: Where does it come from, where does it go to? *Journal of Experimental Botany, 53*, 27–32.

Hertel, R. (1983). The mechanism of auxin transport as a model for auxin action. *Zeitschrift für Pflanzenphysiologie, 112*, 53–67.

Hirose, N., Makita, N., Yamaya, T., & Sakakibara, H. (2005). Functional characterization and expression analysis of a gene, OsENT2, encoding an equilibrative nucleoside transporter in rice suggest a function in cytokinin transport. *Plant Physiology, 138*, 196–206.

Hirose, N., Takei, K., Kuroha, T., Kamada-Nobusada, T., Hayashi, H., & Sakakibara, H. (2007). Regulation of cytokinin biosynthesis, compartmentalization and translocation. *Journal of Experimental Botany, 59*, 75–83.

Hu, B., Cao, J., Ge, K., & Li, L. (2016). The site of water stress governs the pattern of ABA synthesis and transport in peanut. *Scientific Reports, 6*.

Hwang, I., Sheen, J., & Müller, B. (2012). Cytokinin signaling networks. *Annual Review of Plant Biology, 63*, 353–380.

Hwang, J.-U., Song, W.-Y., Hong, D., Ko, D., Yamaoka, Y., Jang, S., et al. (2016). Plant ABC transporters enable many unique aspects of a terrestrial plant's lifestyle. *Molecular Plant, 9*, 338–355.

Ishimaru, Y., Washiyama, K., Oikawa, T., Hamamoto, S., Uozumi, N., & Ueda, M. (2017). Dimerization of GTR1 regulates their plasma membrane localization. *Plant Signaling & Behavior, 12*.

Kaiser, W. M., & Hartung, W. (1981). Uptake and release of abscisic acid by isolated photoautotrophic mesophyll cells, depending on pH gradients. *Plant Physiology, 68*, 202–206.

Kamimoto, Y., Terasaka, K., Hamamoto, M., Takanashi, K., Fukuda, S., Shitan, N., et al. (2012). Arabidopsis ABCB21 is a facultative auxin importer/exporter regulated by cytoplasmic auxin concentration. *Plant & Cell Physiology, 53*, 2090–2100.

Kang, J., Hwang, J.-U., Lee, M., Kim, Y.-Y., Assmann, S. M., Martinoia, E., et al. (2010). PDR-type ABC transporter mediates cellular uptake of the phytohormone abscisic acid. *Proceedings of the National Academy of Sciences of the United States of America, 107*, 2355–2360.

Kang, J., Yim, S., Choi, H., Kim, A., Lee, K. P., Lopez-Molina, L., et al. (2015). Abscisic acid transporters cooperate to control seed germination. *Nature Communications, 6*, 8113.

Kanno, Y., Hanada, A., Chiba, Y., Ichikawa, T., Nakazawa, M., Matsui, M., et al. (2012). Identification of an abscisic acid transporter by functional screening using the receptor complex as a sensor. *Proceedings of the National Academy of Sciences of the United States of America, 109*, 9653–9658.

Kanno, Y., Oikawa, T., Chiba, Y., Ishimaru, Y., Shimizu, T., Sano, N., et al. (2016). AtSWEET13 and AtSWEET14 regulate gibberellin-mediated physiological processes. *Nature Communications*, 7, 13245.

Kim, J.-Y., Henrichs, S., Bailly, A., Vincenzetti, V., Sovero, V., Mancuso, S., et al. (2010). Identification of an ABCB/P-glycoprotein-specific inhibitor of auxin transport by chemical genomics. *The Journal of Biological Chemistry*, *285*, 23309–23317.

Ko, D., Kang, J., Kiba, T., Park, J., Kojima, M., Do, J., et al. (2014). Arabidopsis ABCG14 is essential for the root-to-shoot translocation of cytokinin. *Proceedings of the National Academy of Sciences of the United States of America*, *111*, 7150–7155.

Kohlen, W., Charnikhova, T., Liu, Q., Bours, R., Domagalska, M. A., Beguerie, S., et al. (2011). Strigolactones are transported through the xylem and play a key role in shoot architectural response to phosphate deficiency in nonarbuscular mycorrhizal host Arabidopsis. *Plant Physiology*, *155*, 974–987.

Křeček, P., Skůpa, P., Libus, J., Naramoto, S., Tejos, R., Friml, J., et al. (2009). The PIN-FORMED (PIN) protein family of auxin transporters. *Genome Biology*, *10*, 249.

Kretzschmar, T., Kohlen, W., Sasse, J., Borghi, L., Schlegel, M., Bachelier, J. B., et al. (2012). A petunia ABC protein controls strigolactone-dependent symbiotic signalling and branching. *Nature*, *483*, 341–344.

Krouk, G., Lacombe, B., Bielach, A., Perrine-Walker, F., Malinska, K., Mounier, E., et al. (2010). Nitrate-regulated auxin transport by NRT1.1 defines a mechanism for nutrient sensing in plants. *Developmental Cell*, *18*, 927–937.

Kudo, T., Kiba, T., & Sakakibara, H. (2010). Metabolism and long-distance translocation of cytokinins. *Journal of Integrative Plant Biology*, *52*, 53–60.

Kuromori, T., Miyaji, T., Yabuuchi, H., Shimizu, H., Sugimoto, E., Kamiya, A., et al. (2010). ABC transporter AtABCG25 is involved in abscisic acid transport and responses. *Proceedings of the National Academy of Sciences of the United States of America*, *107*, 2361–2366.

Lacombe, B., & Achard, P. (2016). Long-distance transport of phytohormones through the plant vascular system. *Current Opinion in Plant Biology*, *34*, 1–8.

Lee, S. C., & Luan, S. (2012). ABA signal transduction at the crossroad of biotic and abiotic stress responses. *Plant, Cell & Environment*, *35*, 53–60.

Li, G., Liu, K., Baldwin, S. A., & Wang, D. (2003). Equilibrative nucleoside transporters of Arabidopsis thaliana cDNA cloning, expression pattern, and analysis of transport activities. *The Journal of Biological Chemistry*, *278*, 35732–35742.

Li, R., Wang, M., Wang, Y., Schuman, M. C., Weinhold, A., Schäfer, M., et al. (2017). Flower-specific jasmonate signaling regulates constitutive floral defenses in wild tobacco. *Proceedings of the National Academy of Sciences*, *114*, E7205–E7214.

Luschnig, C., Gaxiola, R. A., Grisafi, P., & Fink, G. R. (1998). EIR1, a root-specific protein involved in auxin transport, is required for gravitropism in Arabidopsis thaliana. *Genes & Development*, *12*, 2175–2187.

Marhava, P., Bassukas, A. E. L., Zourelidou, M., Kolb, M., Moret, B., Fastner, A., et al. (2018). A molecular rheostat adjusts auxin flux to promote root protophloem differentiation. *Nature*, *558*, 297–300.

Merilo, E., Jalakas, P., Laanemets, K., Mohammadi, O., Hõrak, H., Kollist, H., et al. (2015). Abscisic acid transport and homeostasis in the context of stomatal regulation. *Molecular Plant*, *8*, 1321–1333.

Mravec, J., Skůpa, P., Bailly, A., Hoyerová, K., Křeček, P., Bielach, A., et al. (2009). Subcellular homeostasis of phytohormone auxin is mediated by the ER-localized PIN5 transporter. *Nature*, *459*, 1136–1140.

Müller, A., Guan, C., Gälweiler, L., Tänzler, P., Huijser, P., Marchant, A., et al. (1998). AtPIN2 defines a locus of Arabidopsis for root gravitropism control. *The EMBO Journal*, *17*, 6903–6911.

Nguyen, C. T., Martinoia, E., & Farmer, E. E. (2017). Emerging jasmonate transporters. *Molecular Plant, 10*, 659–661.

Nour-Eldin, H. H., Andersen, T. G., Burow, M., Madsen, S. R., Jørgensen, M. E., Olsen, C. E., et al. (2012). NRT/PTR transporters are essential for translocation of glucosinolate defence compounds to seeds. *Nature, 488*, 531–534.

Okada, K., Ueda, J., Komaki, M., Bell, C., & Shimura, Y. (1991). Requirement of the auxin polar transport system in early stages of Arabidopsis floral bud formation. *Plant Cell, 3*, 677–684.

Park, J., Lee, Y., Martinoia, E., & Geisler, M. (2017). Plant hormone transporters: What we know and what we would like to know. *BMC Biology, 15*.

Péret, B., Swarup, K., Ferguson, A., Seth, M., Yang, Y., Dhondt, S., et al. (2012). AUX/LAX genes encode a family of auxin influx transporters that perform distinct functions during Arabidopsis development. *Plant Cell, 24*, 2874–2885.

Petrasek, J. (2006). PIN proteins perform a rate-limiting function in cellular auxin efflux. *Science, 312*, 914–918.

Ranocha, P., Dima, O., Nagy, R., Felten, J., Corratgé-Faillie, C., Novák, O., et al. (2013). Arabidopsis WAT1 is a vacuolar auxin transport facilitator required for auxin homoeostasis. *Nature Communications, 4*, 2625.

Raven, J. A. (1975). Transport of indoleacetic acid in plant cells in relation to pH and electrical potential gradients, and its significance for polar IAA transport. *The New Phytologist, 74*, 163–172.

Regnault, T., Davière, J.-M., Wild, M., Sakvarelidze-Achard, L., Heintz, D., Bergua, E. C., et al. (2015). The gibberellin precursor GA_{12} acts as a long-distance growth signal in *Arabidopsis*. *Nature Plants, 1*, 15073.

Reinhardt, D. (2003). Vascular patterning: More than just auxin? *Current Biology, 13*, R485–R487.

Rojas-Pierce, M., Titapiwatanakun, B., Sohn, E. J., Fang, F., Larive, C. K., Blakeslee, J., et al. (2007). Arabidopsis P-glycoprotein 19 participates in the inhibition of gravitropism by gravacin. *Chemistry & Biology, 14*, 1366–1376.

Rubery, P. H., & Sheldrake, A. R. (1974). Carrier-mediated auxin transport. *Planta, 118*, 101–121.

Ruzicka, K., Strader, L. C., Bailly, A., Yang, H., Blakeslee, J., Langowski, L., et al. (2010). Arabidopsis PIS1 encodes the ABCG37 transporter of auxinic compounds including the auxin precursor indole-3-butyric acid. *Proceedings of the National Academy of Sciences of the United States of America, 107*, 10749–10753.

Saito, H., Oikawa, T., Hamamoto, S., Ishimaru, Y., Kanamori-Sato, M., Sasaki-Sekimoto, Y., et al. (2015). The jasmonate-responsive GTR1 transporter is required for gibberellin-mediated stamen development in Arabidopsis. *Nature Communications, 6*.

Santelia, D., Vincenzetti, V., Azzarello, E., Bovet, L., Fukao, Y., Düchtig, P., et al. (2005). MDR-like ABC transporter AtPGP4 is involved in auxin-mediated lateral root and root hair development. *FEBS Letters, 579*, 5399–5406.

Sauter, A., Dietz, K.-J., & Hartung, W. (2002). A possible stress physiological role of abscisic acid conjugates in root-to-shoot signalling. *Plant, Cell & Environment, 25*, 223–228.

Shin, K., Lee, S., Song, W.-Y., Lee, R.-A., Lee, I., Ha, K., et al. (2015). Genetic identification of ACC-resistant2 reveals involvement of lysine histidine transporter1 in the uptake of 1-aminocyclopropane-1-carboxylic acid in Arabidopsis thaliana. *Plant & Cell Physiology, 56*, 572–582.

Simon, S., Skůpa, P., Viaene, T., Zwiewka, M., Tejos, R., Klíma, P., et al. (2016). PIN6 auxin transporter at endoplasmic reticulum and plasma membrane mediates auxin homeostasis and organogenesis in Arabidopsis. *The New Phytologist, 211*, 65–74.

Strader, L. C., & Bartel, B. (2011). Transport and metabolism of the endogenous auxin precursor Indole-3-butyric acid. *Molecular Plant, 4*, 477–486.

Sun, J., Hirose, N., Wang, X., Wen, P., Xue, L., Sakakibara, H., et al. (2005). Arabidopsis SOI33/AtENT8 gene encodes a putative equilibrative nucleoside transporter that is involved in cytokinin transport in planta. *Journal of Integrative Plant Biology*, *47*, 588–603.

Swarup, R., Friml, J., Marchant, A., Ljung, K., Sandberg, G., Palme, K., et al. (2001). Localization of the auxin permease AUX1 suggests two functionally distinct hormone transport pathways operate in the Arabidopsis root apex. *Genes & Development*, *15*, 2648–2653.

Symons, G. M., Ross, J. J., Jager, C. E., & Reid, J. B. (2007). Brassinosteroid transport. *Journal of Experimental Botany*, *59*, 17–24.

Tal, I., Zhang, Y., Jørgensen, M. E., Pisanty, O., Barbosa, I. C. R., Zourelidou, M., et al. (2016). The Arabidopsis NPF3 protein is a GA transporter. *Nature Communications*, 7.

Tang, J., Han, Z., & Chai, J. (2016). Q&A: What are brassinosteroids and how do they act in plants? *BMC Biology*, *14*, 113.

Tester, M., & Leigh, R. A. (2001). Partitioning of nutrient transport processes in roots. *Journal of Experimental Botany*, *52*, 445–457.

Utsuno, K., Shikanai, T., Yamada, Y., & Hashimoto, T. (1998). AGR, an agravitropic locus of Arabidopsis thaliana, encodes a novel membrane-protein family member. *Plant & Cell Physiology*, *39*, 1111–1118.

Van de Poel, B., & Van Der Straeten, D. (2014). 1-Aminocyclopropane-1-carboxylic acid (ACC) in plants: More than just the precursor of ethylene! *Frontiers in Plant Science*, *5*.

Vukašinović, N., & Russinova, E. (2018). BRexit: Possible Brassinosteroid export and transport routes. *Trends in Plant Science*, *23*, 285–292.

Wang, B., Bailly, A., Zwiewka, M., Henrichs, S., Azzarello, E., Mancuso, S., et al. (2013). Arabidopsis TWISTED DWARF1 functionally interacts with auxin exporter ABCB1 on the root plasma membrane[C][W]. *Plant Cell*, *25*, 202–214.

Wilkinson, S., & Davies, W. J. (1997). ABA-based chemical signalling: The co-ordination of responses to stress in plants. *Plant, Cell & Environment*, *25*, 195–210.

Wiśniewska, J., Xu, J., Seifertová, D., Brewer, P. B., Růžička, K., Blilou, I., et al. (2006). Polar PIN localization directs auxin flow in plants. *Science*, *312*, 883.

Wu, G., Otegui, M. S., & Spalding, E. P. (2010). The ER-localized TWD1 immunophilin is necessary for localization of multidrug resistance-like proteins required for polar auxin transport in Arabidopsis roots. *Plant Cell*, *22*, 3295–3304.

Yang, Y., Hammes, U. Z., Taylor, C. G., Schachtman, D. P., & Nielsen, E. (2006). High-affinity auxin transport by the AUX1 influx carrier protein. *Current Biology*, *16*, 1123–1127.

Yang, S. F., & Hoffman, N. E. (1984). Ethylene biosynthesis and its regulation in higher plants. *Annual Review of Plant Physiology*, *35*, 155–189.

Yang, H., & Murphy, A. S. (2009). Functional expression and characterization of Arabidopsis ABCB, AUX 1 and PIN auxin transporters in Schizosaccharomyces pombe. *The Plant Journal*, *59*, 179–191.

Yao, L., Cheng, X., Gu, Z., Huang, W., Li, S., Wang, L., et al. (2018). The AWPM-19 family protein OsPM1 mediates abscisic acid influx and drought response in rice. *Plant Cell*, *30*, 1258–1276.

Ye, L., Liu, L., Xing, A., & Kang, D. (2013). Characterization of a dwarf mutant allele of Arabidopsis MDR-like ABC transporter AtPGP1 gene. *Biochemical and Biophysical Research Communications*, *441*, 782–786.

Yoneyama, K., Xie, X., Sekimoto, H., Takeuchi, Y., Ogasawara, S., Akiyama, K., et al. (2008). Strigolactones, host recognition signals for root parasitic plants and arbuscular mycorrhizal fungi, from Fabaceae plants. *The New Phytologist*, *179*, 484–494.

Zeevaart, J. A. D., & Creelman, R. A. (1988). Metabolism and physiology of abscisic acid. *Annual Review of Plant Physiology and Plant Molecular Biology*, *39*, 439–473.

Zhang, K., Novak, O., Wei, Z., Gou, M., Zhang, X., Yu, Y., et al. (2014). Arabidopsis ABCG14 protein controls the acropetal translocation of root-synthesized cytokinins. *Nature Communications*, *5*, 3274.
Zhang, H., Zhu, H., Pan, Y., Yu, Y., Luan, S., & Li, L. (2014). A DTX/MATE-type transporter facilitates abscisic acid efflux and modulates ABA sensitivity and drought tolerance in Arabidopsis. *Molecular Plant*, *7*, 1522–1532.
Zourelidou, M., Absmanner, B., Weller, B., Barbosa, I. C., Willige, B. C., Fastner, A., et al. (2014). Auxin efflux by PIN-FORMED proteins is activated by two different protein kinases, D6 protein kinase and pinoid. *eLife*, *19*, 3.
Zürcher, E., Liu, J., di Donato, M., Geisler, M., & Müller, B. (2016). Plant development regulated by cytokinin sinks. *Science*, *353*, 1027–1030.

FURTHER READING

Aryal, B., Laurent, C., & Geisler, M. (2015). Learning from each other: ABC transporter regulation by protein phosphorylation in plant and mammalian systems. *Biochemical Society Transactions*, *43*, 966–974.
Fujioka, S., & Yokota, T. (2003). Biosynthesis and metabolism of Brassinosteroids. *Annual Review of Plant Biology*, *54*, 137–164.
Lefebvre, V., North, H., Frey, A., Sotta, B., Seo, M., Okamoto, M., Nambara, E., and Marion-Poll, A. Functional analysis of Arabidopsis NCED6 and NCED9 genes indicates that ABA synthesized in the endosperm is involved in the induction of seed dormancy. The Plant Journal 45, 309–319.
Went, F. W. (1974). Reflections and speculations. *Annual Review of Plant Physiology*, *25*, 1–27.

CHAPTER SIX

Root Nitrate Uptake

Laurence Lejay[1], Alain Gojon
Biochimie et Physiologie Moléculaire des Plantes (BPMP), Université de Montpellier, CNRS, INRA, SupAgro, Montpellier, France
[1]Corresponding author. e-mail address: laurence.lejay@inra.fr

Contents

Abstract

Nitrogen (N) is the mineral nutrient required in the highest amount in plants, leading to massive use of N fertilizers in agriculture and resulting in pollution problems. It is thus essential to find ways to improve Nitrogen Use Efficiency (NUE) and design biotechnological or breeding strategies for a better N fertilizer use. Since nitrate (NO_3^-) is the primary N source in agricultural soils, root NO_3^- transporters are important targets that have been the purpose of intensive studies. Elucidation of the signalling pathways involved in the regulation of root NO_3^- uptake has led to the identification of several molecular elements regulating both root NO_3^- transporters and root system architecture. The latter determines the volume of soil explored by the plant. The multiple root NO_3^- uptake systems and the molecular signalling pathways involved in their regulation are detailed and discussed in an effort to discriminate how these elements contribute to the regulation of root architecture and/or root NO_3^- uptake. Furthermore, recent reports are discussed to illustrate how NO_3^- transporter genes can be used as molecular tools for improving NUE in crops.

1. INTRODUCTION

Under current agronomic systems nitrogen (N) is one of the key macronutrients in plant nutrition that limits potential yield (Gregory & George, 2011; Tilman, 1999). To date, N inputs have been applied in massive

Advances in Botanical Research, Volume 87
ISSN 0065-2296
https://doi.org/10.1016/bs.abr.2018.09.009

quantities to match crop N demand and thus to maximize crop production. However, lack of accuracy in N fertilization application has both economic and environmental costs (e.g. fossil energy waste in manufacturing N fertilizers and N losses from fields through leaching and volatilizing). Accordingly, improving N use efficiency (NUE) of crops is an absolute requirement for modern agriculture, to face the challenge of enhancing yield for food production in a sustainable, environment friendly way. A classical approach to improve NUE of crops is to exploit the ability of the plants to develop adaptive responses to low N supply, which rely on (i) an up-regulation of specific root membrane transport systems to improve uptake efficiency and (ii) a modulation of root system architecture to increase the soil volume explored by the plant. In this context, soil–plant processes governing root N uptake are of major interest and the elucidation of the plant signalling pathways controlling these processes has a strategic importance to design biotechnological or breeding strategies for improved N fertilizer use (Schroeder et al., 2013).

As nitrate (NO_3^-) is the primary N source in typical aerobic agricultural soils, significant progress has been made, over the last three decades, in understanding the regulatory mechanisms governing the acquisition of NO_3^- both at the level of uptake systems and root architecture. These mechanisms comprise both local signalling pathways acting at the cell level and systemic signalling pathways that communicate internal nutrient status across the plant (Forde, 2002; Gansel, Munos, Tillard, & Gojon, 2001; Gojon, Nacry, & Davidian, 2009; Zhang, Jennings, Barlow, & Forde, 1999). At the molecular level, combinations of transcriptomic approaches and genetic screens have led to the identification of several elements involved essentially in the short-term regulation of root development and root NO_3^- transporters by the local supply of NO_3^-, which is called the primary NO_3^- response (PNR) (Krapp et al., 2014). In contrast, much less is known about molecular components of systemic signalling (Alvarez, Vidal, & Gutierrez, 2012).

This chapter details the discovery of root NO_3^- uptake systems and updates knowledge concerning the molecular signalling pathways in an effort to distinguish how these elements contribute to the regulation of root architecture and/or root NO_3^- uptake. The chapter also discusses recent reports, which illustrate how NO_3^- transporter genes can be used as molecular tools for improving NUE in crops, with a special focus on bread wheat and rice.

2. CHARACTERIZATION OF NO_3^- TRANSPORT SYSTEMS

2.1 Root NO_3^- Uptake

In the soil solution, NO_3^- moves towards the root surface by diffusion and by bulk flow generated by plant transpiration. However, NO_3^- uptake into root cells is largely independent of transpiration, as it relies on the activity of specific membrane transport systems that require energy, even when the external NO_3^- concentration is high, in the mM range (Crawford & Glass, 1998). Thermodynamic considerations and physiological studies have provided evidence that under most situations, root uptake of NO_3^- occurs against its electrochemical potential gradient, and is coupled with H^+ uptake with a stoichiometry of 2(H^+) per NO_3^- anion transported across the membrane (Glass, Shaff, & Kochian, 1992; Meharg & Blatt, 1995). This is also supported by the observation that NO_3^- uptake is faster in acidic conditions and that it leads to an alkalinisation of the external medium (Dijkshoorn, 1962; McClure, Kochian, Spanswick, & Shaff, 1990a, 1990b). In addition, NO_3^- uptake is associated with depolarisation of the plasma membrane (Crawford & Glass, 1998).

Kinetics studies of net NO_3^- uptake or NO_3^- influx as a function of the external NO_3^- concentrations have revealed that roots possess at least three, kinetically distinct, NO_3^- transport systems. First, a constitutive high-affinity transport system (cHATS), with a K_m of 6–20 μM, which mediates a basal level of NO_3^- transport activity in plants grown on medium without NO_3^-. This allows the plants to ensure root NO_3^- influx as soon as NO_3^- appears in the external medium, but the extent of this flux varies among and within species (Glass et al., 2002; King, Siddiqi, Ruth, Warner, & Glass, 1993; Kronzucker, Siddiqi, & Glass, 1995b; Siddiqi, Glass, Ruth, & Fernando, 1989; Zhuo, Okamoto, Vidmar, & Glass, 1999). Second, following a first exposure to NO_3^-, there is a rapid increase in activity of an inducible high-affinity transport system (iHATS), with a K_m of 13–79 μM, which is followed (after several hours) by a downregulation of this activity (Aslam, Travis, & Huffaker, 1992; Glass et al., 2002; Hole, Emran, Fares, & Drew, 1990; Jackson, Flesher, & Hageman, 1973; Kronzucker, Siddiqi, & Glass, 1995a; Lee & Drew, 1986; Siddiqi et al., 1989; Siddiqi, Glass, Ruth, & Rufty, 1990; Zhuo et al., 1999). Third, the low affinity transport system (LATS), which can significantly contribute to NO_3^- uptake at concentrations above 250 μM, and fails to saturate at NO_3^- concentration as high as

50 mM (Aslam et al., 1992; Crawford & Glass, 1998; Glass & Siddiqi, 1995; Kronzucker et al., 1995a; Pace & MacClure, 1986; Siddiqi et al., 1990). Thermodynamic considerations indicate that NO_3^- uptake by LATS is also an active process, in spite of its linear dependency on concentrations (Crawford & Glass, 1998; Glass et al., 1992; Glass & Siddiqi, 1995). Finally, although NO_3^- uptake is associated with a significant expenditure of energy, numerous reports have been published on the occurrence of a NO_3^- efflux that recycles out of the cell a significant proportion of the NO_3^- absorbed. Indeed, upon certain biotic (Garcia-Brugger et al., 2006) or abiotic stresses, such as mechanical or transplant shocks (Aslam, Travis, & Rains, 1996; Delhon, Gojon, Tillard, & Passama, 1995; Macduff & Jackson, 1992; Pearson, Volk, & Jackson, 1981) or medium acidification (Aslam, Travis, & Huffaker, 1995), marked increases in NO_3^- efflux leading to (net) NO_3^- excretion have been reported. However, the biological significance of root NO_3^- efflux remains obscure, especially because in most instances, the changes in net NO_3^- uptake are reflected by changes in NO_3^- influx, while efflux responses are relatively minor.

2.2 Root NO_3^- Transporters

At the molecular level, two families of transporters have been involved in root NO_3^- uptake so far: (i) the NPF (Nitrate Peptide Family) family (previously named NRT1/PTR), which regroups essentially the LATS (Leran et al., 2014) and (ii) the NRT2/NNP (Nitrate-Nitrite Porter) family for the cHATS and iHATS (Krapp et al., 2014; Orsel, Krapp, & Daniel-Vedele, 2002). Two other families participate in NO_3^- transport in the plant, the Chloride Channel Family (CLC) (Barbier-Brygoo et al., 2011) and Slow Anion Associated Channel Homologue (SLAC/SLAH) (Negi et al., 2008) but, to date, none of them was shown to participate in root NO_3^- uptake.

The NPF (NRT1/PTR) and NRT2 proteins share the same structural topology, with 12 transmembrane domains, distributed in two sets of six helices connected by a cytosolic loop. Besides their common structure, no primary sequence homology is found between NPF (NRT1/PTR) and NRT2 transporters.

NPF6.3 (also known as NRT1.1 or CHL1) was the first NO_3^- transporter identified in higher plants using a T-DNA-tagged chlorate resistant mutant (Tsay, Schroeder, Feldmann, & Crawford, 1993). The NO_3^- uptake mutant *chl1* was isolated and characterized 45 years ago (Oostindier-Braaksma & Feenstra, 1973), and the *NPF6.3* (*NRT1.1/CHL1*) gene was identified 20 years after when molecular tools became available in *Arabidopsis* (Tsay et al., 1993). Remarkably, all of the NPF6.3 (NRT1.1/CHL1) homologues identified in bacteria, yeast and animals are di- and tripeptide transporters (Tsay, Chiu, Tsai, Ho, & Hsu, 2007). In contrast, in plants, the NPF (NRT1/PTR) proteins have evolved to mediate the transport of a huge variety of substrates, including dipeptides, NO_3^-, nitrite, chloride, glucosinolates and amino-acids as well as several plant hormones including auxin (IAA), abscisic acid (ABA), jasmonates (JAs) and gibberellins (GAs) (Corratge-Faillie & Lacombe, 2017; Leran et al., 2014). In *Arabidopsis*, the NPF (NRT1/PTR) family consists of 53 members (as predicted by the genome sequence) and molecular genetic studies have revealed that it includes at least 16 NO_3^- transporters expressed in the different parts of the plant. Among them, only NPF6.3 (NRT1.1/CHL1) and NPF4.6 (NRT1.2) are known to be involved in root NO_3^- uptake (Fig. 1) (Wang, Cheng, Chen, & Tsay, 2018).

In contrast to all NO_3^- transporters in the NPF family of *Arabidopsis* characterized to date, which are LATS, NPF6.3 (NRT1.1/CHL1) is a dual-affinity NO_3^- transporter with a K_m of about 50 μM for the high-affinity phase and 5 mM for the low affinity phase (Liu, Huang, & Tsay, 1999;

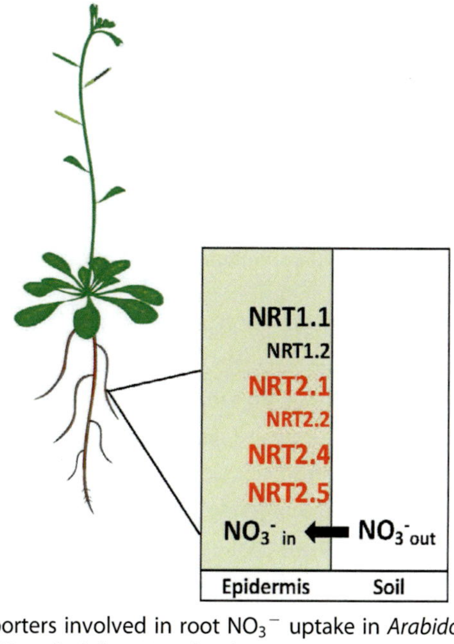

Fig. 1 NO_3^- transporters involved in root NO_3^- uptake in *Arabidopsis thaliana*. Members of the NRT1/PTR and of the NRT2 family are respectively written in black and red. Large fonts are used for the main transporters in each family.

Wang, Liu, & Crawford, 1998). The two action modes of NPF6.3 (NRT1.1/CHL1) are determined by phosphorylation of the threonine 101 residue (Liu & Tsay, 2003). When T101 is phosphorylated, NPF6.3 (NRT1.1/CHL1) functions as a HATS and when dephosphorylated, it functions as a LATS. According to the crystal structures of NPF6.3 (NRT1.1/CHL1), phosphorylation at T101 may decouple its dimer form and/or increase its structural flexibility (Parker & Newstead, 2014; Sun et al., 2014). When external NO_3^- is abundant, dephosphorylated NPF6.3 (NRT1.1/CHL1) forms a dimer and/or acquires a low structural flexibility making it functioning as a LATS. When external NO_3^- is limited, NPF6.3 (NRT1.1/CHL1) is phosphorylated to decouple the dimer and/or increase flexibility, allowing NPF6.3 (NRT1.1/CHL1) to work as a HATS. Interestingly, this dual-affinity mode might not be specific of NPF6.3 (NRT1.1/CHL1) in *Arabidopsis*. Indeed, in *Medicago truncatula*, the NO_3^- transporter MtNPF6.8 (MtNRT1.3) is also a dual-affinity NO_3^- transporter (Morère-Le Paven et al., 2011). Sequence comparison of MtNRT1.3 revealed that its closest *Arabidopsis* homologue, NPF6.4 (NRT1.3), is a putative LATS and even if no dual-affinity has been yet demonstrated for NPF6.4 (NRT1.3), the phosphorylation site RXXT101 responsible for the shift between low and high affinity in NPF6.3 (NRT1.1/CHL1) in *Arabidopsis* is conserved in MtNRT1.3.

Unlike NPF6.3 (NRT1.1/CHL1), NPF4.6 (NRT1.2) mediates only low-affinity NO_3^- uptake in the roots, with a K_m of about 5.9 mM and is constitutively expressed (Huang, Liu, Lo, & Tsay, 1999). This seems to indicate that, like for HATS, there are separate constitutive and NO_3^- inducible low-affinity transport system (cLATS and iLATS, respectively).

Concerning the NRT2/NNP family, the first member to be cloned was the *CRNA* gene from *Aspergillus nidulans* that was identified to be a functional nitrate/nitrite transporter when expressed in *Xenopus* oocytes, with a K_m of about 20 μM (Zhou et al., 2000). The *crna* mutant was isolated, like the *chl1* mutant of *Arabidopsis*, on the basis of its resistance to chlorate (Brownlee & Arst, 1983; Unkles et al., 1991). Later on, the first gene involved in nitrate/nitrite transport in the algae *Chlamydomonas reinhardtii*, *CrNRT2.1*, was identified by complementation of a deletion mutant (Quesada, Galvan, & Fernandez, 1994). Amino acid sequence comparison between CRNA and CrNRT2.1 revealed highly conserved regions. This feature was further exploited to identify and clone *NRT2* genes in different species of land plants and microalgae (Cai et al., 2008; He et al., 2004; Krapp et al., 1998; Orsel et al., 2002). The molecular cloning of *NRT2* genes from a higher plant was first reported in barley (Trueman, Richardson, & Forde, 1996).

Models for the membrane topology of HvNRT2.1 and CRNA polypeptides differ in the size of the predicted central loop (96 amino acids in CRNA versus 32 amino acids in HvNRT2.1) and in the presence in HvNRT2.1 as in CrNRT2.1, of a large C-terminal extension (~70 amino acids), which is absent in CRNA (Forde, 2000). This C-terminal extension seems to be a general feature of NRT2 transporters from algae and higher plants (Forde, 2000; Jacquot, Li, Gojon, Schulze, & Lejay, 2017). The number of high-affinity nitrate/nitrite transporters varies between species from two genes in *Nicotiana plumbaginifolia* to a large multigene family for example in barley (Trueman et al., 1996). However, functional analysis using *Xenopus* oocytes revealed that, except in the yeast *Hansenula polymorpha* and filamentous fungi, most of the NRT2 transporters largely rely on a component partner protein, NAR2 (NRT3), to mediate NO_3^- transport (Orsel et al., 2006; Tong, Zhou, Li, & Miller, 2005; Zhou et al., 2000). This property was first discovered in *C. reinhardtii* where *CrNRT2.1* and *CrNRT2.2* require a second gene, *CrNAR2*, to function in NO_3^- transport (Quesada et al., 1994; Zhou et al., 2000). Similar results were reported for barley, where the product of one of three *NAR2*-like genes, HvNAR2.3, interacts with HvNRT2.1 in the plasma membrane of the roots, and is able to increase NO_3^- transport when co-expressed with HvNRT2.1 in *Xenopus* oocytes (Ishikawa et al., 2009; Tong et al., 2005).

In *Arabidopsis*, seven members have been described in the *NRT2* family along with two *NAR2* genes (Krapp et al., 2014; Okamoto et al., 2006; Orsel et al., 2006; Wang et al., 2018). A yeast two-hybrid assay and biomolecular fluorescence complementation study showed that six NRT2 proteins (except for NRT2.7) interact with NAR2.1 in the plasma membrane (Kotur et al., 2012). Indeed, when NAR2 is co-expressed in *Xenopus* oocytes, the NO_3^- transport activities of NRT2.1, 2.2 and 2.5 are greatly enhanced (Kotur et al., 2012). However, NAR2 does not seem to be required for the NO_3^- transport activity of NRT2.4 (Kiba et al., 2012).

Among the seven NRT2s in *Arabidopsis* four are involved in root NO_3^- uptake (Fig. 1). NRT2.1 appears to be the major component of the HATS since several mutants disrupted for the *NRT2.1* gene have lost up to 75% of the high-affinity NO_3^- uptake activity (Cerezo et al., 2001; Filleur et al., 2001; Li et al., 2007; Little et al., 2005). As a consequence, growth of these mutants is severely impaired at low NO_3^- concentration (Li et al., 2007; Orsel, Eulenburg, Krapp, & Daniel-Vedele, 2004), but not at high NO_3^- concentration when low-affinity transporters, most probably those of the NPF6.3 (NRT1.1/CHL1) family, are active (Crawford & Glass, 1998;

Williams & Miller, 2001). *NRT2.1* is often linked with *NRT2.2,* which is however only weakly expressed compared to *NRT2.1* and does not seem to have a major role in root NO_3^- uptake (Cerezo et al., 2001; Li et al., 2007). More recently, NRT2.4 and NRT2.5 have also been shown to be involved in root NO_3^- uptake. Compared to NRT2.1, NRT2.4 is suspected to be a very high affinity NO_3^- transporter, since uptake at concentrations below 0.025 mM NO_3^- is decreased in the knockout mutant (Kiba et al., 2012). NRT2.5 could, on the other hand, correspond to the cHATS described above since it becomes the major HATS after a long period of N starvation that suppressed expression of the iHATS (Lezhneva et al., 2014).

3. REGULATION OF ROOT NO_3^- ACQUISITION

3.1 Regulation of Root NO_3^- Transporters

Because of the dramatic dependency of plant growth and development on N, plants have evolved to efficiently adapt to environmental or endogenous variations of N availability and use. Root NO_3^- uptake is thus strongly regulated in response to changes in the external NO_3^- availability or in the N demand of the whole plant. Furthermore, NO_3^- uptake is also regulated by photosynthesis to ensure the necessary integration of both N acquisition by roots and C acquisition by shoots. Many of the transport systems detailed above are subjected to such regulations. However, a general conclusion arising from most studies is that the HATS is particularly responsive and displays a much higher flexibility than the corresponding LATS.

There are two major mechanisms that coordinately act to modulate root NO_3^- uptake as a function of external NO_3^- supply or internal N demand (Crawford & Glass, 1998; Gojon et al., 2009). The first one relates to the Primary NO_3^- Response (PNR), which is characterized by a rapid induction of *NRT2.1* and *NPF6.3* (*NRT1.1/CHL1*) shortly after NO_3^- provision (Filleur & Daniel-Vedele, 1999; Lejay et al., 1999; Okamoto, Vidmar, & Glass, 2003; Tsay et al., 1993; Zhuo et al., 1999). The second mechanism is a feedback repression exerted by a high N status of the whole plant and/or high NO_3^-, that down-regulates *NRT2.1*, *NRT2.2*, *NRT2.4* and *NRT2.5* under N satiety conditions (Kiba et al., 2012; Lejay et al., 1999; Lezhneva et al., 2014; Zhuo et al., 1999). This repression is relieved when plants experience N limitation, resulting in a strong increase in HATS capacity that improves N uptake efficiency. For *NRT2.1*, split-root experiments have demonstrated that this regulation relies on systemic signalling pathways. Indeed, *NRT2.1* is up-regulated in roots that are still fed with NO_3^- but that respond to a localized N starvation treatment on another portion of the root system (Gansel et al., 2001). This mechanism was initially attributed to a feedback repression by reduced N metabolites such as ammonium (NH_4^+) and amino acids but recently, shoot-derived glutaredoxin polypeptides named CEPD1 and CEPD2 have been shown to act as descending long-distance mobile signals for the regulation of *NRT2.1* (Lejay et al., 1999; Nazoa et al., 2003; Ohkubo, Tanaka, Tabata, Ogawa-Ohnishi, & Matsubayashi, 2017; Zhuo et al., 1999). On the other hand, feedback repression of *NRT2.1* by high NO_3^- supply is suppressed or strongly attenuated in mutants for the NO_3^- transporter NPF6.3 (NRT1.1/CHL1) (Munos et al., 2004). This results in the overexpression of *NRT2.1* in normally suppressive conditions (e.g. in NH_4NO_3-fed *nrt1.1* plants) along with a lack of stimulation of *NRT2.1* expression by N starvation in *nrt1.1* mutants. Because NH_4NO_3-fed *nrt1.1* mutants were not N deficient, this suggested that *NRT2.1* expression is not only repressed by reduced N metabolites, but also by NO_3^- itself. Indeed, further works revealed that even in wild-type plants, *NRT2.1* is overexpressed under low NO_3^-/high NH_4^+ availability (Krouk, Tillard, & Gojon, 2006). This shows that, at least for *NRT2.1*, the feedback repression exerted by high N status is more complex and involves both feedback regulation by reduced N metabolites and/or CEPD1/CEPD2 and NPF6.3 (NRT1.1/CHL1)-mediated repression by high external NO_3^- (Krouk et al., 2006; Ohkubo et al., 2017). For now, no data are available concerning the regulation of *NRT2.4* and *NRT2.5* by either one of these regulations in response to N starvation.

The regulation of root NO_3^- uptake by photosynthesis has been well described at the physiological level. This regulation is characterized by a diurnal rhythm of NO_3^- uptake, with a peak during the light period and a minimum in the dark. This pattern has been attributed to a stimulatory effect of sugars produced by photosynthesis and transported down to the roots (Delhon, Gojon, Tillard, & Passama, 1996; Lejay et al., 1999). At the molecular level, the regulation of root NO_3^- uptake by light and sugars has been correlated with changes in the expression of *NRT2.1*, *NRT2.4* and *NPF6.3* (*NRT1.1/CHL1*) (Lejay et al., 2003, 1999, 2008). Experiments combining the use of sugar analogues and sugar sensing mutants revealed that the main response mechanisms previously documented for sugar signalling in plants, such as regulation by sucrose (Chiou & Bush, 1998), hexose transport (Lalonde et al., 1999) and hexokinase sensing activity (Moore et al., 2003), are not involved in the induction of *NRT2.1* expression by sugars. However, the lack of *NRT2.1* induction by sugars in a hexokinase antisense line and after addition of the hexokinase inhibitor glucosamine suggested that hexokinase catalytic activity and C metabolism downstream of the hexokinase step were crucial for *NRT2.1* regulation by sugars (Lejay et al., 2003). This result was confirmed by further studies showing that an inhibitor of the Oxidative Pentose Phosphate Pathway (OPPP) or the use of OPPP intermediates was able to impair or mimic, respectively, the induction of *NRT2.1*, *NRT2.4* and *NPF6.3* (*NRT1.1/CHL1*) expression by sugars (de Jong, Thodey, Lejay, & Bevan, 2014; Lejay et al., 2008). It is interesting to note that regulation by a signal coming from OPPP activity does not only concern NO_3^- transporters but also NO_3^- assimilation and the expression of NO_3^- assimilatory genes in response to sucrose in the roots (Bussell, Keech, Fenske, & Smith, 2013; de Jong et al., 2014). Indeed, the link between the OPPP and N metabolism in roots is particularly strong since the OPPP provides the reducing power for nitrite reductase and GOGAT in root plastids (Bowsher, Boulton, Rose, Nayagam, & Emes, 1992; Bowsher, Hucklesby, & Emes, 1989; Oji, Watanabe, Wakiuchi, & Okamoto, 1985). As a consequence, glutamate (Glu) synthesis and the production of reducing power by the OPPP are strictly correlated biochemical processes. However, the signalling pathway involved in this regulation remains as yet unknown.

If most of studies on the regulation of root NO_3^- transporters have focused on the transcriptional level, more and more data indicate that it will be necessary in the future to take into account regulations at the protein level to fully understand the regulatory mechanisms of root NO_3^- uptake. Indeed, transgenic plants expressing *NRT2.1* constitutively showed that regulation by C and N also acts at the post-translational level (Laugier et al., 2012). Furthermore, de Jong et al. (2014) used *gin2-1* plants to show that glucose stimulates NRT2.1 protein levels and transport activity independently of the stimulation of *NRT2.1* expression through OPPP, demonstrating another possible post-transcriptional mechanism influencing NO_3^- uptake in response to C. However, the post-translational modifications of root NO_3^- transporters still need to be characterized and for now, except for the phosphorylation of NPF6.3 (NRT1.1/CHL1) by the protein kinase CIPK23 (Ho, Lin, Hu, & Tsay, 2009), there is no information concerning the molecular elements involved in these regulatory mechanisms.

3.2 Regulation of Root Development

Another important factor of plant's ability to take up mineral nutrients from the soil is the size and architecture of its root system. The development of the root system is genetically determined but also highly plastic, the nutrients playing a major role in its modulation. Indeed, besides the upregulation of root HATS to improve uptake efficiency in response to low NO_3^- availability, root system architecture (RSA) is regulated to increase the volume of soil explored by the plant. Like for root NO_3^- uptake, RSA depends on both external NO_3^- availability and internal N status of the plant, which lead to (i) a systemic repression of lateral root (LR) growth by high N status and (ii) a local stimulation of LR growth by exogenous NO_3^- supply (Zhang et al., 1999). The nature of the signal for the systemic repression of LR growth is unknown but it seems to involve both feedback repression by reduced N metabolites and NO_3^- itself. Indeed, a high shoot NO_3^- status seems important for the inhibitory response, as this effect is enhanced in a nitrate reductase deficient mutant, suggesting that systemic signalling of NO_3^- accumulation in the shoot causes LR inhibition (Zhang et al., 1999). However, reduced N metabolites also trigger repression of LR growth since this repression, as induced by a high NO_3^- concentration, can be alleviated by the inhibition of NH_4^+ assimilation into glutamine, and restored when glutamine is added in the culture medium (Gifford, Dean, Gutierrez, Coruzzi, & Birnbaum, 2008). In contrast with its systemic inhibitory effect on LR growth, NO_3^- locally promotes initiation and/or elongation of LRs. This adaptive response was extensively studied in *Arabidopsis*, in which Zhang and Forde (1998) first demonstrated that a localized patch of NO_3^- in a split agar plate elicited a preferential LR growth in the region of the patch. NO_3^- rather than its downstream metabolites appears to be the signal for the stimulation, as a mutant in nitrate reductase shows the wild-type response (Zhang & Forde, 1998). Linkohr, Williamson, Filter, and Leyser (2002) also found that within the same plant, the uneven distribution of LRs between NO_3^--rich and NO_3^--poor patches is due to repression of LR elongation in the NO_3^--poor patch, which was recently confirmed by Krouk, Lacombe, et al. (2010). Finally, if NO_3^- supply is localized on a highly restricted portion of the total root system, the consequent N limitation alleviates the systemic repression by high N status, further amplifying the local LR proliferation response (Nacry, Bouguyon, & Gojon, 2013).

3.3 Molecular Elements

Important progresses have been made in the last years concerning the molecular bases of the regulation of root NO_3^- acquisition and especially for the PNR. These findings are summarized in Fig. 2.

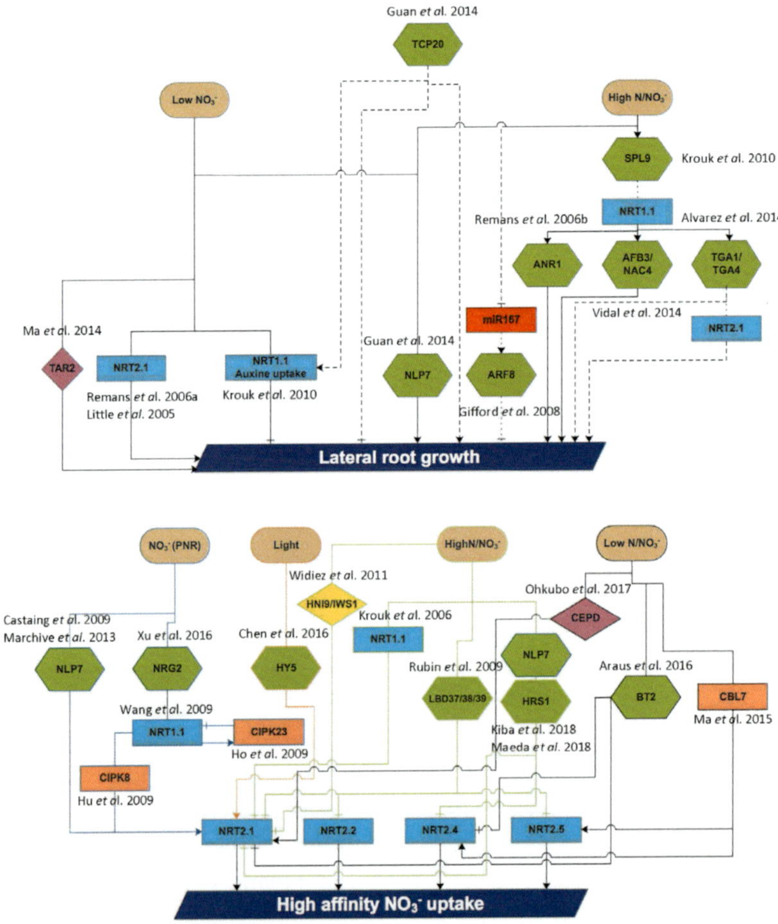

Fig. 2 Known regulatory elements for the regulation of lateral root growth and root high affinity NO_3^- transporters. The extremities of lines, with arrows and horizontal lines, indicate positive and negative signalling steps, respectively. For the sake of clarity, NPF6.3 (NRT1.1/CHL1) was indicated as NRT1.1 and we represented the regulation by N with a box named "high N/NO_3^-" or "low N/NO_3^-" since in most of the studies it is not possible to distinguish whether regulation is mediated by N metabolites and/or by NO_3^- itself.

3.3.1 Common Regulatory Elements for Root NO_3^- Transporters and Root Development

Some regulatory elements control both root NO_3^- transporters and root development. Among them, NPF6.3 (NRT1.1/CHL1) has been shown to act as a NO_3^- transceptor (transporter/sensor) (Gojon, Krouk, Perrine-Walker, & Laugier, 2011), triggering several NO_3^- signalling pathways that control the expression of both *NRT2.1* and the development of LRs (Bouguyon et al., 2015; Ho et al., 2009; Krouk, Lacombe, et al., 2010; Krouk et al., 2006; Remans, Nacry, Pervent, Filleur, et al., 2006; Wang, Xing, Wang, Tran, & Crawford, 2009). The NPF6.3 (NRT1.1/CHL1)-dependent regulation of *NRT2.1* expression was shown to be particularly complex. First, NPF6.3 (NRT1.1/CHL1) can have opposite actions, by either up-regulating *NRT2.1* as a part of PNR, or down-regulating it when NO_3^- is supplied at high concentration for several days (Ho et al., 2009; Krouk et al., 2006; Munos et al., 2004). Second, as for NPF6.3 (NRT1.1/CHL1) NO_3^- transport activity, NPF6.3 (NRT1.1/CHL1)-dependent induction of *NRT2.1* during PNR displays a two-phase pattern, with a high affinity phase (low NO_3^- concentration) leading to a limited *NRT2.1* induction, and a low-affinity phase (high NO_3^- concentration) resulting in full induction of *NRT2.1* (Ho et al., 2009). As for the transport activity, the switch from low- to high-affinity NO_3^- sensing by NPF6.3 (NRT1.1/CHL1) is due to phosphorylation of the T101 residue by CIPK23 (Ho et al., 2009). Conversely, the NPF6.3 (NRT1.1/CHL1)-dependent repression of *NRT2.1* by high NO_3^- requires the phosphorylated form of NPF6.3 (Bouguyon et al., 2015). The CBL-interacting protein kinase, CIPK8, which is induced by NO_3^- at the transcript level, participates in the induction of *NRT2.1* by NO_3^- when NPF6.3 (NRT1.1/CHL1) is not phosphorylated. However, the targets of CIPK8 are still unknown (Hu, Wang, & Tsay, 2009).

The NPF6.3 (NRT1.1/CHL1)-dependent regulation of root development is also complex and several signalling pathways seem to be involved:

(i) Under low NO_3^- concentration in the media, it involves the capacity of NPF6.3 (NRT1.1/CHL1) to transport auxin, which prevents auxin accumulation in LRs and represses their emergence (Krouk, Lacombe, et al., 2010). Consistently, Bouguyon et al. (2015) have shown that the phosphorylated form of NPF6.3 (NRT1.1/CHL1) is the one predominantly active in both auxin transport and regulation of LR growth.

(ii) Besides its role in the repression of LR growth when NO_3^- concentration in the media is low, NPF6.3 (NRT1.1/CHL1) is involved in the signalling pathway triggering root colonization of NO_3^--rich patches by an auxin-independent mechanism (Mounier, Pervent, Ljung, Gojon, & Nacry, 2014; Remans, Nacry, Pervent, Filleur, et al., 2006). This NO_3^- response pathway involves ANR1, a transcription factor from the MADS-box gene family (Zhang & Forde, 1998). Interestingly, MADS-box genes are part of a large family of transcription factors involved in many aspects of plant development (Alvarez-Buylla, Azpeitia, Barrio, Benitez, & Padilla-Longoria, 2010). Several *Arabidopsis* MADS-box genes are expressed in roots (Alvarez-Buylla et al., 2000; Burgeff, Liljegren, Tapia-Lopez, Yanofsky, & Alvarez-Buylla, 2002; Gan, Filleur, Rahman, Gotensparre, & Forde, 2005; Garay-Arroyo et al., 2013; Tapia-Lopez et al., 2008). Gan et al. (2005) identified among them seven MADS-box genes (*AGL16*, *AGL21*, *AGL14*, *AGL19*, *SOC1*, *AGL26* and *AGL56*) with an expression pattern similar to that of *ANR1* in response to different N-treatments (N starvation, NO_3^--resupply). Furthermore, three of these genes (*AGL16*, *AGL21* and *SOC1*) were found to encode proteins that interact with ANR1 in a yeast two-hybrid assay (de Folter et al., 2005). Even though the role of these genes in root development remains as yet elusive, it is very likely that some of these are involved in the regulation of root development and/or NO_3^- uptake by N availability.

(iii) NPF6.3 (NRT1.1/CHL1) NO_3^- transport activity has been shown to be involved in the NO_3^- induction of the *AFB3* auxin receptor and its target, the *NAC4* transcription factor (Vidal, Alvarez, & Gutierrez, 2014). Plants lacking *NAC4* showed altered LR density in response to NO_3^- treatment (Vidal et al., 2010; Vidal, Moyano, Riveras, Contreras-Lopez, & Gutierrez, 2013).

(iv) More recently, two other bZIP transcription factors, TGA1 and TGA4, have been involved in a NPF6.3 (NRT1.1/CHL1)-dependent regulation of root development in response to high NO_3^-. TGA1 and TGA4 have been shown to induce LR growth in high NO_3^- and analysis in various *nrt1.1* mutant alleles positioned these transcription factors in a NO_3^--signalling pathway in roots downstream of NPF6.3 (NRT1.1/CHL1) (Alvarez et al., 2014). Since TGA1 and TGA4 bind to the *NRT2.1* promoter and are involved in its induction by NO_3^-, it is also possible that TGA1 and TGA4 control LR growth by controlling the expression of *NRT2.1*. Indeed, NRT2.1, independently of its role in root NO_3^- uptake, has also been found to be involved in the regulation of LR initiation. However, its role is not clear yet. One study has shown NRT2.1 to stimulate initiation of LR primordia in low NO_3^- conditions (Remans et al., 2006) whereas, in another work, NRT2.1 was found to repress lateral root initiation in low NO_3^- (Little et al., 2005).

Altogether, the multiple roles of NPF6.3 (NRT1.1/CHL1) in the regulation of LR growth by NO_3^- could be determined by the transcription factor SPL9, which has been identified as a potential regulatory hub in response to NO_3^- (Krouk, Mirowski, LeCun, Shasha, & Coruzzi, 2010). SPL9 is predicted to target PNR genes such as *NPF6.3* (*NRT1.1/ CHL1*), and root phenotypic analysis showed that in the presence of NO_3^-, *SPL9* over-accumulators are impaired in LR growth (Krouk, Mirowski, et al., 2010).

Independently of NPF6.3 (NRT1.1/CHL1), another common element for the regulation of root NO_3^- transporters and root development in response to NO_3^- is the transcription factor NLP7 (Castaings et al., 2009; Guan et al., 2014; Marchive et al., 2013; Xu et al., 2016). In *Arabidopsis*, NLP7 was reported as a factor involved in NO_3^- and N starvation responses, as *nlp7* mutants displayed a constitutive N starvation phenotype (Castaings et al., 2009). In addition, the regulation of *NRT2.1* and *NRT2.2* by NO_3^- was severely altered in *nlp7* as compared to wild-type plants (Castaings et al., 2009). The role of NLP7 in LR growth has also been shown using split-root experiments. In that case the *nlp7* mutant shows severely reduced LR growth in both high- and low-NO_3^- media, indicating loss of local stimulation by NO_3^- of root growth (Guan et al., 2014).

Finally, a transcription factor called HY5 has been involved in both the regulation of root NO_3^- transporter and root development in response to illumination of the shoot (Chen, Yao, et al., 2016). Originally, the *hy5* mutant has been described as a photo-morphogenetic mutant involved in a light-dependent signal transduction pathway (Ang & Deng, 1994; Chory, 1992; Oyama, Shimura, & Okada, 1997). More recently, Chen, Yao, et al. (2016) revealed that HY5 is a shoot-to-root mobile signal that mediates light-regulated coupling of shoot growth and C assimilation with root growth and NO_3^- uptake. HY5 loss-of-function mutant is specifically disrupted in shoot illumination promoted root growth. Moreover, ChIPs' experiments demonstrated the in vivo binding of HY5 to the *NRT2.1* promoter in response to light, indicating that HY5 directly regulates *NRT2.1* expression. Interestingly, this binding is also potentiated by increased carbon photo-assimilates in the roots (Chen, Yao, et al., 2016). However, it still remains unknown whether HY5 is part of the signalling pathway linked to the OPPP, which has been involved in the regulation of *NRT2.1* and *NRT2.4* by sugars (de Jong et al., 2014; Lejay et al., 2008).

3.3.2 Regulatory Elements Specific for Root NO_3^- Transporters or Root Development

Some components seem to specifically regulate root development or root NO_3^- transporters. Two elements, TCP20 and miR167/ARF8, appear to be involved in the systemic regulation by N and in the regulation by organic N of root development, respectively. TCP20 is a transcription factor that binds *NRT2.1* and *NPF6.3* (*NRT1.1*/*CHL1*) promoters both in vitro and in vivo (Guan et al., 2014). TCP20 works in systemic NO_3^- signalling for root foraging in NO_3^--rich patches and it may interact with *NPF6.3* (*NRT1.1*/*CHL1*) to repress LR growth on low NO_3^- patches. The miR167/ARF8 module comes from the first of very few studies on NO_3^- response at the tissue-specific level, showing that NO_3^- response is highly cell type-specific (Gifford et al., 2008). Using the glutamine synthetase inhibitor MSX, Gifford et al. (2008) showed that organic N induces *ARF8* through the repression of miR167. This mechanism was shown to regulate the ratio of LR initiation to LR elongation to coordinate rooting under organic N availability. Finally, *TAR2* has been involved, similar to *NRT2.1*, in the induction of LR growth in response to low N conditions (Ma et al., 2014). *TAR2* is an auxin biosynthetic gene, thereby suggesting that the stimulation of LR growth in response to low NO_3^- likely depends on root-synthesized auxin.

For the regulation of root NO_3^- transporters, another component of the PNR, called NRG2, has recently been found using a reverse genetic screen (Xu et al., 2016). Genetic and molecular analysis revealed that NRG2 modulates the expression of *NPF6.3* (*NRT1.1*/*CHL1*) in response to NO_3^- and works upstream of *NPF6.3* (*NRT1.1*/*CHL1*) for the regulation of *NRT2.1* (Xu et al., 2016). Finally, like for the regulation of root development, only few regulatory elements have been found to be involved in the regulation of root HATS by systemic N signalling and/or organic N. First, a forward genetic screen showed that systemic feedback repression of *NRT2.1* promoter under high N conditions is impaired in a *hni9-1* mutant (Girin et al., 2010). HNI9 codes for a component of the RNA polymerase II complex that may increase the accumulation of H3K27me3 histone repressive marks at the *NRT2.1* locus in response to high N, leading to a downregulation of *NRT2.1* (Widiez et al., 2011). More recently, two genes coding for non-secreted polypeptides named CEPD1 and CEPD2 have been involved in the systemic regulation of *NRT2.1* by N starvation (Ohkubo et al., 2017). These shoot-derived polypeptides act as descending long-distance N-demand signals to induce *NRT2.1* in roots. Furthermore, yeast one-hybrid (Y1H) screening, using *NRT2.4* as a N starvation marker gene, identified members of the *Arabidopsis* GARP-type transcription factor NIGT1/HRS1 subfamily. These transcription factors are expressed when N availability is high and they appear to act as direct repressors of *NRT2.4* and *NRT2.1*. They may also act as direct repressors of an array of N-starvation-regulated genes in order to repress N starvation responses under high N availability conditions (Kiba et al., 2018; Maeda et al., 2018). For *NRT2.1*, Maeda et al. (2018) proposed that NO_3^- activates NLP7, which directly upregulates both the *NRT2.1* and NIGT1/HRS1-clade gene promoters. This would explain how *NRT2.1* is initially activated and then repressed by NO_3^- as described by Krouk et al. (2006).

Finally, transcriptomic approaches also enabled the identification of several regulatory transcription factors: (i) *LBD 37/38/39*, which are up-regulated by NO_3^- and to a lesser extent by NH_4^+ and glutamine (Rubin, Tohge, Matsuda, Saito, & Scheible, 2009); they repress the *NRT2.1*, *NRT2.2* and *NRT2.5* genes in N-sufficient conditions but their role in systemic signalling is as yet unknown, and (ii) BT2, a negative regulator of *NRT2.1* and *NRT2.4* under low N conditions (Araus et al., 2016).

4. NITRATE TRANSPORTER-BASED STRATEGIES FOR IMPROVING NUE IN CROPS

Given their pivotal role in nutrient acquisition and utilization, membrane transporters have been considered as important targets for improving nutrient use efficiency, growth and yield in crops (Fan et al., 2017; Schroeder et al., 2013). This is particularly relevant for root NO_3^- uptake, owing to the advances made in the identification of the transporters involved, especially in *Arabidopsis* where all key NRT2 components of the HATS have been isolated. Indeed, quadruple mutants deficient for *NRT2.1*, *NRT2.2*, *NRT2.4* and *NRT2.5* retain less than 5% of the root NO_3^- influx of the wild-type when assayed at 0.2 mM external concentration. In addition, they are unable to grow on 0.5 mM NO_3^- as sole N source (Lezhneva et al., 2014), suggesting that no additional transporter plays a significant role in the HATS. The situation is less clear for the LATS. In *Arabidopsis*, both NPF6.3 (NRT1.1/CHL1) and NPF4.6 (NRT1.2) were shown to contribute to the low-affinity root NO_3^- uptake (Huang, Chiang, Crawford, & Tsay,

1996; Huang et al., 1999), but their relative importance depends on the environmental conditions (Munos et al., 2004; Touraine & Glass, 1997). Furthermore, no mutant was reported to date to be almost totally impaired in the LATS activity, precluding any definite conclusion on the key transporters involved. Since putative NO_3^- transporter genes of both *NPF* (*NRT1*) and *NRT2* families were found in almost all plant species investigated to date (Leran et al., 2014; von Wittgenstein, Le, Hawkins, & Ehlting, 2014), these genes obviously constitute major candidates for improving root acquisition of NO_3^- in crops. The following section provides insights into recent progress made towards this goal, with a focus on bread wheat and rice, two crops of vital importance for human nutrition.

Several studies have delineated the structure of the *NPF* (*NRT1*) and *NRT2* families in wheat and rice (Buchner & Hawkesford, 2014; Fan et al., 2017; Plett et al., 2010). These provided interesting candidate genes for further investigation, but also pointed to significant differences between *Arabidopsis* and cereals in the structure of these gene families and in their expression pattern. This limits the direct translation of knowledge from *Arabidopsis* to wheat and rice, and therefore calls for extensive functional characterization studies in the latter species.

Before determining the relevance of root NO_3^- transporters for improving NUE, the first question to be considered is whether root NO_3^- uptake is a true limiting factor for NUE in crops. Classically, NUE is divided into two components: (i) N uptake efficiency (NUpE), which is calculated as the fraction of total N available in the soil that has been taken up by the plant, and (ii) N utilization efficiency (NUtE), which is calculated as the total biomass production (or grain dry mass) divided by the amount of N taken up by the plant (Good, Shrawat, & Muench, 2004). These concepts are clear, but the experimental quantification of both NUpE and NUtE remains a challenge in field studies (Cormier et al., 2016). This is in particular due to the difficulty in estimating the total N available in the soil, in addition to the amount of N fertilizer applied, and in assessing the amount of N in the below-ground organs. Nevertheless, it is usual to state that NUpE is low for most crops globally, with values often reported to be under 40–50% (Tilman, Cassman, Matson, Naylor, & Polasky, 2002), therefore suggesting that much improvement can be achieved at the root acquisition level. However, this may not be representative of all situations, and recent data convincingly indicate that NUpE can be higher than 80% in modern wheat varieties under temperate conditions (Hawkesford, 2017), leaving a limited margin for further progress. Furthermore, although NUpE displays a significant genetic variability in modern bread wheat, it does not often translate into similar variation of NUE or yield (Hawkesford, 2017). Conversely, genome-wide identification of QTLs associated with variation of NUE components in wheat did not highlight any candidate NO_3^- transporter gene (Cormier et al., 2016; Cormier, Le Gouis, Dubreuil, Lafarge, & Praud, 2014). On the other hand, grain protein concentration (GPC) is directly related to post-flowering root NO_3^- uptake (Taulemesse, Le Gouis, Gouache, Gibon, & Allard, 2016), which appears in turn to be predominantly governed by the TaNRT2.1 transporter (Taulemesse, Le Gouis, Gouache, Gibon, & Allard, 2015). Collectively, the above considerations do not provide a strong case that root NO_3^- transporters are key molecular targets for improving NUE and yield in wheat, although the conclusion may be different for GPC.

The overall picture seems to be pretty different in rice since mounting evidence shows that NO_3^- transporters are key determinants of NUE, growth and yield in this species (Fan et al., 2017; Li, Hu, & Chu, 2017). A striking example is the observation that a significant part of the phenotypic differences in NUE between *japonica* and *indica* rice subspecies can be attributed to a single polymorphism in the *OsNPF6.5* (*NRT1.1B*) transporter gene (Hu et al., 2015). The *indica* subspecies is known to have a higher NUE than *japonica*, and this is associated with a Thr327Met substitution in the *indica* allele of *OsNPF6.5* (*NRT1.1B*) as compared to *japonica*. Accordingly, introgression of the *indica* beneficial allele into the *Nipponbare japonica* variety resulted in both increased NUE and grain yield (Hu et al., 2015). Furthermore, several reports indicate that overexpression of NO_3^- transporter genes in rice is actually a relevant strategy for improving NUE, growth and yield. For instance, transformants for *OsNPF8.9* (*NRT1*), *OsNPF6.5* (*NRT1.1B*), *OsNPF6.3* (*NRT1.1A*), *OsNRT2.3b* or *OsNRT2.1* were all shown to display marked increases (e.g. up to 30–40%) in both NUE and grain yield as compared to the wild-type (Chen, Zhang, et al., 2016; Fan, Feng, et al., 2016; Fan, Tang, et al., 2016; Hu et al., 2015; Wang et al., 2018). Collectively, these observations are very surprising, first because NO_3^- is not considered as a main N source for rice, and second because attempts to overexpress key NO_3^- transporter genes in *Arabidopsis* or in other species did not result in such a dramatic NUE improvement. This is exemplified by the overexpression of *NRT2.1* genes in *Arabidopsis* and *N. plumbaginifolia*, which failed to significantly increase the HATS activity in these species, most probably due to the occurrence of post-transcriptional regulatory mechanisms compensating for enhanced gene expression (Fraisier, Gojon, Tillard, & Daniel-Vedele, 2000; Laugier et al., 2012).

Therefore, the above mentioned studies in rice constitute for sure the most positive outcomes of NO_3^- transporter-based strategies for improving NUE in crops. However, their physiological interpretation remains puzzling because the positive effect of the introgression of the beneficial *OsNPF6.5* (*NRT1.1B*) allele, or of the overexpression of the various transporter genes was not solely attributed to their role in root NO_3^- uptake. Indeed, direct assays of NO_3^- uptake activity (e.g. using ^{15}N labelling) in *OsNPF6.5* (*NRT1.1B*), *OsNPF6.3* (*NRT1.1A*) and *OsNRT2.3b* transformants showed that the increase in NO_3^- uptake remained modest (~10%) as compared to the wild-type, and was not always correlated with the corresponding increases in NUE or yield (Fan, Tang, et al., 2016; Hu et al., 2015; Wang et al., 2018). Furthermore, the effect was not specific of NO_3^-, because NH_4^+ uptake was also stimulated (Fan, Tang, et al., 2016; Wang et al., 2018). These observations suggested additional functions for OsNPF6.5 (NRT1.1B), OsNPF6.3 (NRT1.1A) and OsNRT2.3b that could explain their effects on NUE. Such functions are documented for OsNRT2.3b, which is mainly expressed in the shoot phloem, and have been suggested to act as a pH sensor that contributes to maintaining cytosolic pH homeostasis depending on the form of N (NO_3^- and/or NH_4^+) provided to the plant (Fan, Tang, et al., 2016). On the other hand, OsNPF6.5 (NRT1.1B) and OsNPF6.3 (NRT1.1A) exert, in addition to mediating NO_3^- uptake, a major effect on the expression of a wide range of N-related genes, suggesting they may have a signalling function and act as transceptors (Hu et al., 2015; Wang et al., 2018), as proposed for the *Arabidopsis* NPF6.3 (NRT1.1/CHL1) (Bouguyon et al., 2015; Gojon et al., 2011). In more details, OsNPF6.5 (NRT1.1B) might be a functional homologue of NPF6.3 (NRT1.1/CHL1) as it displays a similar expression pattern, and similar NO_3^- transport and signalling activities. In contrast, OsNPF6.3 (NRT1.1A) is functionally divergent from OsNPF6.5 (NRT1.1B) and NPF6.3 (NRT1.1/CHL1). First, it serves as a tonoplast-localized NO_3^- transporter inducible by NH_4^+ (Wang et al., 2018), whereas OsNPF6.5 (NRT1.1B) and NPF6.3 (NRT1.1/CHL1) are plasma membrane-localized NO_3^- transporters inducible by NO_3^- (Hu et al., 2015; Tsay et al., 1993). Second, OsNPF6.3 (NRT1.1A) controls a different set of N-responsive genes than OsNPF6.5 (NRT1.1B), suggesting that the two proteins are transceptors, but participate in different N signalling pathways (Wang et al., 2018). In addition, OsNPF6.3 (NRT1.1A), but not OsNPF6.5 (NRT1.1B), regulates flowering time. Although the mechanism underlying the transceptor function of OsNPF6.3 (NRT1.1A) requires further investigation, available data indicate that it may act through regulation of cytoplasmic-nuclear shuttling of NLP transcription factors (Wang et al., 2018), which play a pivotal role in NO_3^- signalling in *Arabidopsis* (Liu et al., 2017; Marchive et al., 2013). Interestingly, overexpression of *OsNPF6.3* (*NRT1.1A*) in *Arabidopsis* also resulted in earlier flowering and dramatic increases in total biomass and seed production as compared to the wild-type, pointing to a conservation of OsNPF6.3 (NRT1.1A)-dependent signalling mechanisms between species (Wang et al., 2018).

5. CONCLUSION

Over the past few years, several types of molecular components have been shown to participate in the regulation of root NO_3^- transporters and root development. These include receptors, putative components of signal transduction pathways, and effectors such as transcription factors. Considering the emergent complexity of N signalling pathways in plants (Fig. 2), the challenge will be to find ways to integrate all these data in predictive models and thereby improve our understanding of these mechanisms at the system level. This is crucial for breeding strategies aiming at improving NUE by crops, as these mechanisms dominate most of adaptive responses to N limitation. The idea that these mechanisms are of strategic importance for determining and possibly manipulating NUE in crops has recently received striking support by the study showing that OsNPF6.5 (NRT1.1B) is a major genetic determinant of the NUE difference between *indica* and *japonica* rice subspecies (Hu et al., 2015). Yet, the mechanisms underlying the effect of NO_3^- transporters on NUE remain unclear and data integration remains challenging. It requires in particular the production of time series and tissue specific expression data to construct more reliable gene regulatory networks. Experimental validation is also an essential part through the production of genetically modified plants to validate the models and their predictions. These approaches could be especially challenging depending on the chosen model plants. However, the apparent conservation of the mechanisms between species reinforces the interest of using the model plant *Arabidopsis* to resolve the problems linked to time and space scales studies and develop predictive models based on a refined molecular knowledge.

REFERENCES

Alvarez, J. M., Riveras, E., Vidal, E. A., Gras, D. E., Contreras-Lopez, O., Tamayo, K. P., et al. (2014). Systems approach identifies TGA1 and TGA4 transcription factors as important regulatory components of the nitrate response of Arabidopsis thaliana roots. *Plant Journal*, *80*(1), 1–13. https://doi.org/10.1111/tpj.12618.

Alvarez, J. M., Vidal, E. A., & Gutierrez, R. A. (2012). Integration of local and systemic signaling pathways for plant N responses. *Current Opinion in Plant Biology*, *15*(2), 185–191. https://doi.org/10.1016/j.pbi.2012.03.009.

Alvarez-Buylla, E. R., Azpeitia, E., Barrio, R., Benitez, M., & Padilla-Longoria, P. (2010). From ABC genes to regulatory networks, epigenetic landscapes and flower morphogenesis: Making biological sense of theoretical approaches. *Seminars in Cell and Developmental Biology*, *21*(1), 108–117. https://doi.org/10.1016/j.semcdb.2009.11.010.

Alvarez-Buylla, E. R., Liljegren, S. J., Pelaz, S., Gold, S. E., Burgeff, C., Ditta, G. S., et al. (2000). MADS-box gene evolution beyond flowers: Expression in pollen, endosperm, guard cells, roots and trichomes. *Plant Journal*, *24*(4), 457–466.

Ang, L. H., & Deng, X. W. (1994). Regulatory hierarchy of photomorphogenic loci: Allele-specific and light-dependent interaction between the HY5 and COP1 loci. *The Plant Cell*, *6*(5), 613–628. https://doi.org/10.1105/tpc.6.5.613.

Araus, V., Vidal, E. A., Puelma, T., Alamos, S., Mieulet, D., Guiderdoni, E., et al. (2016). Members of BTB gene family of scaffold proteins suppress nitrate uptake and nitrogen use efficiency. *Plant Physiology*, *171*(2), 1523–1532. https://doi.org/10.1104/pp.15.01731.

Aslam, M., Travis, R. L., & Huffaker, R. C. (1992). Comparative kinetics and reciprocal inhibition of nitrate and nitrite uptake in roots of uninduced and induced barley (*Hordeum vulgare* L.) seedlings. *Plant Physiology*, *99*, 1124–1133.

Aslam, M., Travis, R. L., & Huffaker, R. C. (1995). Effect of pH and calcium on short-term NO_3^- fluxes in roots of barley seedlings. *Plant Physiology*, *108*(2), 727–734.

Aslam, M., Travis, R. L., & Rains, D. W. (1996). Evidence for substrate induction of a nitrate efflux system in barley roots. *Plant Physiology*, *112*(3), 1167–1175.

Barbier-Brygoo, H., De Angeli, A., Filleur, S., Frachisse, J. M., Gambale, F., Thomine, S., et al. (2011). Anion channels/transporters in plants: From molecular bases to regulatory networks. *Annual Review of Plant Biology*, *62*, 25–51. https://doi.org/10.1146/annurev-arplant-042110-103741.

Bouguyon, E., Brun, F., Meynard, D., Kubeš, M., Pervent, M., Leran, S., et al. (2015). Multiple mechanisms of nitrate sensing by Arabidopsis nitrate transceptor NRT1.1. *Nature Plants*, *1*, 15015. https://doi.org/10.1038/nplants.2015.15. http://www.nature.com/articles/nplants201515#supplementary-information.

Bowsher, C. G., Boulton, E. L., Rose, J., Nayagam, S., & Emes, M. J. (1992). Reductant for glutamate synthase is generated by the oxidative pentose phosphate pathway in non-photosynthetic root plastids. *Plant Journal*, *2*, 893–898.

Bowsher, C. G., Hucklesby, D. P., & Emes, M. J. (1989). Nitrite reduction and carbohydrate metabolism in plastids purified from roots of *Pisum sativum* L. *Planta*, *177*, 359–366.

Brownlee, A. G., & Arst, H. N. (1983). Nitrate uptake in *Aspergillus nidulans* and involvement of the third gene of the nitrate assimilation gene cluster. *Journal of Bacteriology*, *155*, 1138–1146.

Buchner, P., & Hawkesford, M. J. (2014). Complex phylogeny and gene expression patterns of members of the NITRATE TRANSPORTER 1/PEPTIDE TRANSPORTER family (NPF) in wheat. *Journal of Experimental Botany*, *65*(19), 5697–5710. https://doi.org/10.1093/jxb/eru231.

Burgeff, C., Liljegren, S. J., Tapia-Lopez, R., Yanofsky, M. F., & Alvarez-Buylla, E. R. (2002). MADS-box gene expression in lateral primordia, meristems and differentiated tissues of Arabidopsis thaliana roots. *Planta*, *214*(3), 365–372.

Bussell, J. D., Keech, O., Fenske, R., & Smith, S. M. (2013). Requirement for the plastidial oxidative pentose phosphate pathway for nitrate assimilation in Arabidopsis. *Plant Journal, 75*(4), 578–591. https://doi.org/10.1111/tpj.12222.

Cai, C., Wang, J. Y., Zhu, Y. G., Shen, Q. R., Li, B., Tong, Y. P., et al. (2008). Gene structure and expression of the high-affinity nitrate transport system in rice roots. *Journal of Integrative Plant Biology, 50*, 443–451. https://doi.org/10.1111/j.1744-7909.2008.00642.x.

Castaings, L., Camargo, A., Pocholle, D., Gaudon, V., Texier, Y., Boutet-Mercey, S., et al. (2009). The nodule inception-like protein 7 modulates nitrate sensing and metabolism in Arabidopsis. *Plant Journal, 57*(3), 426–435. https://doi.org/10.1111/j.1365-313X.2008.03695.x.

Cerezo, M., Tillard, P., Filleur, S., Munos, S., Daniel-Vedele, F., & Gojon, A. (2001). Major alterations of the regulation of root NO(3)(-) uptake are associated with the mutation of Nrt2.1 and Nrt2.2 genes in Arabidopsis. *Plant Physiology, 127*(1), 262–271.

Chen, X., Yao, Q., Gao, X., Jiang, C., Harberd, N. P., & Fu, X. (2016). Shoot-to-root mobile transcription factor HY5 coordinates plant carbon and nitrogen acquisition. *Current Biology, 26*(5), 640–646. https://doi.org/10.1016/j.cub.2015.12.066.

Chen, J., Zhang, Y., Tan, Y., Zhang, M., Zhu, L., Xu, G., et al. (2016). Agronomic nitrogen-use efficiency of rice can be increased by driving OsNRT2.1 expression with the OsNAR2.1 promoter. *Plant Biotechnology Journal, 14*(8), 1705–1715. https://doi.org/10.1111/pbi.12531.

Chiou, T. J., & Bush, D. R. (1998). Sucrose is a signal molecule in assimilate partitioning. *Proceedings of the National Academy of Sciences of the United States of America, 95*(8), 4784–4788.

Chory, J. (1992). A genetic model for light-regulated seedling Arabidopsis. *Development, 115*(1), 337–354.

Cormier, F., Foulkes, J., Hirel, B., Gouache, D., Moënne-Loccoz, Y., & Le Gouis, J. (2016). Breeding for increased nitrogen-use efficiency: A review for wheat (T. aestivum L.). *Plant Breeding, 135*(3), 255–278. https://doi.org/10.1111/pbr.12371.

Cormier, F., Le Gouis, J., Dubreuil, P., Lafarge, S., & Praud, S. (2014). A genome-wide identification of chromosomal regions determining nitrogen use efficiency components in wheat (Triticum aestivum L.). *Theoretical and Applied Genetics, 127*(12), 2679–2693. https://doi.org/10.1007/s00122-014-2407-7.

Corratge-Faillie, C., & Lacombe, B. (2017). Substrate (un)specificity of Arabidopsis NRT1/PTR FAMILY (NPF) proteins. *Journal of Experimental Botany, 68*(12), 3107–3113. https://doi.org/10.1093/jxb/erw499.

Crawford, N. M., & Glass, A. D. M. (1998). Molecular and physiological aspects of nitrate uptake in plants. *Trends in Plant Science, 3*(10), 389–395.

de Folter, S., Immink, R. G., Kieffer, M., Parenicova, L., Henz, S. R., Weigel, D., et al. (2005). Comprehensive interaction map of the Arabidopsis MADS box transcription factors. *The Plant Cell, 17*(5), 1424–1433. https://doi.org/10.1105/tpc.105.031831.

de Jong, F., Thodey, K., Lejay, L. V., & Bevan, M. W. (2014). Glucose elevates NITRATE TRANSPORTER2.1 protein levels and nitrate transport activity independently of its HEXOKINASE1-mediated stimulation of NITRATE TRANSPORTER2.1 expression. *Plant Physiology, 164*(1), 308–320. https://doi.org/10.1104/pp.113.230599.

Delhon, P., Gojon, A., Tillard, P., & Passama, L. (1995). Diurnal regulation of NO_3^- uptake in soybean plants. I. Changes in NO_3^- influx, efflux, and N utilization in the plant during the day/night cycle. *Journal of Experimental Botany, 46*, 1585–1594.

Delhon, P., Gojon, A., Tillard, P., & Passama, L. (1996). Diurnal regulation of NO_3^- uptake in soybean plants IV. Dependence on current photosynthesis and sugar availability to the roots. *Journal of Experimental Botany, 47*, 893–900.

Dijkshoorn, W. (1962). Metabolic regulation of the alkaline effect of nitrate utilization by plants. *Nature, 194*, 165–167.

Fan, X., Feng, H., Tan, Y., Xu, Y., Miao, Q., & Xu, G. (2016). A putative 6-transmembrane nitrate transporter OsNRT1.1b plays a key role in rice under low nitrogen. *Journal of Integrative Plant Biology, 58*(6), 590–599. https://doi.org/10.1111/jipb.12382.

Fan, X., Naz, M., Fan, X., Xuan, W., Miller, A. J., & Xu, G. (2017). Plant nitrate transporters: From gene function to application. *Journal of Experimental Botany, 68*(10), 2463–2475. https://doi.org/10.1093/jxb/erx011.

Fan, X., Tang, Z., Tan, Y., Zhang, Y., Luo, B., Yang, M., et al. (2016). Overexpression of a pH-sensitive nitrate transporter in rice increases crop yields. *Proceedings of the National Academy of Sciences of the United States of America, 113*(26), 7118–7123.

Filleur, S., & Daniel-Vedele, F. (1999). Expression analysis of a high-affinity nitrate transporter isolated from Arabidopsis thaliana by differential display. *Planta, 207*, 461–469. https://doi.org/10.1007/s004250050505.

Filleur, S., Dorbe, M. F., Cerezo, M., Orsel, M., Granier, F., Gojon, A., et al. (2001). An arabidopsis T-DNA mutant affected in Nrt2 genes is impaired in nitrate uptake. *FEBS Letters, 489*(2–3), 220–224.

Forde, B. G. (2000). Nitrate transporters in plants: Structure, function and regulation. *Biochimica et Biophysica Acta, 1465*(1–2), 219–235.

Forde, B. G. (2002). Local and long-range signaling pathways regulating plant responses to nitrate. *Annual Review of Plant Biology, 53*, 203–224. https://doi.org/10.1146/annurev.arplant.53.100301.135256.

Fraisier, V., Gojon, A., Tillard, P., & Daniel-Vedele, F. (2000). Constitutive expression of a putative high-affinity nitrate transporter in Nicotiana plumbaginifolia: Evidence for post-transcriptional regulation by a reduced nitrogen source. *Plant Journal, 23*(4), 489–496.

Gan, Y., Filleur, S., Rahman, A., Gotensparre, S., & Forde, B. G. (2005). Nutritional regulation of ANR1 and other root-expressed MADS-box genes in Arabidopsis thaliana. *Planta, 222*(4), 730–742. https://doi.org/10.1007/s00425-005-0020-3.

Gansel, X., Munos, S., Tillard, P., & Gojon, A. (2001). Differential regulation of the NO_3^- and NH_4^+ transporter genes AtNrt2.1 and AtAmt1.1 in Arabidopsis: Relation with long-distance and local controls by N status of the plant. *Plant Journal, 26*(2), 143–155.

Garay-Arroyo, A., Ortiz-Moreno, E., de la Paz Sanchez, M., Murphy, A. S., Garcia-Ponce, B., Marsch-Martinez, N., et al. (2013). The MADS transcription factor XAL2/AGL14 modulates auxin transport during Arabidopsis root development by regulating PIN expression. *The EMBO Journal, 32*(21), 2884–2895. https://doi.org/10.1038/emboj.2013.216.

Garcia-Brugger, A., Lamotte, O., Vandelle, E., Bourque, S., Lecourieux, D., Poinssot, B., et al. (2006). Early signaling events induced by elicitors of plant defenses. *Molecular Plant-Microbe Interactions, 19*(7), 711–724. https://doi.org/10.1094/MPMI-19-0711.

Gifford, M. L., Dean, A., Gutierrez, R. A., Coruzzi, G. M., & Birnbaum, K. D. (2008). Cell-specific nitrogen responses mediate developmental plasticity. *Proceedings of the National Academy of Sciences of the United States of America, 105*(2), 803–808.

Girin, T., El-Kafafi el, S., Widiez, T., Erban, A., Hubberten, H. M., Kopka, J., et al. (2010). Identification of Arabidopsis mutants impaired in the systemic regulation of root nitrate uptake by the nitrogen status of the plant. *Plant Physiology, 153*(3), 1250–1260. https://doi.org/10.1104/pp.110.157354.

Glass, A. D., Britto, D. T., Kaiser, B. N., Kinghorn, J. R., Kronzucker, H. J., Kumar, A., et al. (2002). The regulation of nitrate and ammonium transport systems in plants. *Journal of Experimental Botany, 53*(370), 855–864.

Glass, A. D., Shaff, J. E., & Kochian, L. V. (1992). Studies of the uptake of nitrate in barley: IV. Electrophysiology. *Plant Physiology, 99*(2), 456–463.

Glass, A. D. M., & Siddiqi, M. Y. (1995). Nitrogen absorption by plant roots. In H. S. Srivastava & R. P. Singh (Eds.), *Nitrogen Nutrition in Higher Plants* (pp. 21–56). New Delhi: Associated Publishing Company.

Gojon, A., Krouk, G., Perrine-Walker, F., & Laugier, E. (2011). Nitrate transceptor(s) in plants. *Journal of Experimental Botany*, *62*(7), 2299–2308. https://doi.org/10.1093/jxb/erq419.

Gojon, A., Nacry, P., & Davidian, J. C. (2009). Root uptake regulation: A central process for NPS homeostasis in plants. *Current Opinion in Plant Biology*, *12*(3), 328–338. https://doi.org/10.1016/j.pbi.2009.04.015.

Good, A. G., Shrawat, A. K., & Muench, D. G. (2004). Can less yield more? Is reducing nutrient input into the environment compatible with maintaining crop production? *Trends in Plant Science*, *9*(12), 597–605. https://doi.org/10.1016/j.tplants.2004.10.008.

Gregory, P. J., & George, T. S. (2011). Feeding nine billion: The challenge to sustainable crop production. *Journal of Experimental Botany*, *62*(15), 5233–5239. https://doi.org/10.1093/jxb/err232.

Guan, P., Wang, R., Nacry, P., Breton, G., Kay, S. A., Pruneda-Paz, J. L., et al. (2014). Nitrate foraging by Arabidopsis roots is mediated by the transcription factor TCP20 through the systemic signaling pathway. *Proceedings of the National Academy of Sciences of the United States of America*, *111*(42), 15267–15272. https://doi.org/10.1073/pnas.1411375111.

Hawkesford, M. J. (2017). Genetic variation in traits for nitrogen use efficiency in wheat. *Journal of Experimental Botany*, *68*(10), 2627–2632. https://doi.org/10.1093/jxb/erx079.

He, Q., Qiao, D., Zhang, Q., Li, Y., Xu, H., Wei, L., et al. (2004). Cloning and expression study of a putative high-affinity nitrate transporter gene from Dunaliella salina. *Journal of Applied Phycology*, *16*(5), 395–400. https://doi.org/10.1023/B:JAPH.0000047950.76549.ce.

Ho, C. H., Lin, S. H., Hu, H. C., & Tsay, Y. F. (2009). CHL1 functions as a nitrate sensor in plants. *Cell*, *138*, 1184–1194. https://doi.org/10.1016/j.cell.2009.07.004.

Hole, D. J., Emran, A. M., Fares, Y., & Drew, M. C. (1990). Induction of nitrate transport in maize roots, and kinetics of influx, measured with nitrogen-13. *Plant Physiology*, *93*, 642–647.

Hu, B., Wang, W., Ou, S., Tang, J., Li, H., Che, R., et al. (2015). Variation in NRT1.1B contributes to nitrate-use divergence between rice subspecies. *Nature Genetics*, *47*(7), 834–838. https://doi.org/10.1038/ng.3337.

Hu, H. C., Wang, Y. Y., & Tsay, Y. F. (2009). AtCIPK8, a CBL-interacting protein kinase, regulates the low-affinity phase of the primary nitrate response. *Plant Journal*, *57*(2), 264–278.

Huang, N. C., Chiang, C. S., Crawford, N. M., & Tsay, Y. F. (1996). CHL1 encodes a component of the low-affinity nitrate uptake system in Arabidopsis and shows cell type-specific expression in roots. *The Plant Cell*, *8*(12), 2183–2191.

Huang, N. C., Liu, K. H., Lo, H. J., & Tsay, Y. F. (1999). Cloning and functional characterization of an Arabidopsis nitrate transporter gene that encodes a constitutive component of low-affinity uptake. *The Plant Cell*, *11*, 1381–1392. https://doi.org/10.1105/tpc.11.8.1381.

Ishikawa, S., Ito, Y., Sato, Y., Fukaya, Y., Takahashi, M., Morikawa, H., et al. (2009). Two-component high-affinity nitrate transport system in barley: Membrane localization, protein expression in roots and a direct protein–protein interaction. *Plant Biotechnology Journal*, *26*, 197–205.

Jackson, W. A., Flesher, D., & Hageman, R. H. (1973). Nitrate uptake by dark-grown corn seedlings: Some characteristics of apparent induction. *Plant Physiology*, *51*(1), 120–127.

Jacquot, A., Li, Z., Gojon, A., Schulze, W., & Lejay, L. (2017). Post-translational regulation of nitrogen transporters in plants and microorganisms. *Journal of Experimental Botany*, *68*(10), 2567–2580. https://doi.org/10.1093/jxb/erx073.

Kiba, T., Feria-Bourrellier, A. B., Lafouge, F., Lezhneva, L., Boutet-Mercey, S., Orsel, M., et al. (2012). The Arabidopsis nitrate transporter NRT2.4 plays a double role in roots and shoots of nitrogen-starved plants. *The Plant Cell*, *24*(1), 245–258. https://doi.org/10.1105/tpc.111.092221.

Kiba, T., Inaba, J., Kudo, T., Ueda, N., Konishi, M., Mitsuda, N., et al. (2018). Repression of nitrogen starvation responses by members of the Arabidopsis GARP-type transcription factor NIGT1/HRS1 subfamily. *The Plant Cell*, *30*(4), 925–945. https://doi.org/10.1105/tpc.17.00810.

King, B. J., Siddiqi, M. Y., Ruth, T. J., Warner, R. L., & Glass, A. (1993). Feedback regulation of nitrate influx in barley roots by nitrate, nitrite, and ammonium. *Plant Physiology*, *102*(4), 1279–1286.

Kotur, Z., Mackenzie, N., Ramesh, S., Tyerman, S. D., Kaiser, B. N., & Glass, A. D. (2012). Nitrate transport capacity of the Arabidopsis thaliana NRT2 family members and their interactions with AtNAR2.1. *The New Phytologist*, *194*(3), 724–731. https://doi.org/10.1111/j.1469-8137.2012.04094.x.

Krapp, A., David, L. C., Chardin, C., Girin, T., Marmagne, A., Leprince, A. S., et al. (2014). Nitrate transport and signalling in Arabidopsis. *Journal of Experimental Botany*, *65*(3), 789–798. https://doi.org/10.1093/jxb/eru001.

Krapp, A., Fraisier, V., Scheible, W. R., Quesada, A., Gojon, A., Stitt, M., et al. (1998). Expression studies of Nrt2:1Np, a putative high-affinity nitrate transporter: Evidence for its role in nitrate uptake. *Plant Journal*, *14*(6), 723–731.

Kronzucker, H. J., Siddiqi, M. Y., & Glass, A. (1995a). Kinetics of NO_3^- influx in spruce. *Plant Physiology*, *109*(1), 319–326.

Kronzucker, H. J., Siddiqi, M. Y., & Glass, A. D. M. (1995b). Compartmentation and flux characteristics of nitrate in spruce. *Planta*, *196*, 674–682.

Krouk, G., Lacombe, B., Bielach, A., Perrine-Walker, F., Malinska, K., Mounier, E., et al. (2010). Nitrate-regulated auxin transport by NRT1.1 defines a mechanism for nutrient sensing in plants. *Developmental Cell*, *18*(6), 927–937. https://doi.org/10.1016/j.devcel.2010.05.008.

Krouk, G., Mirowski, P., LeCun, Y., Shasha, D. E., & Coruzzi, G. M. (2010). Predictive network modeling of the high-resolution dynamic plant transcriptome in response to nitrate. *Genome Biology*, *11*(12), R123. https://doi.org/10.1186/gb-2010-11-12-r123.

Krouk, G., Tillard, P., & Gojon, A. (2006). Regulation of the high-affinity NO_3^- uptake system by NRT1.1-mediated NO_3^- demand signaling in Arabidopsis. *Plant Physiology*, *142*(3), 1075–1086.

Lalonde, S., Boles, E., Hellmann, H., Barker, L., Patrick, J. W., Frommer, W. B., et al. (1999). The dual function of sugar carriers. Transport and sugar sensing. *The Plant Cell*, *11*(4), 707–726.

Laugier, E., Bouguyon, E., Mauries, A., Tillard, P., Gojon, A., & Lejay, L. (2012). Regulation of high-affinity nitrate uptake in roots of Arabidopsis depends predominantly on posttranscriptional control of the NRT2.1/NAR2.1 transport system. *Plant Physiology*, *158*(2), 1067–1078. https://doi.org/10.1104/pp.111.188532.

Lee, R. B., & Drew, M. C. (1986). Nitrogen-13 studies of nitrate fluxes in barley roots. II. Effect of plant N-status on the kinetic parameters of nitrate influx. *Journal of Experimental Botany*, *185*, 1768–1779.

Lejay, L., Gansel, X., Cerezo, M., Tillard, P., Muller, C., Krapp, A., et al. (2003). Regulation of root ion transporters by photosynthesis: Functional importance and relation with hexokinase. *The Plant Cell*, *15*(9), 2218–2232.

Lejay, L., Tillard, P., Lepetit, M., Olive, F., Filleur, S., Daniel-Vedele, F., et al. (1999). Molecular and functional regulation of two NO3- uptake systems by N- and C-status of Arabidopsis plants. *Plant Journal*, *18*(5), 509–519.

Lejay, L., Wirth, J., Pervent, M., Cross, J. M., Tillard, P., & Gojon, A. (2008). Oxidative pentose phosphate pathway-dependent sugar sensing as a mechanism for regulation of root ion transporters by photosynthesis. *Plant Physiology*, *146*(4), 2036–2053. https://doi.org/10.1104/pp.107.114710.

Leran, S., Varala, K., Boyer, J. C., Chiurazzi, M., Crawford, N., Daniel-Vedele, F., et al. (2014). A unified nomenclature of NITRATE TRANSPORTER 1/PEPTIDE TRANSPORTER family members in plants. *Trends in Plant Science*, *19*(1), 5–9. https://doi.org/10.1016/j.tplants.2013.08.008.

Lezhneva, L., Kiba, T., Feria-Bourrellier, A. B., Lafouge, F., Boutet-Mercey, S., Zoufan, P., et al. (2014). The Arabidopsis nitrate transporter NRT2.5 plays a role in nitrate acquisition and remobilization in nitrogen-starved plants. *Plant Journal*, *80*(2), 230–241. https://doi.org/10.1111/tpj.12626.

Li, H., Hu, B., & Chu, C. (2017). Nitrogen use efficiency in crops: Lessons from Arabidopsis and rice. *Journal of Experimental Botany*, *68*, 2477–2488.

Li, W., Wang, Y., Okamoto, M., Crawford, N. M., Siddiqi, M. Y., & Glass, A. D. (2007). Dissection of the AtNRT2.1:AtNRT2.2 inducible high-affinity nitrate transporter gene cluster. *Plant Physiology*, *143*(1), 425–433.

Linkohr, B. I., Williamson, L. C., Filter, A. H., & Leyser, H. M. (2002). Nitrate and phosphate availability and distribution have different effects on root system architecture of Arabidopsis. *Plant Journal*, *29*(6), 751–760.

Little, D. Y., Rao, H., Oliva, S., Daniel-Vedele, F., Krapp, A., & Malamy, J. E. (2005). The putative high-affinity nitrate transporter NRT2.1 represses lateral root initiation in response to nutritional cues. *Proceedings of the National Academy of Sciences of the United States of America*, *102*(38), 13693–13698. https://doi.org/10.1073/pnas.0504219102.

Liu, K. H., Huang, C. Y., & Tsay, Y. F. (1999). CHL1 is a dual-affinity nitrate transporter of Arabidopsis involved in multiple phases of nitrate uptake. *The Plant Cell*, *11*(5), 865–874.

Liu, K. H., Niu, Y., Konishi, M., Wu, Y., Du, H., Sun Chung, H., et al. (2017). Discovery of nitrate-CPK-NLP signalling in central nutrient-growth networks. *Nature*, *545*(7654), 311–316. https://doi.org/10.1038/nature22077.

Liu, K. H., & Tsay, Y. F. (2003). Switching between the two action modes of the dual-affinity nitrate transporter CHL1 by phosphorylation. *The EMBO Journal*, *22*(5), 1005–1013.

Ma, W., Li, J., Qu, B., He, X., Zhao, X., Li, B., et al. (2014). Auxin biosynthetic gene TAR2 is involved in low nitrogen-mediated reprogramming of root architecture in Arabidopsis. *Plant Journal*, *78*(1), 70–79. https://doi.org/10.1111/tpj.12448.

Macduff, J. H., & Jackson, S. B. (1992). Influx and efflux of nitrate and ammonium in italian ryegrass and white clover roots: Comparisons between effects of darkness and defoliation. *Journal of Experimental Botany*, *43*(249), 525–535.

Maeda, Y., Konishi, M., Kiba, T., Sakuraba, Y., Sawaki, N., Kurai, T., et al. (2018). A NIGT1-centred transcriptional cascade regulates nitrate signalling and incorporates phosphorus starvation signals in Arabidopsis. *Nature Communications*, *9*(1), 1376. https://doi.org/10.1038/s41467-018-03832-6.

Marchive, C., Roudier, F., Castaings, L., Brehaut, V., Blondet, E., Colot, V., et al. (2013). Nuclear retention of the transcription factor NLP7 orchestrates the early response to nitrate in plants. *Nature Communications*, *4*, 1713. https://doi.org/10.1038/ncomms2650.

McClure, P. R., Kochian, L. V., Spanswick, R. M., & Shaff, J. E. (1990a). Evidence for cotransport of nitrate and protons in maize roots. I. Effects of nitrate on the membrane potential. *Plant Physiology*, *93*, 271–289.

McClure, P. R., Kochian, L. V., Spanswick, R. M., & Shaff, J. E. (1990b). Evidence for cotransport of nitrate and protons in maize roots. II. Measurements of NO_3^- and H^+ fluxes with ion-selective microelectrodes. *Plant Physiology*, *93*, 290–294.

Meharg, A. A., & Blatt, M. R. (1995). NO_3^- transport across the plasma membrane of Arabidopsis thaliana root hairs: Kinetic control by pH and membrane voltage. *Journal of Membrane Biology*, *145*(1), 49–66.

Moore, B., Zhou, L., Rolland, F., Hall, Q., Cheng, W. H., Liu, Y. X., et al. (2003). Role of the Arabidopsis glucose sensor HXK1 in nutrient, light, and hormonal signaling. *Science*, *300*(5617), 332–336.

Morère-Le Paven, M.-C., Viau, L., Hamon, A., Vandecasteele, C., Pellizzaro, A., Bourdin, C., et al. (2011). Characterization of a dual-affinity nitrate transporter MtNRT1.3 in the model legume Medicago truncatula. *Journal of Experimental Botany*, *62*(15), 5595–5605. https://doi.org/10.1093/jxb/err243.

Mounier, E., Pervent, M., Ljung, K., Gojon, A., & Nacry, P. (2014). Auxin-mediated nitrate signalling by NRT1.1 participates in the adaptive response of Arabidopsis root architecture to the spatial heterogeneity of nitrate availability. *Plant, Cell & Environment*, *37*(1), 162–174. https://doi.org/10.1111/pce.12143.

Munos, S., Cazettes, C., Fizames, C., Gaymard, F., Tillard, P., Lepetit, M., et al. (2004). Transcript profiling in the chl1-5 mutant of Arabidopsis reveals a role of the nitrate transporter NRT11 in the regulation of another nitrate transporter, NRT2.1. *The Plant Cell*, *16*, 2433–2447. https://doi.org/10.1105/tpc.104.024380.

Nacry, P., Bouguyon, E., & Gojon, A. (2013). Nitrogen acquisition by roots: Physiological and developmental mechanisms ensuring plant adaptation to a fluctuating resource. *Plant and Soil*, *370*, 1–29.

Nazoa, P., Vidmar, J. J., Tranbarger, T. J., Mouline, K., Damiani, I., Tillard, P., et al. (2003). Regulation of the nitrate transporter gene AtNRT2.1 in Arabidopsis thaliana: Responses to nitrate, amino acids and developmental stage. *Plant Molecular Biology*, *52*(3), 689–703.

Negi, J., Matsuda, O., Nagasawa, T., Oba, Y., Takahashi, H., Kawai-Yamada, M., et al. (2008). CO2 regulator SLAC1 and its homologues are essential for anion homeostasis in plant cells. *Nature*, *452*(7186), 483–486. https://doi.org/10.1038/nature06720.

Ohkubo, Y., Tanaka, M., Tabata, R., Ogawa-Ohnishi, M., & Matsubayashi, Y. (2017). Shoot-to-root mobile polypeptides involved in systemic regulation of nitrogen acquisition. *Nature Plants*, *3*, 17029. https://doi.org/10.1038/nplants.2017.29.

Oji, Y., Watanabe, M., Wakiuchi, N., & Okamoto, S. (1985). Nitrite reduction in barley-root plastids: Dependence on NADPH coupled with glucose-6-phoshate and 6-phosphogluconate dehydrogenases, and possible involvement of an electron carrier and a diaphorase. *Planta*, *165*, 85–90.

Okamoto, M., Kumar, A., Li, W., Wang, Y., Siddiqi, M. Y., Crawford, N. M., et al. (2006). High-affinity nitrate transport in roots of Arabidopsis depends on expression of the NAR2-like gene AtNRT3.1. *Plant Physiology*, *140*(3), 1036–1046.

Okamoto, M., Vidmar, J. J., & Glass, A. D. (2003). Regulation of NRT1 and NRT2 gene families of Arabidopsis thaliana: Responses to nitrate provision. *Plant and Cell Physiology*, *44*(3), 304–317.

Oostindier-Braaksma, F. J., & Feenstra, W. J. (1973). Isolation and characterization of chlorate-resistant mutants of *Arabidopsis thaliana*. *Mutation Research*, *19*, 175–185.

Orsel, M., Chopin, F., Leleu, O., Smith, S. J., Krapp, A., Daniel-Vedele, F., et al. (2006). Characterization of a two-component high-affinity nitrate uptake system in Arabidopsis. Physiology and protein-protein interaction. *Plant Physiology*, *142*(3), 1304–1317.

Orsel, M., Eulenburg, K., Krapp, A., & Daniel-Vedele, F. (2004). Disruption of the nitrate transporter genes AtNRT2.1 and AtNRT2.2 restricts growth at low external nitrate concentration. *Planta*, *219*, 714–721. https://doi.org/10.1007/s00425-004-1266-x.

Orsel, M., Krapp, A., & Daniel-Vedele, F. (2002). Analysis of the NRT2 nitrate transporter family in Arabidopsis. Structure and gene expression. *Plant Physiology*, *129*(2), 886–896.

Oyama, T., Shimura, Y., & Okada, K. (1997). The Arabidopsis HY5 gene encodes a bZIP protein that regulates stimulus-induced development of root and hypocotyl. *Genes and Development*, *11*(22), 2983–2995.

Pace, G. M., & MacClure, P. R. (1986). Comparison of nitrate uptake parameters across maize inbreed lines. *Journal of Plant Nutrition*, *9*, 1095–1111.

Parker, J. L., & Newstead, S. (2014). Molecular basis of nitrate uptake by the plant nitrate transporter NRT1.1. *Nature*, *507*(7490), 68–72. https://doi.org/10.1038/nature13116.

Pearson, C. J., Volk, R. J., & Jackson, W. A. (1981). Daily changes in nitrate influx, efflux and metabolism in maize and pearl millet. *Planta*, *152*(4), 319–324. https://doi.org/10.1007/BF00388256.

Plett, D., Toubia, J., Garnett, T., Tester, M., Kaiser, B. N., & Baumann, U. (2010). Dichotomy in the NRT gene families of dicots and grass species. *PLoS One*, *5*(12), e15289. https://doi.org/10.1371/journal.pone.0015289.

Quesada, A., Galvan, A., & Fernandez, E. (1994). Identification of nitrate transporter genes in Chlamydomonas reinhardtii. *Plant Journal*, *5*(3), 407–419.

Remans, T., Nacry, P., Pervent, M., Filleur, S., Diatloff, E., Mounier, E., et al. (2006). The Arabidopsis NRT1.1 transporter participates in the signaling pathway triggering root colonization of nitrate-rich patches. *Proceedings of the National Academy of Sciences of the United States of America*, *103*(50), 19206–19211. https://doi.org/10.1073/pnas.0605275103.

Remans, T., Nacry, P., Pervent, M., Girin, T., Tillard, P., Lepetit, M., et al. (2006). A central role for the nitrate transporter NRT2.1 in the integrated morphological and physiological responses of the root system to nitrogen limitation in Arabidopsis. *Plant Physiology*, *140*(3), 909–921.

Rubin, G., Tohge, T., Matsuda, F., Saito, K., & Scheible, W. R. (2009). Members of the LBD family of transcription factors repress anthocyanin synthesis and affect additional nitrogen responses in Arabidopsis. *The Plant Cell*, *21*(11), 3567–3584. https://doi.org/10.1105/tpc.109.067041.

Schroeder, J. I., Delhaize, E., Frommer, W. B., Guerinot, M. L., Harrison, M. J., Herrera-Estrella, L., et al. (2013). Using membrane transporters to improve crops for sustainable food production. *Nature*, *497*(7447), 60–66. https://doi.org/10.1038/nature11909.

Siddiqi, M. Y., Glass, A. D., Ruth, T. J., & Fernando, M. (1989). Studies of the regulation of nitrate influx by barley seedlings using NO(3). *Plant Physiology*, *90*(3), 806–813.

Siddiqi, M. Y., Glass, A. D., Ruth, T. J., & Rufty, T. W. (1990). Studies of the uptake of nitrate in barley: I. kinetics of NO(3) influx. *Plant Physiology*, *93*(4), 1426–1432.

Sun, J., Bankston, J. R., Payandeh, J., Hinds, T. R., Zagotta, W. N., & Zheng, N. (2014). Crystal structure of the plant dual-affinity nitrate transporter NRT1.1. *Nature*, *507*(7490), 73–77. https://doi.org/10.1038/nature13074.

Tapia-Lopez, R., Garcia-Ponce, B., Dubrovsky, J. G., Garay-Arroyo, A., Perez-Ruiz, R. V., Kim, S. H., et al. (2008). An AGAMOUS-related MADS-box gene, XAL1 (AGL12), regulates root meristem cell proliferation and flowering transition in Arabidopsis. *Plant Physiology*, *146*(3), 1182–1192. https://doi.org/10.1104/pp.107.108647.

Taulemesse, F., Le Gouis, J., Gouache, D., Gibon, Y., & Allard, V. (2015). Post-flowering nitrate uptake in wheat is controlled by N status at flowering, with a putative major role of root nitrate transporter NRT2.1. *PLoS One*, *10*(3), e0120291. https://doi.org/10.1371/journal.pone.0120291.

Taulemesse, F., Le Gouis, J., Gouache, D., Gibon, Y., & Allard, V. (2016). Bread wheat (Triticum aestivum L.) grain protein concentration is related to early post-flowering nitrate uptake under putative control of plant satiety level. *PLoS One*, *11*(2), e0149668. https://doi.org/10.1371/journal.pone.0149668.

Tilman, D. (1999). Global environmental impacts of agricultural expansion: The need for sustainable and efficient practices. *Proceedings of the National Academy of Sciences of the United States of America, 96*(11), 5995–6000.

Tilman, D., Cassman, K. G., Matson, P. A., Naylor, R., & Polasky, S. (2002). Agricultural sustainability and intensive production practices. *Nature, 418*(6898), 671–677. https://doi.org/10.1038/nature01014.

Tong, Y., Zhou, J. J., Li, Z., & Miller, A. J. (2005). A two-component high-affinity nitrate uptake system in barley. *Plant Journal, 41*(3), 442–450.

Touraine, B., & Glass, A. D. (1997). NO_3^- and ClO_3^- fluxes in the *chl1-5* mutant of *Arabidopsis thaliana*. Does the *CHL1-5* gene encode a low-affinity NO_3^- transporter? *Plant Physiology, 114*(1), 137–144.

Trueman, L. J., Richardson, A., & Forde, B. G. (1996). Molecular cloning of higher plant homologues of the high-affinity nitrate transporters of Chlamydomonas reinhardtii and Aspergillus nidulans. *Gene, 175*(1–2), 223–231.

Tsay, Y. F., Chiu, C. C., Tsai, C. B., Ho, C. H., & Hsu, P. K. (2007). Nitrate transporters and peptide transporters. *FEBS Letters, 581*(12), 2290–2300.

Tsay, Y. F., Schroeder, J. I., Feldmann, K. A., & Crawford, N. M. (1993). The herbicide sensitivity gene *CHL1* of *Arabidopsis* encodes a nitrate-inducible nitrate transporter. *Cell, 72*(5), 705–713.

Unkles, S. E., Hawker, K. L., Grieve, C., Campbell, E. I., Montague, P., & Kinghorn, J. R. (1991). crnA encodes a nitrate transporter in Aspergillus nidulans. *Proceedings of the National Academy of Sciences of the United States of America, 88*(1), 204–208.

Vidal, E. A., Alvarez, J. M., & Gutierrez, R. A. (2014). Nitrate regulation of AFB3 and NAC4 gene expression in Arabidopsis roots depends on NRT1.1 nitrate transport function. *Plant Signaling & Behavior, 9*(3), e28501.

Vidal, E. A., Araus, V., Lu, C., Parry, G., Green, P. J., Coruzzi, G. M., et al. (2010). Nitrate-responsive miR393/AFB3 regulatory module controls root system architecture in Arabidopsis thaliana. *Proceedings of the National Academy of Sciences of the United States of America, 107*(9), 4477–4482. https://doi.org/10.1073/pnas.0909571107.

Vidal, E. A., Moyano, T. C., Riveras, E., Contreras-Lopez, O., & Gutierrez, R. A. (2013). Systems approaches map regulatory networks downstream of the auxin receptor AFB3 in the nitrate response of Arabidopsis thaliana roots. *Proceedings of the National Academy of Sciences of the United States of America, 110*(31), 12840–12845. https://doi.org/10.1073/pnas.1310937110.

von Wittgenstein, N. J., Le, C. H., Hawkins, B. J., & Ehlting, J. (2014). Evolutionary classification of ammonium, nitrate, and peptide transporters in land plants. *BMC Evolutionary Biology, 14*(1), 11. https://doi.org/10.1186/1471-2148-14-11.

Wang, Y. Y., Cheng, Y. H., Chen, K. E., & Tsay, Y. F. (2018). Nitrate transport, signaling, and use efficiency. *Annual Review of Plant Biology, 69*, 85–122. https://doi.org/10.1146/annurev-arplant-042817-040056.

Wang, R., Liu, D., & Crawford, N. M. (1998). The Arabidopsis CHL1 protein plays a major role in high-affinity nitrate uptake. *Proceedings of the National Academy of Sciences of the United States of America, 95*(25), 15134–15139.

Wang, R., Xing, X., Wang, Y., Tran, A., & Crawford, N. M. (2009). A genetic screen for nitrate regulatory mutants captures the nitrate transporter gene NRT1.1. *Plant Physiology, 151*(1), 472–478. https://doi.org/10.1104/pp.109.140434.

Widiez, T., El Kafafi el, S., Girin, T., Berr, A., Ruffel, S., Krouk, G., et al. (2011). High nitrogen insensitive 9 (HNI9)-mediated systemic repression of root NO_3^- uptake is associated with changes in histone methylation. *Proceedings of the National Academy of Sciences of the United States of America, 108*(32), 13329–13334. https://doi.org/10.1073/pnas.1017863108.

Williams, L., & Miller, A. (2001). Transporters responsible for the uptake and partitioning of nitrogenous solutes. *Annual Review of Plant Physiology and Plant Molecular Biology*, *52*, 659–688.

Xu, N., Wang, R., Zhao, L., Zhang, C., Li, Z., Lei, Z., et al. (2016). The Arabidopsis NRG2 protein mediates nitrate signaling and interacts with and regulates key nitrate regulators. *The Plant Cell*, *28*(2), 485–504. https://doi.org/10.1105/tpc.15.00567.

Zhang, H., & Forde, B. G. (1998). An Arabidopsis MADS box gene that controls nutrient-induced changes in root architecture. *Science*, *279*(5349), 407–409.

Zhang, H., Jennings, A., Barlow, P. W., & Forde, B. G. (1999). Dual pathways for regulation of root branching by nitrate. *Proceedings of the National Academy of Sciences of the United States of America*, *96*(11), 6529–6534.

Zhou, J. J., Trueman, L. J., Boorer, K. J., Theodoulou, F. L., Forde, B. G., & Miller, A. J. (2000). A high affinity fungal nitrate carrier with two transport mechanisms. *Journal of Biological Chemistry*, *275*(51), 39894–39899. https://doi.org/10.1074/jbc.M004610200.

Zhuo, D., Okamoto, M., Vidmar, J. J., & Glass, A. D. (1999). Regulation of a putative high-affinity nitrate transporter (*Nrt2;1At*) in roots of *Arabidopsis thaliana*. *Plant Journal*, *17*(5), 563–568.

CHAPTER SEVEN

The Regulation of Ion Channels and Transporters in the Guard Cell

Brice Roux, Nathalie Leonhardt[1]
Laboratoire de Biologie du Développement des Plantes, Commissariat à l'Energie Atomique et aux Energies Alternatives (CEA), CNRS, Aix-Marseille Université, UMR 7265, Institut de Biosciences et Biotechnologies d'Aix-Marseille, CEA Cadarache, Saint-Paul-lez-Durance, France
[1]Corresponding author: e-mail address: nathalie.leonhardt@cea.fr

Contents

Abstract

Stomatal guard cells delimit pores in the epidermis of plant leaves and modulate plant gas exchange through the regulation of aperture of the pore. The opening and closure of stomatal pores control the trade-off between efficient photosynthetic CO_2 fixation and water loss via transpiration, playing a critical role in plant growth and stress responses. Stomata open in response to light and close in response to drought stress, elevated CO_2, ozone, low humidity and plant pathogens. Stomatal movement relies on transport of solutes across channels and transporters in the plasma membrane and vacuolar membrane of guard cells. The fine-tuned regulation of this transport requires an integrated convergence of multiple endogenous signals and environmental stimuli. Current knowledge of the plasma membrane and vacuolar membrane ion channels and transporters of guard cell model systems and their function in guard cell signalling is reviewed in this chapter.

Advances in Botanical Research, Volume 87
ISSN 0065-2296
https://doi.org/10.1016/bs.abr.2018.09.013

1. INTRODUCTION

When plants left the aquatic environment and invaded the land almost 400 million years ago, they had to develop special features called stomata to prevent excessive water loss while allowing CO_2 to enter the plant body for photosynthesis (Edwards, Fanning, & Richardson, 1986). Stomata are pores formed by pairs of specialized epidermal cells, the guard cells, which are found in the surface of aerial parts of most land plants. The stomatal pore called ostiole can open and close through dynamic changes in turgor of the guard cells to control gas exchange between the plant and its environment. Thus, their major function is to optimize photosynthetic CO_2 fixation with respect to transpiration water loss. Stomatal pores occupy less than 5% of the leaf surface, but can account for over 90% of the CO_2 entering the leaf and over 70% of water loss from the plant (Lawson & Blatt, 2014). The opening and closing of stomata are regulated by the integration of environmental and endogenous signals such as light, plant hormones including abscisic acid (ABA) and auxin, Ca^{2+}, CO_2, air humidity, pathogens and air pollutants. These signals are transduced into intracellular messages, which ultimately trigger the stomatal movements (Schroeder, Allen, Hugouvieux, Kwak, & Waner, 2001; Shimazaki, Doi, Assmann, & Kinoshita, 2007). Since several decades, guard cells have emerged as a key model system to decipher plant ion channel functions and signalling. Indeed, these cells are located in the epidermis and can be easily isolated. In particular, they are not connected to adjacent cells by plasmodesmata (Pandey, Zhang, & Assmann, 2007; Wille & Lucas, 1984). Thus, they can respond cell-autonomously to physiological stimulation, allowing cell biological analyses of stomatal opening and closing in response to various stimuli. In addition, the development of powerful approaches in Arabidopsis combining molecular and cell biology, genetics, electrophysiology, biochemistry and whole plant biology has contributed to gaining substantial new insights regarding ion transporters and their regulation during stomatal responses to signals.

During stomatal opening, plasma membrane H^+-ATPases mediate H^+ efflux from the cytosol, which hyperpolarize the membrane potential above the equilibrium potential for K^+. Accordingly, K^+ uptake through inward-rectifying K^+ channels is activated. In addition, K^+ charge is balanced by malate^{2-} production from starch breakdown and the uptake of inorganic anions including Cl^- and NO_3^-, presumably through proton-coupled uptake transporters. This lowers the inner guard cell water potential and drives osmotic water uptake increasing guard cell volume by approximately 40% (Gao et al., 2005; Shope, DeWald, & Mott, 2003). Most of cell volume change is accounted for by increased vacuolar volume, with multiple small vacuoles fusing to form a large central vacuole (Gao et al., 2005). K^+ uptake into vacuoles during stomatal opening is likely due to a H^+/K^+ antiporter activity (Padmanaban, Lin, Perera, Kawamura, & Sze, 2004), whereas anion uptake into vacuoles may occur through a combination of low-affinity anion channels (Hafke, Hafke, Smith, Lüttge, & Thiel, 2003; Kovermann et al., 2007; Pei, Ward, Harper, & Schroeder, 1996) and an active H^+/anion exchange system (De Angeli et al., 2006; Eisenach & De Angeli, 2017). During stomatal closing, membrane depolarization occurs due to the inhibition of plasma membrane H^+-ATPases and the activation of plasma membrane anion channels, which mediate passive efflux of Cl^-, malate^{2-}, and NO_3^-. Plasma membrane depolarization below the equilibrium potential for K^+ drives K^+ efflux through outward-rectifying K^+ channels (Keller, Hedrich, & Raschke, 1989; Schroeder & Hagiwara, 1990; Schroeder, Raschke, & Neher, 1987). The resulting cellular export of K^+, Cl^-, and organic ions such as malate^{2-} drives water efflux. Elevation of cytosolic Ca^{2+} through Ca^{2+}-permeable channels at the plasma membrane and intracellular Ca^{2+}-release channels is observed to precede stomatal closure. At the vacuolar membrane, Ca^{2+}-activated vacuolar K^+ channels are required for K^+ release to the cytoplasm (Gobert, Isayenkov, Voelker, Czempinski, & Maathuis, 2007; MacRobbie, 1998; Ward & Schroeder, 1994).

One of the primary roles of stomatal closure is to prevent water loss, thereby inducing tolerance to water deficit. Under drought conditions, ABA is produced or/and accumulated in guard cells, thereby inducing stomatal closure and conserving water (Bauer et al., 2013; Lee & Luan, 2012). The cellular and molecular regulation of stomatal opening and closure under drought conditions has been extensively investigated and reviewed (Assmann, 2003; Cutler, Rodriguez, Finkelstein, & Abrams, 2010; Hubbard, Nishimura, Hitomi, Getzoff, & Schroeder, 2010; Lim, Baek, Jung, Kim, & Lee, 2015; Popko, Hänsch, Mendel, Polle, & Teichmann, 2010; Sirichandra, Wasilewska, Vlad, Valon, & Leung, 2009; Wasilewska et al., 2008; Wilkinson & Davies, 2010; Zhang, Chen, & Harmon, 2016). ABA-controlled processes are necessary for plant survival, and ABA-deficient mutants are more susceptible to water stress (Finkelstein, Gampala, & Rock, 2002; Frey et al., 2012; Kang, Choi, Im, & Kim, 2002). Based on the genetic identification of some of its components and the knowledge of protein-protein interactions in Arabidopsis, the core pathway of ABA signalling has been successfully reconstituted. In the absence of ABA, the protein kinase open stomata1 [OST1, named after the phenotype of the null mutant upon ABA treatment, also called sucrose nonfermenting-1 (Snf1)-related protein kinase 2.6 or SnRK2.6] is bound to a type A protein phosphatase 2C. Several of these have been identified such as abscisic acid insensitive 1 or 2, ABI1 or ABI2, or homology to ABI1, HAB1. The PP2C binds to the activation loop of OST1 blocking its active site. In the presence of ABA, ABA binds to soluble ABA receptors, called pyrabactin resistance/pyrabactin-like/regulatory component of ABA receptor (PYR/PYL/RCAR), and induces their conformational change, which in turn promotes binding to PP2C. This binding inhibits the PP2C activity and releases OST1 inhibition, which can phosphorylate downstream targets including ion channels to promote stomatal closure.

Various foliar pathogens such as bacteria, fungi, and viruses are known to disrupt stomatal movement in order to successfully infect plants (Arnaud & Hwang, 2015; McLachlan, Kopischke, & Robatzek, 2014; Melotto, Underwood, & He, 2008). The first line of plant defence is the recognition of evolutionary conserved pathogen materials named the pathogen-associated molecular patterns (PAMP) by plant pattern recognition receptors (PRRS), thereby leading to PAMP-triggered immunity (PTI) (Jones & Dangl, 2006; Monaghan & Zipfel, 2012; Schwessinger & Zipfel, 2008). The second line of defence is the recognition of effectors through plant resistance (R) proteins, thereby leading to effector-triggered immunity (ETI) (Chisholm, Coaker, Day, & Staskawicz, 2006; Jones & Dangl, 2006). Melotto, Underwood, Koczan, Nomura, and He (2006) showed that stomata constitute not only a major path for transpiration but also an active gate limiting pathogen entry. Thus, stomata play an important role in the plant immune response, and the regulation of their movements is an early stage of defence mechanism against pathogens. The present chapter reviews our current knowledge and understanding of the properties and regulation of ion channels and transporters at the plasma membrane and tonoplast of guard cells, and their roles in stomatal movements and guard cell signalling pathways. Models highlighting the role in stomatal movement of transport systems at the plasma membrane and tonoplast are illustrated in Figs 1 and 2, respectively. Table 1 summarizes the major proteins involved in regulation of guard cell transporters and channels.

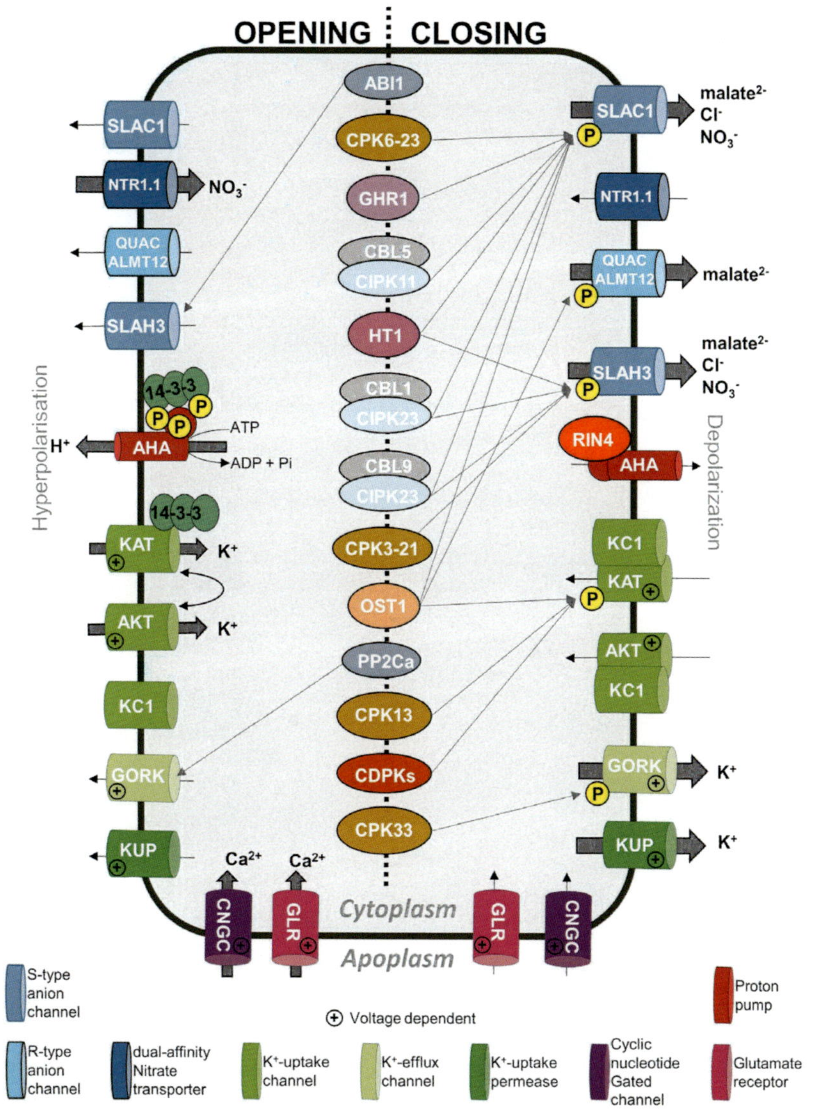

Fig. 1 Ion channels, transporters and pumps localized at the plasma membrane and their regulation during stomatal opening (left) and closure (right).

2. PROTON PUMPS

2.1 Plasma Membrane H^+-ATPases

In plants and fungi, the energetics at the plasma membrane is provided by a large proton motive force generated by a family of P-type ATPases,

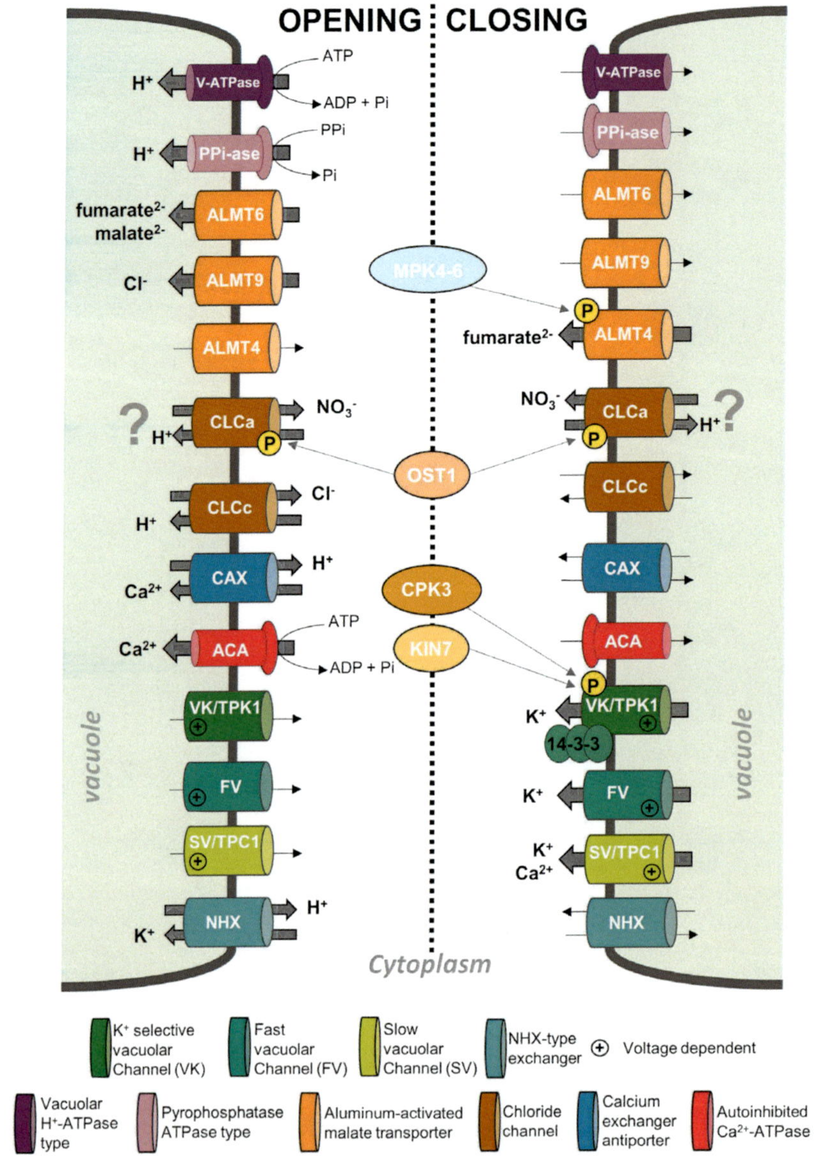

Fig. 2 Ion channels, transporters and pumps localized at vacuolar membrane and their regulation during stomatal opening (left) and closure (right).

commonly called plasma membrane H^+-ATPases or, in Arabidopsis, AHAs for Arabidopsis H^+-ATPases. These plasma membrane H^+-ATPases act as primary proton transporters and drive a large number of secondary transporters by enhancing the membrane potential and pH gradient across the

Table 1 Major Proteins Involved in Regulation of Guard Cell Transporters and Channels

Name	Type	Subcellular Localisation	Post Translational Modification	Protein	References
AHA1	H^{+}-ATPase	Plasma membrane		RIN4	Mackey, Holt, Wiig, and Dangl (2002); Liu, Elmore, and Coaker (2009)
				14-3-3	Kinoshita and Shimazaki (2001); Baekgaard, Fuglsang, and Palmgren (2005); Palmgren (2001)
AHA2	H^{+}-ATPase	Plasma membrane		RIN4	Mackey et al. (2002); Liu et al. (2009)
				14-3-3	Kinoshita and Shimazaki (2001); Bækgaard et al. (2005); Palmgren (2001)
KAT1	K^{+} inward rectifier channel	Plasma membrane	Phosphorylation	CDPK	Li, Lee, and Assmann (1998); Berkowitz, Zhang, Mercie, Leng, and Lawton (2000); Sato et al. (2009); Sato, Gambale, Dreyer, and Uozumi (2010)
				OST1	Sato et al. (2010, 2009)
				CPK13	Ronzier et al. (2014)
				14-3-3	Sottocornola et al. (2006)
				KC1	Duby et al. (2008)
KAT2	K^{+} inward rectifier channel	Plasma membrane	Phosphorylation	CPK13	Ronzier et al. (2014)
				KC1	Duby et al. (2008)
AKT1	K^{+} inward rectifier channel	Plasma membrane		KC1	Jeanguenin et al. (2011); Wang, He, Li, Xu, and Wu (2010); Wang, Hu, et al. (2016); Wang, Chen, et al. (2016)

Continued

Table 1 Major Proteins Involved in Regulation of Guard Cell Transporters and Channels—cont'd

Name	Type	Subcellular Localisation	Post Translational Modification	Protein	References
AKT2	K^+ inward rectifier channel	Plasma membrane		KC1	Jeanguenin et al. (2011)
GORK	K^+ outward rectifier channel	Plasma membrane	Phosphorylation	CPK33	Corratgé-Faillie et al. (2017)
			Dephosphorylation	PP2CA	Lefoulon et al. (2016)
TPK1	K^+ selective vacuolar	Vacuole membrane	Phosphorylation	CPK3	Latz et al. (2013)
				KIN7	Isner, Begum, Nuehse, Hetherington, and Maathuis (2018)
				14-3-3 (GRF6)	Latz, Becker, et al. (2007); Latz, Ivashikina, et al. (2007); Latz et al. (2013)
SLAC1	S-type anion channel	Plasma membrane	Phosphorylation	OST1	Geiger et al. (2009); Lee, Lan, Buchanan, and Luan (2009); Vahisalu et al. (2010)
				GHR1	Hua et al. (2012)
				CPK3	Geiger et al. (2010); Scherzer, Maierhofer, Al-Rasheid, Geiger, and Hedrich (2012); Brandt et al. (2015)
				CPK6	Geiger et al. (2010); Scherzer et al. (2012); Brandt et al. (2015)

				CPK21	Geiger et al. (2010); Scherzer et al. (2012)
				CPK23	Geiger et al. (2010); Scherzer et al. (2012)
				CIPK11plusCBL5	Saito et al. (2018)
				CIPK23plusCBL1	Saito et al. (2018); Maierhofer et al. (2014)
				CIPK23plusCBL9	Saito et al. (2018); Maierhofer et al. (2014)
			Dephosphorylation	HT1	Tian et al. (2015)
SLAH3	S-type anion channel	Plasma membrane	Phosphorylation	CPK21	Geiger et al. (2011)
				CIPK23	Maierhofer et al. (2014)
				CIPK23plusCBL1	Saito et al. (2018); Maierhofer et al. (2014)
				CIPK23plusCBL9	Saito et al. (2018); Maierhofer et al. (2014)
				HT1	Hashimoto et al. (2006); Merilo et al. (2013); Tian et al. (2015); Xue et al. (2011)
			Dephosphorylation	ABI1	Geiger et al. (2011); Maierhofer et al. (2014)
QUAC1/ ALMT12	R-type anion channel	Plasma membrane	Phosphorylation	OST1	Imes et al. (2013)
ALMT4	R-type anion channel	Vacuole membrane	Phosphorylation	MPK4	Eisenach et al. (2017); Imes et al. (2013)
				MPK6	Eisenach et al. (2017); Imes et al. (2013)
CLCa	Chloride channel	Vacuole membrane	Phosphorylation	OST1	Wege et al. (2014)

plasma membrane (Palmgren, 2001). H^+-ATPases are activated in response to various signals including blue light, osmotic stress, and the phytohormone auxin (Kerkeb, Venema, Donaire, & Rodríguez-Rosales, 2002; Kinoshita & Shimazaki, 1999; Niittylä, Fuglsang, Palmgren, Frommer, & Schulze, 2007; Okumura, Inoue, Kuwata, & Kinoshita, 2016; Takahashi, Hayashi, & Kinoshita, 2012). In the Arabidopsis genome, there are 11 H^+-ATPase encoding genes and important progress has been made in recent years, concerning the identification and organization of these genes, their expression, and also the kinetics and regulation of individual H^+-ATPase isoforms.

In the guard cell, plasma membrane H^+-ATPases play a key role in stomatal opening, particularly in response to blue light (Assmann, Simoncini, & Schroeder, 1985; Hayashi, Inoue, Takahashi, & Kinoshita, 2011; Kinoshita & Shimazaki, 1999; Shimazaki et al., 2007; Shimazaki, Tominaga, & Shigenaga, 1997). H^+-ATPases drive H^+ efflux, thereby acidifying the guard cell wall space and generating a membrane potential, negative inside, which triggers K^+ uptake through voltage-gated K^+ channels during stomatal opening (Assmann & Shimazaki, 1999; Schroeder et al., 2001). In guard cell protoplasts, plasma membrane H^+-ATPAses drive voltage-dependent outward currents that reverse at −230 mV (Lohse & Hedrich, 1992). Besides, free-running membrane potentials as negative as −210 mV have been found for guard cells in epidermal peels (Blatt, 1987; Roelfsema & Prins, 1997). In the guard cells, the H^+-ATPase activity is positively regulated by blue light and auxin, whereas Ca^{2+} and ABA act as negative regulators (Brault et al., 2004). The plasma membrane H^+-ATPase is self-inhibited by binding of its C-terminal domain to its catalytic domain (Morsomme et al., 1996; Palmgren, 2001; Palmgren, Larsson, & Sommarin, 1990; Palmgren, Sommarin, Serrano, & Larsson, 1991). Conversely, activation of H^+-ATPases occurs by phosphorylation of the C-terminal penultimate threonine followed by binding of 14-3-3 proteins, which in turn causes a conformational change and results in a release of the C-terminal domain and de-repression of the catalytic domain (Kinoshita & Shimazaki, 1999, 2002). The 11 Arabidopsis AHA isoforms are thought to function redundantly in stomatal opening, since they are all expressed in guard cells with higher expression levels for *AHA1*, *AHA2*, and *AHA5* (Leonhardt et al., 2004; Ueno, Kinoshita, Inoue, Emi, & Shimazaki, 2005; Yamauchi et al., 2016). The importance of Arabidopsis AHA1 for stomatal opening has been demonstrated by dominant mutations *ost2-1D* and *ost2-2D* (*Open Stomata 2*), which lead to expression of constitutively active forms of AHA1 and defects in stomatal closure (Merlot et al., 2007). Recently, analyses of *aha1*, *aha2* and *aha5* loss-of-function mutants provided additional genetic evidence for the important role of AHA1 in blue light-dependent stomatal opening (Yamauchi et al., 2016). Overexpression of H^+-ATPase (AHA2) in guard cells or increasing the accumulation of H^+-ATPase at the plasma membrane of guard cells resulted in the promotion of light-induced stomatal opening and thus higher photosynthetic activity (Hashimoto-Sugimoto et al., 2013; Wang et al., 2014). In tobacco, suppression of one H^+-ATPase isoform (PMA4) impaired stomata opening and caused stomatal insensitivity to fusicoccin, a fungal toxin that activates H^+-ATPases (Zhao, Dielen, Kinet, & Boutry, 2000).

Blue-light-activation of plasma membrane H^+-ATPases has been intensively investigated in guard cells. Blue light is mainly perceived by two phototropins, PHOT1 and PHOT2, which are serine/threonine protein kinases with two LOV (light, oxygen and voltage) domains (Briggs & Christie, 2002; Kinoshita et al., 2001; Shimazaki et al., 2007). Activated phototropins undergo autophosphorylation, bind to 14-3-3 proteins, and ultimately activate the plasma membrane H^+-ATPase (Kinoshita et al., 2001, 2003; Ueno et al., 2005). It is still unknown whether phototropin excitation induces direct phosphorylation of the H^+-ATPase or whether there are intermediate signalling elements. Furthermore, the proton pump activation requires the phosphorylation of its penultimate threonine residue at the C-terminus end. This phosphorylation event alone is not sufficient to activate H^+ pumping, as subsequent binding of 14-3-3 proteins is also needed (Bækgaard et al., 2005; Palmgren, 2001) Recent work also indicates how 14-3-3 proteins can activate H^+-ATPases. In *Vicia faba* guard cells, a 32kDa 14-3-3 protein is able to bind to the phosphorylated C-terminus of the H^+-ATPase, but not to the nonphosphorylated one (Emi, Kinoshita, & Shimazaki, 2001; Kinoshita & Shimazaki, 2001). Moreover, blue light increases the amount of bound 14-3-3 proteins, which activates H^+-ATPase activity in proportion (Kinoshita & Shimazaki, 2002). Finally, the H^+-ATPase/14-3-3 complex is stabilized by fusicoccin (Palmgren, 2001). Fusicoccin binds to 14-3-3 proteins, thereby increasing their affinity for the autoinhibitory C-terminal end of the plasma membrane H^+-ATPase. This prevents interaction of the C-terminal end with the H^+-ATPase catalytic domain, thereby increasing proton pump activity and causing an irreversible opening of stomata (Assmann & Schwartz, 1992; Kinoshita & Shimazaki, 2001). Arabidopsis possesses 13 genes encoding 14-3-3 proteins, defining two major subclasses, the ε and non-ε isoforms. Interestingly, Pallucca et al. (2014) showed that H^+-ATPases preferentially interact with the latter. However, further studies will be needed to identify which 14-3-3 isoforms interact with guard cell-expressed proton pumps.

The protein kinase(s) that is critical for phosphorylation of the penultimate threonine residue has not been identified yet. However, in *V. faba*, this kinase(s) appears to be insensitive to general protein kinase inhibitors (Kinoshita & Shimazaki, 2001), and in spinach it seems to be associated with the plasma membrane (Svennelid et al., 1999).

Blue light-induced phosphorylation of the guard cell H^+-ATPase is counteracted by ABA (Hayashi & Kinoshita, 2011). Knowledge of the mechanism involved remains fragmentary, but part of the effects might be exerted indirectly through nitric oxide and H_2O_2 (Zhang, Takemiya, Kinoshita, & Shimazaki, 2007). Indeed, the inhibitory effect of ABA on proton pumping, H^+-ATPase phosphorylation, and stomatal opening can be rescued by 2-carboxyphenyl-4,4,5,5-tetramethylimidazoline-1-oxyl-3-oxide (c-PTIO), a scavenger of NO, but not by H_2O_2 (Zhang et al., 2007). In addition, the inhibitory effect of ABA on the guard cell plasma membrane H^+-ATPase was not observed in ABA signalling mutants *abi1-1*, *abi2-1* and *ost1-2*, suggesting that these effects are mediated via the classical PP2C-OST1 pathway (Hayashi et al., 2011; Roelfsema, Staal, & Prins, 1998). It is not known, however, whether OST1 and the PP2Cs directly inactivate proton pumping. The only negative regulator of H^+-ATPases identified to date is a calcium-stimulated protein kinase, PKS5. This member of the SnRK3 kinase family phosphorylates the serine residue at position 931 of AHA2, interfering with binding of 14-3-3 proteins to the penultimate threonine (Fuglsang et al., 2007). In addition, several leucine-rich repeat receptor kinases (LRR-RK) including PSY1R and FERONIA interact with plasma membrane H^+-ATPases (Fuglsang et al., 2014; Pertl-Obermeyer, Schulze, & Obermeyer, 2014). For example, AHA1 and AHA2 were shown to interact directly *in vitro* and *in planta* with PSY1R, that serves as a plasma membrane receptor for the peptide growth hormone PSY1. The intracellular protein kinase domain of PSY1R phosphorylates the two proton pumps at the autoinhibitory region I of their C-terminal domain (Fuglsang et al., 2014). However, no stomatal behaviour phenotype has been reported for loss-of-function mutants in PSY1R and FERONIA.

Interestingly, AHA1 and AHA2 play also an important role in PAMP-induced stomatal closure (Liu et al., 2009). The *ost2* mutant (which exhibits constitutive activation of AHA1) does not close stomata in response to bacteria and PAMPs such as flg22 and lipopolysaccharides (Liu et al., 2009). AHA1 and AHA2 interact with RIN4 (RPM1-INTERACTING PROTEIN4), which is a well-known regulator of multiple plant immune responses (Liu et al., 2009). *RIN4* is expressed in guard cells. It has been shown to play a major role in PAMP-triggered immunity, effector-triggered immunity, and stomatal opening upon pathogen perception (Liu et al., 2009; Mackey et al., 2002). Overexpression of *RIN4* increased the activation of H^+-ATPases, while a mutant affected in RIN4 had reduced H^+-ATPase activity and was resistant to *Pseudomonas syringae* pv. tomato, DC3000 (Liu et al., 2009).

2.2 Vacuolar V-Type ATPases and Pyrophosphatases

Guard cell vacuoles are major storage compartments for protons. The vacuolar accumulation of protons essentially depends on a type of vacuolar H^+-ATPases, the V-type ATPases, and on inorganic pyrophosphatases, the H^+-PPases. This accumulation generates a gradient of proton concentration and membrane potential that can be used to transport compounds against their concentration or electrochemical potential gradients (Gaxiola, Fink, & Hirschi, 2002). The movement of ions, such as K^+, Cl^- or malate between vacuoles and the cytoplasm causes changes in the guard cell water potential, which drive reversible water flows across the tonoplast (Schroeder et al., 2001). Such processes result in changes in turgor pressure, thereby driving stomatal movement. To date, the resting value of the guard cell vacuolar membrane potential is still unknown. Yet, the modelling platform OnGuard points to tonoplast potential values between -15 and -35 mV using vacuolar transport parameters and predicts a vacuolar membrane potential depolarization during stomatal opening and an hyperpolarization during closure (Chen et al., 2012; Hills, Chen, Amtmann, Blatt, & Lew, 2012). Nonetheless, no variations of the vacuolar membrane potential in response to extracellular or intracellular stimuli have been reported so far.

The V-type ATPases are composed of multiple subunits that form a large protein complex (Sze, Schumacher, Müller, Padmanaban, & Taiz, 2002). Genes for at least 26 different V-type ATPase subunits have been identified in the Arabidopsis genome, with some subunits encoded by small gene families (Padmanaban et al., 2004). ATP hydrolysis within this protein complex induces conformational changes that contribute to a rotary mechanism that drives H^+ transport across the vacuolar membrane (Schumacher & Krebs, 2010). Interestingly, the number of protons transported per consumed ATP molecule can vary significantly (Davies, Hunt, & Sanders, 1994; Rienmüller et al., 2012). Under low pH gradients, V-type ATPases transport several H^+ ions per ATP, but the coupling ratio drops if H^+ has to be transported against a higher pH gradient. V-type ATPases are also strongly inhibited by electrical potentials (Rienmüller et al., 2012), which explains why the vacuolar membrane does not become more hyperpolarized than −30 mV (Felle, Brummer, Bertl, & Parish, 1986; Walker, Leigh, & Miller, 1996). Finally, guard cell V-type ATPases are likely regulated by the redox state and cytoplasmic nucleotide concentrations (Kluge et al., 2003), but no data are available about these types of regulation. In complement to V-type ATPases, H^+-pumping pyrophosphatases were shown to be active in the vacuolar membrane of guard cells (Darley, Skiera, Northrop, Sanders, & Davies, 1998). These pumps are large homodimeric enzymes, which utilize the energy of pyrophosphate (PPi) hydrolysis to drive H^+ uptake into the vacuole (Gaxiola, Palmgren, & Schumacher, 2007; Lin et al., 2012). To date, several genetic evidences support the importance of V-type ATPases and PPi-ases in guard cell physiology. Arabidopsis has three isoforms of the membrane-integral subunit VHA-a that are differentially localized: VHA-a1 is localized in the TGN/early endosome, whereas VHA-a2 and VHA-a3 sit in the tonoplast. In a *vha-a2 vha-a3* double loss-of-function mutant that has no V-type ATPase activity and exhibits a vacuolar pH of 6.4 (instead of 5.9 in wild-type plants), ABA-induced stomatal closure is delayed (Bak et al., 2013; Krebs et al., 2010). These results demonstrate that the V-type ATPase is critical for rapid stomatal closure in response to ABA. A *avp1* mutant (*Vacuolar proton pyrophosphatase1*) also exhibited substantial delays in ABA-induced vacuolar acidification and stomatal closure, which indicates that, in mature stomatal guard cells, the V-PPase also contributes significantly to these functions (Bak et al., 2013). Altogether, these observations confirm the initial presumption that, by contributing to guard cell cytosol alkalization, vacuolar acidification is essential for ABA-induced stomatal closure (Bak et al., 2013; Gonugunta, Srivastava, Puli, & Raghavendra, 2008; Irving, Gehring, & Parish, 1992; Suhita, Raghavendra, Kwak, & Vavasseur, 2004). The det*3* mutant, which is affected in the C subunit of the V-type ATPase and only displays 70% of wild-type V-type ATPase activity, is also impaired in stomatal closure in response to an increase in extracellular Ca^{2+} concentration or to exogenous H_2O_2 (Allen et al., 2000; Brüx et al., 2008; Schumacher et al., 1999). These defects are linked to an altered Ca^{2+} homeostasis in det*3* guard cells. Interestingly, none of these mutants has been reported to be affected in stomatal opening. Thus, in contrast to stomatal closure, the role of a vacuolar proton gradient during stomatal opening remains elusive and needs further investigations.

3. K^+ CHANNELS AND TRANSPORTERS

3.1 Plasma Membrane K^+ Channels

Stomatal movements are associated with large K^+ fluxes across the guard cell plasma membrane, facilitated by K^+ inward- and outward-rectifying ion channels (Blatt, 1992; Blatt & Clint, 1989; Schroeder, 1988; Schroeder et al., 1987). These movements are passive and their direction depends on the free-running membrane potential (Blatt, 1992; Thiel, MacRobbie, & Blatt, 1992). The K^+ equilibrium potential, E_k, as determined by impalement of intact guard cells, is between −50 and −70 mV (Roelfsema, Steinmeyer, Staal, & Hedrich, 2001). When the plasma membrane electrical potential is higher than (positive to) E_k, the K^+ flux is directed out of the cell. In contrast, when the plasma membrane electrical potential is lower (more negative) than E_k, inward-rectifying K^+ channels are activated allowing K^+ influx. Thus, changes in the guard cell membrane potential determine the direction of K^+ fluxes, thereby acting on guard cell turgor and volume. In line with these ideas, several forms of plasma membrane K^+ channels with different voltage-dependent activation properties have been identified (Dreyer & Blatt, 2009; Hedrich, 2012; Lebaudy, Véry, & Sentenac, 2007). Yet, all these channels display significant structural similarities with voltage-dependent K^+ channels of the Shaker family. For instance, Arabidopsis guard cells express six Shaker-like K^+ channel subunits including four K^+-uptake channels (KAT1, KAT2, AKT1, AKT2), a regulatory subunit (AtKC1) and a K^+-efflux channel (GORK) (Szyroki et al., 2001).

The K^+-uptake channels mediate hyperpolarization-induced K^+ influx, resulting in water influx and stomatal opening in response to signals such as light. They can form heteromeric K^+ channels with each other, to form hetero-tetrameric channels (Baizabal-Aguirre, Clemens, Uozumi, & Schroeder, 1999; Ivashikina, Deeken, Fischer, Ache, & Hedrich, 2005; Lebaudy et al., 2007, 2010; Pilot et al., 2001; Urbach, Chérel, Sentenac, & Gaymard, 2000; Xicluna et al., 2007). The regulatory subunit AtKC1 can also assemble with these pore forming subunits to reduce the conductance and negatively shift the activation threshold of the functional channels (Duby et al., 2008; Jeanguenin et al., 2011; Wang, Chen, et al., 2016; Wang et al., 2010). Functional redundancy probably exists between K^+-uptake channel isoforms, as stomatal functioning remained intact in single loss-of-function mutants (Lebaudy, Hosy, et al., 2008; Lebaudy, Vavasseur, et al., 2008; Szyroki et al., 2001). However, the expression of dominant-negative KAT1 and KAT2 variants suppressed light- and low CO_2-induced K^+ uptake in guard cells and reduced stomatal opening suggesting that, as homo- or heteromers, KAT1 and KAT2 play a key role in K^+ uptake during stomatal opening (Kwak et al., 2001; Lebaudy, Hosy, et al., 2008; Lebaudy, Vavasseur, et al., 2008). In accordance with such important role, K^+ uptake channel activity is tightly regulated at various levels. During stomatal opening, the acidification of the extracellular space due to the proton pumping activity of H^+-ATPases increases the voltage-sensitivity of K^+-uptake channels leading to an increased number of open channels (Hoth et al., 1997; Hoth, Geiger, Becker, & Hedrich, 2001; Müller-Röber et al., 1995). Extracellular but also cytosolic protons are known to act on amino acid residues (including histidines) of plant Shaker-like K^+ channels to regulate their voltage-dependent gating properties (González et al., 2012; Hoth et al., 1997; Hoth & Hedrich, 1999). Another important component of K^+ channel regulation is protein trafficking. The number of active KAT1 channels in the plasma membrane seems to be regulated by turgor controlled *endo-* and *exo-*cytosis events (Hurst, Meckel, Tayefeh, Thiel, & Homann, 2004; Meckel, Hurst, Thiel, & Homann, 2005). Upon stomatal opening, when the guard cell surface area increases by up to 25% (Meckel et al., 2005), a pre-plasma membrane pool carrying clusters of K^+-uptake channels is ready to be incorporated into the plasma membrane. In contrast, endocytic retrieval of membrane vesicles containing K^+-uptake channels can be observed during stomatal closure, when the guard cell volume and surface area decrease (Sutter et al., 2007). The use of *Xenopus* oocyte expression further indicated that KAT1 can be activated by external acidification (Véry, Gaymard, Bosseux, Sentenac, & Thibaud, 1995) or co-expression with 14-3-3 proteins (Saponaro et al., 2017; Sottocornola et al., 2008, 2006) and is inhibited by co-expression with the regulatory subunit KC1 (Duby et al., 2008). KAT1 channel activity is also impaired by calcium-dependent protein kinases (CDPKs) from *V. faba* guard cells and soybean (Berkowitz et al., 2000; Li et al., 1998). Interestingly, studies on an Arabidopsis CDPK, CPK13, which exerts inhibiting effects on both KAT1 and KAT2 channel subunits, revealed that this regulation is Ca^{2+} independent (Ronzier et al., 2014). Thus, Ca^{2+}-dependent regulation of Shaker channels remains uncovered. Furthermore, a guard cell calcium-independent, ABA-activated protein kinase/ABA-responsive kinase (AAPK/ABR kinase) from *V. faba* was shown to phosphorylate the C-terminal region of KAT1 (Mori, Uozumi, & Muto, 2000). This AAPK is an ortholog of Arabidopsis OST1, described above as an essential element of ABA signalling in guard cells (Li et al., 1998; Li & Assmann, 1996). Residues of KAT1 involved in this regulation have been identified by site-directed mutagenesis. Single point mutation of a threonine at position 306, which is targeted by OST1, reduced the K^+ transport uptake activity of KAT1 upon expression in *Xenopus* oocytes and yeast (Sato et al., 2010, 2009). These studies indicate how K^+ uptake channels such as KAT1 can be downregulated by protein kinases known to promote stomatal closure.

In contrast to different K^+ uptake channels, GORK represents the only K^+-efflux channel expressed in guard cells and contributes to stomatal closure (Ache et al., 2000; Becker et al., 2003). Depolarization of the guard cell by anion efflux activates GORK, and the resulting K^+ efflux promotes water loss and stomatal closure. Interestingly, a full loss of function of *GORK* in *gork1* mutants resulted in no measurable K^+-efflux channel currents in guard cells but only partially impaired stomatal closure in response to darkness or ABA, suggesting the involvement of additional channels or transporters (Hosy et al., 2003). Accordingly, a triple mutant lacking two members of the HAK/KUP transporter family, KUP6 and KUP8, together with GORK, was impaired in ABA-mediated stomatal closure indicating that KUP6 and KUP8 are also involved in ABA-induced stomatal closure (Osakabe et al., 2013). GORK channel activity is modulated in response to different signalling factors, such as reactive oxygen species (Garcia-Mata et al., 2010), external K^+ (Ache et al., 2000), and internal and external pH (Ache et al., 2000; Lacombe, Pilot, Gaymard, Sentenac, & Thibaud, 2000). In addition, clustering of GORK channels was correlated with changes in their gating properties (Eisenach, Papanatsiou, Hillert, & Blatt, 2014). *GORK* expression is upregulated during drought, salt and cold stresses and this regulation was impaired in the PP2C mutants, *abi1* and *abi2* (Becker et al., 2003). This finding is supported by physiological studies in guard cells of *Nicotiana benthamiana* which showed that outward K^+ fluxes were not simply driven by membrane depolarization and could be controlled by PP2Cs: overexpression in this cell system of a ABI1-1 dominant mutant form, showing ABA-independent constitutive activity, led to a strong and specific decrease of the amplitude of guard cell outward K^+ currents. This process was independent of the cell membrane potential and was not observed with K^+ inward and anionic currents suggesting that K^+-efflux channel can be inhibited by ABI1-1 independently of the membrane potential and ABA signalling pathway (Armstrong et al., 1995). More recently, the channel activity of GORK was shown to be directly inhibited by another PP2C, AtPP2CA (Lefoulon et al., 2016). GORK also shows a calcium-dependent activation: its channel activity was stimulated upon co-expression with the calcium-dependent protein kinase, CPK33, in

Xenopus oocytes (Corratgé-Faillie et al., 2017). In addition, stomatal closure during diurnal cycles was delayed in *cpk33* loss-of-function mutants suggesting that CPK33 acts as positive regulator of GORK channel to promote stomatal closure at dusk (Corratgé-Faillie et al., 2017).

3.2 Vacuolar K^+ Transport

Stomatal closure is mediated by a release of K^+ from the vacuole into the cytoplasm, which is followed by the extrusion of K^+ across the guard cell plasma membrane (Becker et al., 2003). In guard cells, three types of channels allowing K^+ fluxes across the tonoplast have been characterized: the "fast vacuolar" (FV), "K^+ selective vacuolar" (VK) and "slow vacuolar" (SV) channels (Allen & Sanders, 1996; Hedrich & Neher, 1987; Ward & Schroeder, 1994).

FV channels display an instantaneous activation and are inhibited by elevated cytosolic Ca^{2+} concentrations. These channels are voltage-dependent with lower open probabilities at membrane potentials close to 0 mV suggesting that FV channel activity promotes K^+ efflux from the vacuole to the cytoplasm for stomatal closure in a Ca^{2+}-independent manner (Hedrich & Neher, 1987; Pei, Ward, & Schroeder, 1999). The molecular identification of FV channels will definitely provide a better understanding of their role in guard cell functioning.

Early patch-clamp experiments on guard cell vacuoles from different plant species revealed a voltage-independent K^+ channel conductance attributed to VK channels (Pottosin & Martínez-Estévez, 2003; Ward & Schroeder, 1994). These currents are highly selective for K^+, activated by elevated cytosolic Ca^{2+}, and stimulated by low cytosolic pH (Allen & Sanders, 1996; Ward & Schroeder, 1994). In *Arabidopsis thaliana*, the ubiquitously expressed two-pore-domain K^+ channel, TPK1, was shown to encode this VK channel. In particular, expression in yeast revealed that the ion selectivity features and activation properties of TPK1 are consistent with guard cell VK channel currents (Gobert et al., 2007). While wild-type plants and TPK1-overexpressing plants showed a normal decrease in leaf conductance in response to an ABA treatment, indicating that their stomata were properly closing, *tpk1* loss-of-function mutant showed a slower response (Gobert et al., 2007). Altogether, these observations point to a vacuolar K^+ release function of TPK1 in guard cells that promotes stomatal closure. The TPK1 activity is regulated by interaction with the 14-3-3 protein GRF6. GRF6 binding depends on phosphorylation of a serine residue at position 42 (Ser42) in the cytosolic N-terminus of TPK1 increasing the open probability of the channel (Latz, Becker, et al., 2007; Latz, Ivashikina, et al., 2007). Ser42 phosphorylation is mediated by CPK3 (Latz et al., 2013). More recently, the LRR kinase, KIN7, was shown to interact with TPK1 and to stimulate its activity in response to ABA (Isner et al., 2018). In agreement with activating effects of elevated cytosolic calcium concentrations on VK channels, TPK1 possesses EF-hands in its C-terminal end. Thus, Ca^{2+} binding can directly activate the channel to promote K^+ release from the vacuole (Gobert et al., 2007; Ward & Schroeder, 1994).

In addition to VK and FV channels, a slow activating cation-selective channel, the SV channel, is also present in the guard cell vacuolar membrane (Ward & Schroeder, 1994). Elevated cytosolic calcium concentrations and positive trans-tonoplast electrical potentials (e.g. positive voltages on the cytoplasmic membrane side) activate the SV channel current (Allen & Sanders, 1996; Ward & Schroeder, 1994). In guard cells, SV channels exhibit a complex ion selectivity and transport monovalent as well as divalent cations, such as K^+, Na^+ and Ca^{2+} (Hedrich & Marten, 2011). This dual permeability has generated a controversy on SV channel function, and its respective roles in controlling K^+ and/or Ca^{2+} fluxes. In Arabidopsis, the SV channel is encoded by the unique tandem-pore channel gene, *TPC1*, which consists of two interconnected Shaker-type α-subunits associated with two EF-hands in the cytosolic interconnection loop (Peiter et al., 2005). The *tpc1-2* loss-of-function mutant is impaired in stomatal closure in response to high external Ca^{2+} (Peiter et al., 2005). However, this phenotype does not reflect a function of TPC1 in Ca^{2+}-induced Ca^{2+} release (CICR). It can rather be indirectly explained by an activation of plasma membrane S-type channels (Islam et al., 2010; Rienmüller et al., 2010). Moreover, no stomatal behaviour phenotype has been reported for the *fou2* mutant, which has a point mutation in the vacuolar loop of TPC1 (D454N) making the channel constitutively active (Beyhl et al., 2009). Overall, the stomatal function of the SV/TPC1 channels is still uncertain and further experiments are needed to determine the guard cell-specific SV channel features.

During stomatal opening, K^+ enters into the cytoplasm and is accumulated within the vacuole, thereby increasing turgor and promoting guard cell swelling. Since the vacuolar uptake of K^+ occurs against electrochemical and concentration gradients, it has to be mediated by secondary active carriers such as H^+/K^+ antiporters which are energized by the pH gradient across the tonoplast (Roelfsema & Hedrich, 2005; Walker et al., 1996).

The function of Na^+,K^+/H^+ antiporter (NHX)-type exchangers in guard cell vacuoles was investigated in this context. In Arabidopsis, the two major isoforms, NHX1 and NHX2, are highly expressed in guard cells (Barragán et al., 2012). A detailed study indicated that light-induced stomatal opening was impaired in *nhx1 nhx2* double mutants. Unexpectedly, stomatal closure was also affected in this double mutant, suggesting that reduced K^+ concentrations in the vacuole, in combination with K^+ accumulation in the cytosol, can compromise depolarization of the tonoplast by vacuolar K^+ efflux, and thereby stomatal closure (Andrés et al., 2014). In addition, the vacuolar pH in guard cell of *nhx1 nhx2* double mutant was found to be more acidic than in wild-type plants (Andrés et al., 2014). Ion transport by NHX proteins also supports the intense remodelling of vacuoles and associated endosomes that accompany stomatal movements (Andrés et al., 2014). Altogether, these results indicate that vacuolar accumulation of K^+ during stomatal opening is mediated by via NHX-type H^+ antiporters and relies on the proton motive force established by vacuolar proton pumps.

4. ANION TRANSPORT

4.1 Plasma Membrane Anion Channels

Activation of anion channels at the guard cell plasma membrane is a key step of stomatal closure (Pei, Kuchitsu, Ward, Schwarz, & Schroeder, 1997; Schroeder, Schmidt, & Sheaffer, 1993). The resulting efflux of anion induces a depolarization of the guard cell plasma membrane, which becomes more positive than E_k, and thereby promotes the activation of outwardly rectifying K^+ channels (Kim, Böhmer, Hu, Nishimura, & Schroeder, 2010; Kollist, Nuhkat, & Roelfsema, 2014). The overall ion efflux reduces the cell osmotic potential and as a consequence triggers water efflux, shrinkage of the guard cells, and closure of the pore. Patch-clamp experiments in *V. faba* guard cell protoplasts have led to the identification of two different plasma membrane anion channel currents classified as R-type or QUAC for quick activating anion channels, and S-type or SLAC for slow activating anion channels (Keller et al., 1989; Linder & Raschke, 1992; Raschke, 2003; Schroeder & Hagiwara, 1990; Schroeder & Keller, 1992). R-type anion currents are voltage dependent and exhibit rapid activation and deactivation kinetics within the millisecond range. In contrast, activation and deactivation time constants of S-type anion currents are in the range of tens of seconds. In addition, these currents show a weak voltage dependence and do not inactivate over time (Roelfsema, Levchenko, & Hedrich, 2004;

Schmidt, Schelle, Liao, & Schroeder, 1995). Both R-type and S-type channels are permeable to a range of physiologically important anions including malate^{2-}, Cl^- and NO_3^- and can be activated by an elevation of cytosolic Ca^{2+} (Hedrich, Busch, & Raschke, 1990; Marten, Konrad, Dietrich, Roelfsema, & Hedrich, 2007; Schroeder & Hagiwara, 1990). Thus, both channels must participate in stomatal closure, but with distinct contributions depending on physiological context. For instance, ABA activates both S- and R-type anion currents, whereas an increase in atmospheric CO_2 consistently activates S-type currents only, while R-type channels can be either activated or inactivated (Raschke, 2003).

The molecular identity of the guard cell S-type anion channel was genetically determined from independent Arabidopsis mutant screens for defects in stomatal responses to ozone or atmospheric CO_2 (Negi et al., 2008; Vahisalu et al., 2008). Plants homozygous for the *slac1* recessive mutation were found to be insensitive to a wide range of endogenous or environmental stimuli promoting stomatal closure, such as ABA, Ca^{2+}, NO, CO_2, darkness, high humidity or ozone (Negi et al., 2008; Vahisalu et al., 2008). Accordingly, Ca^{2+}- and ABA-induced S-type anion currents were dramatically impaired in guard cells of *slac1* mutants, whereas activation of R-type channels remained intact (Vahisalu et al., 2008). Altogether, these results provide strong genetic evidence of the crucial role of S-type channels in stomatal closure. When expressed alone in *Xenopus* oocytes, SLAC1 did not generate significant anion currents, whereas co-expression with a protein kinase such as OST1 consistently induced S-type anion channel currents (Geiger et al., 2009; Lee et al., 2009). In addition, the concomitant presence of PP2C phosphatase ABI1 prevented SLAC1 activation. In *Xenopus* oocytes, the functionally expressed SLAC1 anion channels showed properties that were similar to those of guard cell S-type anion channels, such as a weak voltage dependence, slow deactivation kinetics and a higher permeability for nitrate and chloride than for malate (Chen, Hills, Lim, & Blatt, 2010; Geiger et al., 2009; Schmidt & Schroeder, 1994). Interestingly, multiple serines in the hydrophilic N-terminal domain of SLAC1 are targeted by OST1 and serve in phosphorylation-dependent activation (Vahisalu et al., 2010). Mutation of one of the phosphorylated serines (Ser120) affects plant stomatal responses to ozone indicating that OST1-dependent phosphorylation at this site is physiologically relevant for induction of stomatal closure (Vahisalu et al., 2010). In addition, OST1 can phosphorylate the C-terminal tail of SLAC1 *in vitro* (Lee et al., 2009). The crucial role of reversible phosphorylation in the regulation of S-type anion channels was previously pointed out using protein kinase and protein phosphatase inhibitors (Schmidt et al., 1995). In addition, ABA activation of S-type anion currents is dramatically impaired in the ABA-insensitive *abi1-1*, *abi2-1*, and *ost1* mutants (Acharya, Jeon, Zhang, & Assmann, 2013; Finkelstein & Somerville, 1990; Koornneef, Reuling, & Karssen, 1984; Leung et al., 1994; Meyer, Leube, & Grill, 1994).

Patch-clamp on isolated guard cell protoplasts and microelectrode analyses on intact guard cells have shown that elevation of cytosolic Ca^{2+} activates guard cell S-type currents (Schroeder & Hagiwara, 1990; Stange, Hedrich, & Roelfsema, 2010). Moreover, ABA activates these currents in a calcium-independent as well as -dependent manner (Hetherington & Brownlee, 2004; Levchenko, Konrad, Dietrich, Roelfsema, & Hedrich, 2005; Marten et al., 2007; Schroeder & Hagiwara, 1990). These properties explain why S-type anion channel currents of *abi1-1* guard cells do not respond to ABA, whereas they are still activated by calcium (Allen, Kuchitsu, Chu, Murata, & Schroeder, 1999; Pei et al., 1997). In addition, guard cells from single and double mutants of Arabidopsis CPK3 and CPK6 showed impaired ABA- and Ca^{2+}-induced S-type anion currents suggesting that CDPK-mediated signalling pathway functions in the ABA response (Mori et al., 2006). Accordingly, phosphorylation and activation of SLAC1 channels in *Xenopus* oocytes can be mediated by OST1 and several other protein kinases including CDPKs such as CPK3, CPK6, CPK21 and CPK23 (Geiger et al., 2010; Scherzer et al., 2012). While CPK3 and CPK21 operate in the high-affinity calcium range, CPK6 and CPK23 display a prominent calcium-independent SLAC1 phosphorylation activity (Brandt et al., 2012; Geiger et al., 2010; Scherzer et al., 2012). Interestingly, activation of SLAC1 S-type anion currents by CDPKs or OST1 requires phosphorylation of the Ser59 or Ser120 residues, respectively (Brandt et al., 2015). These results suggest that CDPKs and OST1 have distinct affinities for two phospho-sites, which could explain the interdependence of the Ca^{2+}-dependent and -independent pathways of the ABA signalling network. In addition to CDPKs, SLAC1 is also activated by calcium sensor kinases of the calcineurin B-like/CBL-interacting protein kinase type (CBL/CIPK). Co-expression of CBL5 with CIPK11strongly activates SLAC1 in *Xenopus* oocytes (Saito et al., 2018). However, co-expression of CBL5 with CIPK23 has no activation effects, in contrast to co-expression of CIPK23 with CBL1 or CBL9 (Maierhofer et al., 2014; Saito et al., 2018). These results highlight the remarkable specificity of distinct CBL-CIPK complexes towards their cognate substrate proteins. In addition, SLAC1 activation by the CBL5-CIPK11 complex also depended on the dual lipid modification of CBL5. Thus, it can be proposed that Ca^{2+}-mediated regulation of SLAC1 also depends on lipid modifications (Batistic, Sorek, Schültke, Yalovsky, & Kudla, 2008; Geiger et al., 2010; Maierhofer et al., 2014; Scherzer et al., 2012). Overall, these observations suggest that a complex CDPK-mediated signalling pathway contributes to the response of guard cells to ABA. A leucine-rich repeat (LRR) RLK, named GUARD CELL HYDROGEN PEROXIDE-RESISTANT1, GHR1, can directly interact with, phosphorylate, and regulate SLAC1. GHR1 is negatively regulated by ABI2 but not ABI1, and localized in the plasma membrane. The *ghr1* mutation impaired H_2O_2 activation of the Ca^{2+} channel and induction of stomatal closure by ABA and H_2O_2, suggesting that GHR1 might also activate Ca^{2+} channels, indirectly or by direct phosphorylation (Hua et al., 2012). It will possible to test this hypothesis, once guard cell Ca^{2+} channels are identified. More generally, the interaction between GHR1 and SLAC1 provides a simple model for the regulation of downstream targets by an RLK in plants.

High CO_2 also leads to activation of S-type anion currents. It has been proposed that CO_2 influx across the guard cell plasma membrane is enhanced by aquaporins, an associated carbonic anhydrase accelerating the production of intracellular bicarbonate. Cytoplasmic bicarbonate then acts as an intracellular signalling molecule to activate S-type anion channels (Hu et al., 2010; Tian et al., 2015; Wang, Chen, et al., 2016; Wang, Hu, et al., 2016; Xue et al., 2011). Bicarbonate-induced activation of S-type anion currents is positively and negatively controlled by OST1 and the HIGH LEAF TEMPERATURE1 (HT1) protein kinase, respectively (Hashimoto et al., 2006; Merilo et al., 2013; Tian et al., 2015; Xue et al., 2011). In addition, HT1 counteracts the activation of SLAC1 by GHR1 in *Xenopus* oocytes (Hõrak et al., 2016). Moreover, elevated intracellular bicarbonate concentration strongly enhances SLAC1 anion currents when SLAC1 is co-expressed in *Xenopus* oocytes with OST1, CPK6 or CPK23 (Wang, Chen, et al., 2016, Wang, Hu, et al., 2016). Despite this information, the mode of SLAC1 activation by bicarbonate is not fully understood yet.

Four SLAC1 homologs are recognized in *A. thaliana* (SLAH1-4; Negi et al., 2008) of which, only SLAH3 is functionally characterized so far (Geiger et al., 2011). Co-expression of SLAH3 with CPK21 or CIPK23 in *Xenopus* oocytes resulted in nitrate-induced anion currents through phosphorylation of the N-terminal domain of SLAH3 (Geiger et al., 2011). In contrast, SLAH3 is not activated by OST1. Similar to SLAC1, SLAH3 phosphorylation and activation are blocked by the PP2C-type protein phosphatase, ABI1. In comparison with SLAC1, SLAH3 exhibits a higher preference for nitrate and requires the presence of nitrate at its extracellular face to be activated (Geiger et al., 2011; Maierhofer et al., 2014).

To date, no stomatal phenotype has been reported for *slah3* loss-of-function mutant. This suggests that SLAC1 and SLAH3 release chloride and nitrate in a concerted action upon stomatal closure (Geiger et al., 2011; Negi et al., 2008; Vahisalu et al., 2008).

R-type or QUAC anion channels seem to be related to the aluminium activated malate transporters (ALMTs) family. The ALMT12 isoform was identified as preferentially localized in the guard cell plasma membrane (Meyer et al., 2010; Sasaki et al., 2010) and the corresponding *atmt12* loss-of-function mutant is impaired by 40% in guard cell malate-induced R-type currents and in stomatal closure (Meyer et al., 2010). However, in the absence of extracellular malate, R-type currents are similar to those observed in wild-type (Meyer et al., 2010; Sasaki et al., 2010). When expressed in *Xenopus* oocytes, ALMT12 operates as a channel rather than a transporter and exhibits the hallmark electrical properties of guard cell R-type anion channels (Meyer et al., 2010). Interestingly, extracellular Al^{3+} does not activate ALMT12. According to the nomenclature proposed by Raschke (Linder & Raschke, 1992), ALMT12 was renamed QUAC1. Patch-clamp studies on guard cells showed that ABA activates R-type currents in wild-type plants, but to a much lesser extent in *abi1-1* or *ost1-2* mutant plants. Co-expression of QUAC1 and OST1 in *Xenopus* oocytes resulted in a pronounced activation of the R-type anion channel (Imes et al., 2013). Thus, OST1 targets both S-type and R-type guard cell anion channels. Overall, these results indicate that, although other transport systems are surely involved, ALMT12 belongs to a channel complex that mediates the release of anions during stomatal closure.

Finally, we note that stomatal opening requires an influx of anions against their concentration gradient. Such transport can be mediated by secondary active carriers such as the H^+/anion symporters or OH^-/anion antiporters. Interestingly, the NRT1.1 nitrate transporter, which mediates both high- and low-affinity nitrate uptake, is highly expressed in guard cells. Moreover, the *chl1* loss-of function mutant, which lacks NRT1.1 function, shows reductions in stomatal opening, guard cell nitrate content, and transpiration rate in the light (Guo, Young, & Crawford, 2003). Unfortunately, transport proteins for uptake of other ions including chloride have not yet been identified.

4.2 Vacuolar Anion Transport During Stomatal Movement

During stomatal opening, the K^+ is charge-balanced by Cl^-, NO_3^-, and Mal^{2-}, which act as essential counterions as well as osmotica

(Eisenach & De Angeli, 2017; Roelfsema & Hedrich, 2005). Because vacuolar V-type H^+-ATPases and H^+-pyrophosphatases cause a negative potential on the cytosolic membrane side of vacuoles, anion accumulation into vacuoles is essentially passive and mediated by ion channels. Chloride and malate conductances activated by a CDPK were initially identified in guard cell vacuoles of *V. faba* (Pei et al., 1996). To date, two major families of anion transporters, the AtALMT and the AtCLC families, have been identified for their role in stomatal movement in Arabidopsis.

Among the 13 members of the Arabidopsis AtALMT gene family, 3 have been identified and functionally characterized as being involved in malate transport at the guard cell tonoplast. ALMT6 is specifically expressed in guard cells, and patch-clamp analyses of vacuoles have shown that this channel mediates the transport of malate and fumarate, but not of NO_3^- or Cl^-, in a Ca^{2+}- and pH-dependent manner (Meyer et al., 2011). The physiological role of ALMT6 *in planta* is still unclear since no defect in stomatal response to different stimuli could be observed in *atlm6* loss-of-function mutant plants (Meyer et al., 2011). ALMT9 has been shown to mediate malate-activated Cl^- currents and promotes chloride fluxes into the guard cell vacuole (De Angeli, Zhang, Meyer, & Martinoia, 2013). *atml9* loss-of-function mutant plants exhibit reduced stomatal aperture and slower opening kinetics. These defects result in reduced transpiration and enhanced drought tolerance of *atml9* as compared to wild type (De Angeli et al., 2013). More recently, ALMT4 was described to mediate an efflux of malate from the guard cell vacuole (Eisenach et al., 2017). Genetic and biochemical studies indicated that ALMT4 is phosphorylated *in vitro* by mitogen-activated protein kinases, MPK4 and MPK6, but not by OST1 as previously reported for QUAC1 (Eisenach et al., 2017; Imes et al., 2013). A single point mutation of Ser382, a putative target site for ALMT4 phosphorylation, affected current activity. In addition, a *almt4* loss-of-function mutant was impaired in stomatal closure in response to ABA but displayed normal stomatal opening responses (Eisenach et al., 2017). Altogether, these results suggest that ALMT4 is responsible for malate efflux during stomatal closure as part of a Ca^{2+}-independent ABA-specific signalling pathway. Further research will be necessary to increase our understanding of the functions and redundancy of ALMT family members in the overall transport machinery in guard cells.

The other family of anion transport systems identified in guard cell vacuole is the CLC family. Of the seven members of Arabidopsis CLC family, two of them, AtCLCa and AtCLCc, have been shown to be involved in guard cell physiology. AtCLCa was the first NO_3^- transport system identified in the tonoplast that acts as 2 anions/1H$^+$ exchanger driving the uptake of anions into vacuoles (De Angeli et al., 2006). This protein is highly expressed in guard cells, where it plays a dual role (Wege et al., 2014). Firstly, it is involved in stomatal opening in response to light indicating that it contributes to anion accumulation in the vacuole. Unexpectedly, AtCLCa is also required for ABA-induced stomata closure suggesting a second role in anion release from the vacuole. Interestingly, ABA-dependent protein kinase OST1 phosphorylates the N-terminal domain of AtCLCa, enhancing its anion release activity (Wege et al., 2014) and mutation of the corresponding phosphorylation site impacts stomata closure. Thus, phosphorylation of AtCLCa would enhance both the release of anions from vacuoles and their acidification during ABA-induced stomata closure (Wege et al., 2014). AtCLCc, the other *Arabidopsis* CLC member expressed in guard cell, was described as the first putative vacuolar anion transporter involved in guard cell chloride homeostasis (Jossier et al., 2010). Based on amino acid sequence, AtCLCc is predicted to be an anion/H$^+$ exchanger with a preference for Cl$^-$. Unfortunately, no data are currently available on the transport properties on this channel. Nonetheless, *atclc-c* loss-of-function mutant plants were impaired in light-induced stomatal opening and stomatal closure in response to ABA and showed a dramatic decrease in guard cell chloride content. These defects suggest an important coordination between anion transport at the plasma and vacuolar membrane during stomatal movements (Jossier et al., 2010). Since the efflux of anion from vacuole into the cytosol can be passive and therefore mediated by ion channels, the roles of AtCLCa and AtCLCc in stomatal closure are unexpected. However, it cannot be excluded that the lack of these transporters can affect vacuolar pH and/or vacuolar membrane potential, thereby modifying other transport system required for stomatal closure.

5. Ca^{2+} TRANSPORTERS

5.1 Plasma Membrane Ca^{2+} Transporters

Electrophysiological analyses have revealed non-selective Ca^{2+}-permeable channel activity in the plasma membrane of guard cells (Hamilton, Hills, Kohler, & Blatt, 2000; Hua et al., 2012; Köhler & Blatt, 2002; Miao et al., 2006; Mori et al., 2006; Murata, Pei, Mori, & Schroeder, 2001; Pei et al., 2000; Schroeder & Hagiwara, 1990; Suh et al., 2007; Vahisalu et al., 2008). These channels are activated by hyperpolarization and can be stimulated with H_2O_2 and ABA (Mori et al., 2006; Pei et al., 2000).

They are thought to play a crucial role in the regulation of cytoplasmic Ca^{2+} concentration, a key element involved in guard cell signalling (McAinsh, Gray, Hetherington, Leckie, & Ng, 2000). At the molecular level, two gene families have been proposed to encode Ca^{2+}-permeable channels: the cyclic nucleotide gated channel (CNGC) family and the glutamate receptor family (GLR). To date, Arabidopsis CNGC5 and CNGC6 have been shown to be required for cGMP-activated non-selective Ca^{2+}-permeable cation channel activity in the guard cell plasma membrane (Wang et al., 2013). Functional characterization of corresponding loss-of-function mutants showed that CNGC5 and CNGC6 encode a new type of non-selective cation channel in the plasma membrane of Arabidopsis guard cells (Wang et al., 2013). Yet, ABA-activated Ca^{2+}-permeable channels were still present in the guard cell plasma membrane of the mutant plants, suggesting an overlapping role of additional CNGCs (Wang et al., 2013).

At least three GLR isoforms, GLR3.4, GLR3.1 and GLR3.5, are expressed in guard cells (Kong et al., 2016; Meyerhoff et al., 2005). GLR3.1 and GLR3.5 are able to form Ca^{2+} channels activated by L-methionine at physiological concentrations (Kong et al., 2016). In addition, *glr3.1/3.5* loss-of-function mutant plants show a lower cytosolic Ca^{2+} level and a defect in Ca^{2+}-induced stomatal closure. The GLR3.1/3.5 Ca^{2+} channels are distinct from previously characterized ROS-activated Ca^{2+} channels (Kong et al., 2016). Thus, additional Ca^{2+}-permeable channels at the guard cell plasma membrane remain to be molecularly identified.

5.2 Vacuolar Membrane Ca^{2+} Transporters

As previously mentioned for the plasma membrane, molecular knowledge of the Ca^{2+} transporters in the guard cell vacuolar membrane is very poor. Two Ca^{2+} uptake transport systems have been described in plant vacuoles, the H^+/Ca^{2+} antiporter (CAX) family and the Ca^{2+}-ATPase (ACA) family. Among the six members of the CAX family, CAX1, CAX2, CAX3 and CAX5 are expressed in guard cells (Cho et al., 2012; Leonhardt et al., 2004). CAX1 and CAX3 are tonoplast-localized and play an important role in controlling tissue Ca^{2+} homeostasis (Conn et al., 2011; Hirschi, Zhen, Cunningham, Rea, & Fink, 1996; Pittman & Hirschi, 2016; Shigaki, Cheng, Pittman, & Hirschi, 2001). Genetic analyses of *cax1*, *cax3*, and *cax1 cax3* loss-of-function mutants suggested that a putative CAX1-CAX3 complex may influence guard cell closure and apoplastic pH (Cho et al., 2012). More recently, it was shown that these mutants are also non-responsive to increased Ca^{2+} concentrations, indicating that each isoform is required in guard cell response to apoplastic Ca^{2+} (Hocking et al., 2017). Altogether these observations suggest that the CAX1-CAX3 complex may function downstream of apoplastic Ca^{2+} signalling and is required for maintenance of normal stomatal aperture in response to changes in external Ca^{2+} concentration (Hocking et al., 2017). Nevertheless, further work is required to fully understand the role of CAXs in guard cell signalling. The role of ACA family members in guard cell Ca^{2+} signalling is also unclear.

Concerning vacuolar release of Ca^{2+}, the SV channel TCP1, which mediates K^+ efflux through the tonoplast, is also permeable to Ca^{2+} and activated by cations. As previously discussed, TPC1 could be involved in intracellular Ca^{2+} signalling (Hedrich & Marten, 2011).

6. SUMMARY

Guard cells provide an attractive single-cell type model system for the study of ion channels and transporters and their regulation. Guard cell ion channels and transporters mediate transmembrane ion fluxes, which generate cell volume changes. Thus, regulation of ion channels and transporters is essential to adjust stomatal movement in response to environmental stimuli including biotic and abiotic stresses. Over the last few decades, the development and combination of multidisciplinary approaches including molecular and cell biology, genetics, electrophysiology, biochemistry and whole plant physiology have allowed the molecular identification of an important array of ions channels and transporters and elucidation of their function in guard cell signalling pathways. These pathways involve a number of ubiquitous signalling molecules such as Ca^{2+} and ROS, regulatory proteins such as kinases, phosphatases and 14-3-3 proteins as well as hormones including ABA and auxin. Nethertheless, there are still significant gaps in our understanding of the molecular mechanisms that connect signalling pathways to ion channels and transporter activity and abundance. Another important challenge is to identify genes and proteins responsible for orphan electrophysiological activities. Integration of our knowledge on the function of guard cell ion channels and transporters, on their regulation and integration into guard cell metabolism networks remains a challenge for the future and may help us improve plant water use and productivity.

ACKNOWLEDGEMENT

We would like to apologize to all investigators whose research could not be cited appropriately in this chapter.

REFERENCES

Acharya, B. R., Jeon, B. W., Zhang, W., & Assmann, S. M. (2013). Open stomata 1 (OST1) is limiting in abscisic acid responses of Arabidopsis guard cells. *The New Phytologist, 200*, 1049–1063. https://doi.org/10.1111/nph.12469.

Ache, P., Becker, D., Ivashikina, N., Dietrich, P., Roelfsema, M. R., & Hedrich, R. (2000). GORK, a delayed outward rectifier expressed in guard cells of Arabidopsis thaliana, is a K (+)-selective, K(+)-sensing ion channel. *FEBS Letters, 486*, 93–98.

Allen, G. J., Chu, S. P., Schumacher, K., Shimazaki, C. T., Vafeados, D., Kemper, A., et al. (2000). Alteration of stimulus-specific guard cell calcium oscillations and stomatal closing in Arabidopsis det3 mutant. *Science, 289*, 2338–2342.

Allen, G. J., Kuchitsu, K., Chu, S. P., Murata, Y., & Schroeder, J. I. (1999). Arabidopsis abi1-1 and abi2-1 phosphatase mutations reduce abscisic acid-induced cytoplasmic calcium rises in guard cells. *Plant Cell, 11*, 1785–1798.

Allen, G. J., & Sanders, D. (1996). Control of ionic currents in guard cell vacuoles by cytosolic and luminal calcium. *The Plant Journal, 10*, 1055–1069.

Andrés, Z., Pérez-Hormaeche, J., Leidi, E. O., Schlücking, K., Steinhorst, L., McLachlan, D. H., et al. (2014). Control of vacuolar dynamics and regulation of stomatal aperture by tonoplast potassium uptake. *Proceedings of the National Academy of Sciences of the United States of America, 111*, E1806–E1814. https://doi.org/10.1073/pnas.1320421111.

Armstrong, F., Leung, J., Grabov, A., Brearley, J., Giraudat, J., & Blatt, M. R. (1995). Sensitivity to abscisic acid of guard-cell K+ channels is suppressed by abi1-1, a mutant Arabidopsis gene encoding a putative protein phosphatase. *Proceedings of the National Academy of Sciences of the United States of America, 92*, 9520–9524.

Arnaud, D., & Hwang, I. (2015). A sophisticated network of signaling pathways regulates stomatal defenses to bacterial pathogens. *Molecular Plant, 8*, 566–581. https://doi.org/10.1016/j.molp.2014.10.012.

Assmann, S. M. (2003). OPEN STOMATA1 opens the door to ABA signaling in Arabidopsis guard cells. *Trends in Plant Science, 8*, 151–153. https://doi.org/10.1016/S1360-1385(03)00052-9.

Assmann, S. M., & Schwartz, A. (1992). Synergistic effect of light and fusicoccin on stomatal opening: Epidermal peel and patch clamp experiments. *Plant Physiology, 98*, 1349–1355.

Assmann, S. M., & Shimazaki, C. T. (1999). The multisensory guard cell. Stomatal responses to blue light and abscisic acid. *Plant Physiology, 119*, 809–816.

Assmann, S. M., Simoncini, L., & Schroeder, J. I. (1985). Blue light activates electrogenic ion pumping in guard cell protoplasts of Vicia faba. *Nature, 318*, 285–287. https://doi.org/10.1038/318285a0.

Baekgaard, L., Fuglsang, A. T., & Palmgren, M. G. (2005). Regulation of plant plasma membrane H+- and Ca2+-ATPases by terminal domains. *Journal of Bioenergetics and Biomembranes, 37*, 369–374. https://doi.org/10.1007/s10863-005-9473-0.

Baizabal-Aguirre, V. M., Clemens, S., Uozumi, N., & Schroeder, J. I. (1999). Suppression of inward-rectifying K+ channels KAT1 and AKT2 by dominant negative point mutations in the KAT1 alpha-subunit. *The Journal of Membrane Biology, 167*, 119–125.

Bak, G., Lee, E.-J., Lee, Y., Kato, M., Segami, S., Sze, H., et al. (2013). Rapid structural changes and acidification of guard cell vacuoles during stomatal closure require phosphatidylinositol 3,5-bisphosphate. *Plant Cell, 25*, 2202–2216. https://doi.org/10.1105/tpc.113.110411.

Barragán, V., Leidi, E. O., Andrés, Z., Rubio, L., De Luca, A., Fernández, J. A., et al. (2012). Ion exchangers NHX1 and NHX2 mediate active potassium uptake into vacuoles to regulate cell turgor and stomatal function in Arabidopsis. *Plant Cell, 24*, 1127–1142. https://doi.org/10.1105/tpc.111.095273.

Batistic, O., Sorek, N., Schültke, S., Yalovsky, S., & Kudla, J. (2008). Dual fatty acyl modification determines the localization and plasma membrane targeting of CBL/CIPK Ca2+ signaling complexes in Arabidopsis. *Plant Cell*, *20*, 1346–1362. https://doi.org/10.1105/tpc.108.058123.

Bauer, H., Ache, P., Lautner, S., Fromm, J., Hartung, W., Al-Rasheid, K. A. S., et al. (2013). The stomatal response to reduced relative humidity requires guard cell-autonomous ABA synthesis. *Current Biology*, *23*, 53–57. https://doi.org/10.1016/j.cub.2012.11.022.

Becker, D., Hoth, S., Ache, P., Wenkel, S., Roelfsema, M. R. G., Meyerhoff, O., et al. (2003). Regulation of the ABA-sensitive Arabidopsis potassium channel gene GORK in response to water stress. *FEBS Letters*, *554*, 119–126.

Berkowitz, G., Zhang, X., Mercie, R., Leng, Q., & Lawton, M. (2000). Co-expression of calcium-dependent protein kinase with the inward rectified guard cell K+ channel KAT1 alters current parameters in Xenopus laevis oocytes. *Plant & Cell Physiology*, *41*, 785–790.

Beyhl, D., Hörtensteiner, S., Martinoia, E., Farmer, E. E., Fromm, J., Marten, I., et al. (2009). The fou2 mutation in the major vacuolar cation channel TPC1 confers tolerance to inhibitory luminal calcium. *The Plant Journal*, *58*, 715–723. https://doi.org/10.1111/j.1365-313X.2009.03820.x.

Blatt, M. R. (1987). Electrical characteristics of stomatal guard cells: The ionic basis of the membrane potential and the consequence of potassium chlorides leakage from microelectrodes. *Planta*, *170*, 272–287. https://doi.org/10.1007/BF00397898.

Blatt, M. R. (1992). K+ channels of stomatal guard cells. Characteristics of the inward rectifier and its control by pH. *The Journal of General Physiology*, *99*, 615–644.

Blatt, M. R., & Clint, G. M. (1989). Mechanisms of fusicoccin action: Kinetic modification and inactivation of K(+) channels in guard cells. *Planta*, *178*, 509–523. https://doi.org/10.1007/BF00963821.

Brandt, B., Brodsky, D. E., Xue, S., Negi, J., Iba, K., Kangasjärvi, J., et al. (2012). Reconstitution of abscisic acid activation of SLAC1 anion channel by CPK6 and OST1 kinases and branched ABI1 PP2C phosphatase action. *Proceedings of the National Academy of Sciences of the United States of America*, *109*, 10593–10598. https://doi.org/10.1073/pnas.1116590109.

Brandt, B., Munemasa, S., Wang, C., Nguyen, D., Yong, T., Yang, P. G., et al. (2015). Calcium specificity signaling mechanisms in abscisic acid signal transduction in Arabidopsis guard cells. *eLife*, *4*. https://doi.org/10.7554/eLife.03599.

Brault, M., Amiar, Z., Pennarun, A.-M., Monestiez, M., Zhang, Z., Cornel, D., et al. (2004). Plasma membrane depolarization induced by abscisic acid in Arabidopsis suspension cells involves reduction of proton pumping in addition to anion channel activation, which are both Ca^{2+} dependent. *Plant Physiology*, *135*, 231–243. https://doi.org/10.1104/pp.103.039255.

Briggs, W. R., & Christie, J. M. (2002). Phototropins 1 and 2: Versatile plant blue-light receptors. *Trends in Plant Science*, 7, 204–210.

Brüx, A., Liu, T. Y., Krebs, M., Stierhof, Y.-D., Lohmann, J. U., Miersch, O., et al. (2008). Reduced V-ATPase activity in the trans-Golgi network causes oxylipin-dependent hypocotyl growth inhibition in Arabidopsis. *Plant Cell*, *20*, 1088–1100. https://doi.org/10.1105/tpc.108.058362.

Chen, Z.-H., Hills, A., Bätz, U., Amtmann, A., Lew, V. L., & Blatt, M. R. (2012). Systems dynamic modeling of the stomatal guard cell predicts emergent behaviors in transport, signaling, and volume control. *Plant Physiology*, *159*, 1235–1251. https://doi.org/10.1104/pp.112.197350.

Chen, Z.-H., Hills, A., Lim, C. K., & Blatt, M. R. (2010). Dynamic regulation of guard cell anion channels by cytosolic free Ca2+ concentration and protein phosphorylation. *The Plant Journal*, *61*, 816–825. https://doi.org/10.1111/j.1365-313X.2009.04108.x.

Chisholm, S. T., Coaker, G., Day, B., & Staskawicz, B. J. (2006). Host-microbe interactions: Shaping the evolution of the plant immune response. *Cell*, *124*, 803–814. https://doi.org/10.1016/j.cell.2006.02.008.

Cho, D., Villiers, F., Kroniewicz, L., Lee, S., Seo, Y. J., Hirschi, K. D., et al. (2012). Vacuolar CAX1 and CAX3 influence auxin transport in guard cells via regulation of apoplastic pH. *Plant Physiology*, *160*, 1293–1302. https://doi.org/10.1104/pp.112.201442.

Conn, S. J., Gilliham, M., Athman, A., Schreiber, A. W., Baumann, U., Moller, I., et al. (2011). Cell-specific vacuolar calcium storage mediated by CAX1 regulates apoplastic calcium concentration, gas exchange, and plant productivity in Arabidopsis. *Plant Cell*, *23*, 240–257. https://doi.org/10.1105/tpc.109.072769.

Corratgé-Faillie, C., Ronzier, E., Sanchez, F., Prado, K., Kim, J.-H., Lanciano, S., et al. (2017). The Arabidopsis guard cell outward potassium channel GORK is regulated by CPK33. *FEBS Letters*, *591*, 1982–1992. https://doi.org/10.1002/1873-3468.12687.

Cutler, S. R., Rodriguez, P. L., Finkelstein, R. R., & Abrams, S. R. (2010). Abscisic acid: Emergence of a core signaling network. *Annual Review of Plant Biology*, *61*, 651–679. https://doi.org/10.1146/annurev-arplant-042809-112122.

Darley, C. P., Skiera, L. A., Northrop, F. D., Sanders, D., & Davies, J. M. (1998). Tonoplast inorganic pyrophosphatase in Vicia faba guard cells. *Planta*, *206*, 272–277. https://doi.org/10.1007/s004250050400.

Davies, J. M., Hunt, I., & Sanders, D. (1994). Vacuolar H(+)-pumping ATPase variable transport coupling ratio controlled by pH. *Proceedings of the National Academy of Sciences of the United States of America*, *91*, 8547–8551.

De Angeli, A., Monachello, D., Ephritikhine, G., Frachisse, J. M., Thomine, S., Gambale, F., et al. (2006). The nitrate/proton antiporter AtCLCa mediates nitrate accumulation in plant vacuoles. *Nature*, *442*, 939–942. https://doi.org/10.1038/nature05013.

De Angeli, A., Zhang, J., Meyer, S., & Martinoia, E. (2013). AtALMT9 is a malate-activated vacuolar chloride channel required for stomatal opening in Arabidopsis. *Nature Communications*, *4*, 1804. https://doi.org/10.1038/ncomms2815.

Dreyer, I., & Blatt, M. R. (2009). What makes a gate? The ins and outs of Kv-like K+ channels in plants. *Trends in Plant Science*, *14*, 383–390. https://doi.org/10.1016/j.tplants.2009.04.001.

Duby, G., Hosy, E., Fizames, C., Alcon, C., Costa, A., Sentenac, H., et al. (2008). AtKC1, a conditionally targeted shaker-type subunit, regulates the activity of plant K+ channels. *The Plant Journal*, *53*, 115–123. https://doi.org/10.1111/j.1365-313X.2007.03324.x.

Edwards, D., Fanning, U., & Richardson, J. B. (1986). Stomata and sterome in early land plants. *Nature*, *323*, 438–440. https://doi.org/10.1038/323438a0.

Eisenach, C., Baetz, U., Huck, N. V., Zhang, J., De Angeli, A., Beckers, G. J. M., et al. (2017). ABA-induced stomatal closure involves ALMT4, a phosphorylation-dependent vacuolar anion channel of Arabidopsis. *Plant Cell*, *29*, 2552–2569. https://doi.org/10.1105/tpc.17.00452.

Eisenach, C., & De Angeli, A. (2017). Ion transport at the vacuole during stomatal movements. *Plant Physiology*, *174*, 520–530. https://doi.org/10.1104/pp.17.00130.

Eisenach, C., Papanatsiou, M., Hillert, E.-K., & Blatt, M. R. (2014). Clustering of the K+ channel GORK of Arabidopsis parallels its gating by extracellular K+. *The Plant Journal*, *78*, 203–214. https://doi.org/10.1111/tpj.12471.

Emi, T., Kinoshita, T., & Shimazaki, K. (2001). Specific binding of vf14-3-3a isoform to the plasma membrane H+-ATPase in response to blue light and fusicoccin in guard cells of broad bean. *Plant Physiology*, *125*, 1115–1125.

Felle, H., Brummer, B., Bertl, A., & Parish, R. W. (1986). Indole-3-acetic acid and fusicoccin cause cytosolic acidification of corn coleoptile cells. *Proceedings of the National Academy of Sciences of the United States of America*, *83*, 8992–8995.

Finkelstein, R. R., Gampala, S. S. L., & Rock, C. D. (2002). Abscisic acid signaling in seeds and seedlings. *Plant Cell, 14*(Suppl), S15–S45.

Finkelstein, R. R., & Somerville, C. R. (1990). Three classes of abscisic acid (ABA)-insensitive mutations of Arabidopsis define genes that control overlapping subsets of ABA responses. *Plant Physiology, 94*, 1172–1179.

Frey, A., Effroy, D., Lefebvre, V., Seo, M., Perreau, F., Berger, A., et al. (2012). Epoxycarotenoid cleavage by NCED5 fine-tunes ABA accumulation and affects seed dormancy and drought tolerance with other NCED family members: Functional analysis of the NCED5 gene. *The Plant Journal, 70*, 501–512. https://doi.org/10.1111/j.1365-313X.2011.04887.x.

Fuglsang, A. T., Guo, Y., Cuin, T. A., Qiu, Q., Song, C., Kristiansen, K. A., et al. (2007). Arabidopsis protein kinase PKS5 inhibits the plasma membrane H+-ATPase by preventing interaction with 14-3-3 protein. *Plant Cell, 19*, 1617–1634. https://doi.org/10.1105/tpc.105.035626.

Fuglsang, A. T., Kristensen, A., Cuin, T. A., Schulze, W. X., Persson, J., Thuesen, K. H., et al. (2014). Receptor kinase-mediated control of primary active proton pumping at the plasma membrane. *The Plant Journal, 80*, 951–964. https://doi.org/10.1111/tpj.12680.

Gao, X.-Q., Li, C.-G., Wei, P.-C., Zhang, X.-Y., Chen, J., & Wang, X.-C. (2005). The dynamic changes of tonoplasts in guard cells are important for stomatal movement in Vicia faba. *Plant Physiology, 139*, 1207–1216. https://doi.org/10.1104/pp.105.067520.

Garcia-Mata, C., Wang, J., Gajdanowicz, P., Gonzalez, W., Hills, A., Donald, N., et al. (2010). A minimal cysteine motif required to activate the SKOR K+ channel of Arabidopsis by the reactive oxygen species H2O2. *The Journal of Biological Chemistry, 285*, 29286–29294. https://doi.org/10.1074/jbc.M110.141176.

Gaxiola, R. A., Fink, G. R., & Hirschi, K. D. (2002). Genetic manipulation of vacuolar proton pumps and transporters. *Plant Physiology, 129*, 967–973. https://doi.org/10.1104/pp.020009.

Gaxiola, R. A., Palmgren, M. G., & Schumacher, K. (2007). Plant proton pumps. *FEBS Letters, 581*, 2204–2214. https://doi.org/10.1016/j.febslet.2007.03.050.

Geiger, D., Maierhofer, T., Al-Rasheid, K. A. S., Scherzer, S., Mumm, P., Liese, A., et al. (2011). Stomatal closure by fast abscisic acid signaling is mediated by the guard cell anion channel SLAH3 and the receptor RCAR1. *Science Signaling, 4*, ra32. https://doi.org/10.1126/scisignal.2001346.

Geiger, D., Scherzer, S., Mumm, P., Marten, I., Ache, P., Matschi, S., et al. (2010). Guard cell anion channel SLAC1 is regulated by CDPK protein kinases with distinct Ca2+ affinities. *Proceedings of the National Academy of Sciences of the United States of America, 107*, 8023–8028. https://doi.org/10.1073/pnas.0912030107.

Geiger, D., Scherzer, S., Mumm, P., Stange, A., Marten, I., Bauer, H., et al. (2009). Activity of guard cell anion channel SLAC1 is controlled by drought-stress signaling kinase-phosphatase pair. *Proceedings of the National Academy of Sciences of the United States of America, 106*, 21425–21430. https://doi.org/10.1073/pnas.0912021106.

Gobert, A., Isayenkov, S., Voelker, C., Czempinski, K., & Maathuis, F. J. M. (2007). The two-pore channel TPK1 gene encodes the vacuolar K+ conductance and plays a role in K+ homeostasis. *Proceedings of the National Academy of Sciences of the United States of America, 104*, 10726–10731. https://doi.org/10.1073/pnas.0702595104.

Gonugunta, V. K., Srivastava, N., Puli, M. R., & Raghavendra, A. S. (2008). Nitric oxide production occurs after cytosolic alkalinization during stomatal closure induced by abscisic acid. *Plant, Cell & Environment, 31*, 1717–1724. https://doi.org/10.1111/j.1365-3040.2008.01872.x.

González, W., Riedelsberger, J., Morales-Navarro, S. E., Caballero, J., Alzate-Morales, J. H., González-Nilo, F. D., et al. (2012). The pH sensor of the plant K+-uptake channel KAT1 is built from a sensory cloud rather than from single key amino acids. *The Biochemical Journal, 442*, 57–63. https://doi.org/10.1042/BJ20111498.

Guo, F.-Q., Young, J., & Crawford, N. M. (2003). The nitrate transporter AtNRT1.1 (CHL1) functions in stomatal opening and contributes to drought susceptibility in Arabidopsis. *Plant Cell, 15*, 107–117.

Hafke, J. B., Hafke, Y., Smith, J. A. C., Lüttge, U., & Thiel, G. (2003). Vacuolar malate uptake is mediated by an anion-selective inward rectifier. *The Plant Journal, 35*, 116–128.

Hamilton, D. W., Hills, A., Kohler, B., & Blatt, M. R. (2000). Ca2+ channels at the plasma membrane of stomatal guard cells are activated by hyperpolarization and abscisic acid. *Proceedings of the National Academy of Sciences of the United States of America, 97*, 4967–4972. https://doi.org/10.1073/pnas.080068897.

Hashimoto, M., Negi, J., Young, J., Israelsson, M., Schroeder, J. I., & Iba, K. (2006). Arabidopsis HT1 kinase controls stomatal movements in response to CO2. *Nature Cell Biology, 8*, 391–397. https://doi.org/10.1038/ncb1387.

Hashimoto-Sugimoto, M., Higaki, T., Yaeno, T., Nagami, A., Irie, M., Fujimi, M., et al. (2013). A Munc13-like protein in Arabidopsis mediates H+-ATPase translocation that is essential for stomatal responses. *Nature Communications, 4*, 2215. https://doi.org/10.1038/ncomms3215.

Hayashi, M., & Kinoshita, T. (2011). Crosstalk between blue-light- and ABA-signaling pathways in stomatal guard cells. *Plant Signaling & Behavior, 6*, 1662–1664. https://doi.org/10.4161/psb.6.11.17800.

Hayashi, M., Inoue, S.-I., Takahashi, K., & Kinoshita, T. (2011). Immunohistochemical detection of blue light-induced phosphorylation of the plasma membrane H+-ATPase in stomatal guard cells. *Plant & Cell Physiology, 52*, 1238–1248. https://doi.org/10.1093/pcp/pcr072.

Hedrich, R. (2012). Ion channels in plants. *Physiological Reviews, 92*, 1777–1811. https://doi.org/10.1152/physrev.00038.2011.

Hedrich, R., Busch, H., & Raschke, K. (1990). Ca2+ and nucleotide dependent regulation of voltage dependent anion channels in the plasma membrane of guard cells. *The EMBO Journal, 9*, 3889–3892.

Hedrich, R., & Marten, I. (2011). TPC1-SV channels gain shape. *Molecular Plant, 4*, 428–441. https://doi.org/10.1093/mp/ssr017.

Hedrich, R., & Neher, E. (1987). Cytoplasmic calcium regulates voltage-dependent ion channels in plant vacuoles. *Nature, 329*, 833–836. https://doi.org/10.1038/329833a0.

Hetherington, A. M., & Brownlee, C. (2004). The generation of Ca(2+) signals in plants. *Annual Review of Plant Biology, 55*, 401–427. https://doi.org/10.1146/annurev.arplant.55.031903.141624.

Hills, A., Chen, Z.-H., Amtmann, A., Blatt, M. R., & Lew, V. L. (2012). OnGuard, a computational platform for quantitative kinetic modeling of guard cell physiology. *Plant Physiology, 159*, 1026–1042. https://doi.org/10.1104/pp.112.197244.

Hirschi, K. D., Zhen, R. G., Cunningham, K. W., Rea, P. A., & Fink, G. R. (1996). CAX1, an H+/Ca2+ antiporter from Arabidopsis. *Proceedings of the National Academy of Sciences of the United States of America, 93*, 8782–8786.

Hocking, B., Conn, S. J., Manohar, M., Xu, B., Athman, A., Stancombe, M. A., et al. (2017). Heterodimerization of Arabidopsis calcium/proton exchangers contributes to regulation of guard cell dynamics and plant defense responses. *Journal of Experimental Botany, 68*, 4171–4183. https://doi.org/10.1093/jxb/erx209.

Hõrak, H., Sierla, M., Tõldsepp, K., Wang, C., Wang, Y.-S., Nuhkat, M., et al. (2016). A dominant mutation in the HT1 kinase uncovers roles of MAP kinases and GHR1 in CO2-induced stomatal closure. *The Plant Cell, 28*, 2493–2509. https://doi.org/10.1105/tpc.16.00131.

Hosy, E., Vavasseur, A., Mouline, K., Dreyer, I., Gaymard, F., Porée, F., et al. (2003). The Arabidopsis outward K+ channel GORK is involved in regulation of stomatal movements and plant transpiration. *Proceedings of the National Academy of Sciences of the United States of America, 100*, 5549–5554. https://doi.org/10.1073/pnas.0733970100.

Hoth, S., Dreyer, I., Dietrich, P., Becker, D., Müller-Röber, B., & Hedrich, R. (1997). Molecular basis of plant-specific acid activation of K+ uptake channels. *Proceedings of the National Academy of Sciences of the United States of America, 94*, 4806–4810.

Hoth, S., Geiger, D., Becker, D., & Hedrich, R. (2001). The pore of plant K(+) channels is involved in voltage and pH sensing: Domain-swapping between different K(+) channel alpha-subunits. *Plant Cell, 13*, 943–952.

Hoth, S., & Hedrich, R. (1999). Distinct molecular bases for pH sensitivity of the guard cell K+ channels KST1 and KAT1. *The Journal of Biological Chemistry, 274*, 11599–11603.

Hu, H., Boisson-Dernier, A., Israelsson-Nordström, M., Böhmer, M., Xue, S., Ries, A., et al. (2010). Carbonic anhydrases are upstream regulators of CO2-controlled stomatal movements in guard cells. *Nature Cell Biology, 12*, 87–93. sup pp 1–18. https://doi.org/10.1038/ncb2009.

Hua, D., Wang, C., He, J., Liao, H., Duan, Y., Zhu, Z., et al. (2012). A plasma membrane receptor kinase, GHR1, mediates abscisic acid- and hydrogen peroxide-regulated stomatal movement in Arabidopsis. *Plant Cell, 24*, 2546–2561. https://doi.org/10.1105/tpc.112.100107.

Hubbard, K. E., Nishimura, N., Hitomi, K., Getzoff, E. D., & Schroeder, J. I. (2010). Early abscisic acid signal transduction mechanisms: Newly discovered components and newly emerging questions. *Genes & Development, 24*, 1695–1708. https://doi.org/10.1101/gad.1953910.

Hurst, A. C., Meckel, T., Tayefeh, S., Thiel, G., & Homann, U. (2004). Trafficking of the plant potassium inward rectifier KAT1 in guard cell protoplasts of Vicia faba. *The Plant Journal, 37*, 391–397.

Imes, D., Mumm, P., Böhm, J., Al-Rasheid, K. A. S., Marten, I., Geiger, D., et al. (2013). Open stomata 1 (OST1) kinase controls R-type anion channel QUAC1 in Arabidopsis guard cells. *The Plant Journal, 74*, 372–382. https://doi.org/10.1111/tpj.12133.

Irving, H. R., Gehring, C. A., & Parish, R. W. (1992). Changes in cytosolic pH and calcium of guard cells precede stomatal movements. *Proceedings of the National Academy of Sciences of the United States of America, 89*, 1790–1794.

Islam, M. M., Munemasa, S., Hossain, M. A., Nakamura, Y., Mori, I. C., & Murata, Y. (2010). Roles of AtTPC1, vacuolar two pore channel 1, in Arabidopsis stomatal closure. *Plant & Cell Physiology, 51*, 302–311. https://doi.org/10.1093/pcp/pcq001.

Isner, J. C., Begum, A., Nuehse, T., Hetherington, A. M., & Maathuis, F. J. M. (2018). KIN7 kinase regulates the vacuolar TPK1 K+ channel during stomatal closure. *Current Biology, 28*. 466–472.e4. https://doi.org/10.1016/j.cub.2017.12.046.

Ivashikina, N., Deeken, R., Fischer, S., Ache, P., & Hedrich, R. (2005). AKT2/3 subunits render guard cell K^+ channels Ca^{2+} sensitive. *The Journal of General Physiology, 125*, 483–492. https://doi.org/10.1085/jgp.200409211.

Jeanguenin, L., Alcon, C., Duby, G., Boeglin, M., Chérel, I., Gaillard, I., et al. (2011). AtKC1 is a general modulator of Arabidopsis inward shaker channel activity. *The Plant Journal, 67*, 570–582. https://doi.org/10.1111/j.1365-313X.2011.04617.x.

Jones, J. D. G., & Dangl, J. L. (2006). The plant immune system. *Nature, 444*, 323–329. https://doi.org/10.1038/nature05286.

Jossier, M., Kroniewicz, L., Dalmas, F., Le Thiec, D., Ephritikhine, G., Thomine, S., et al. (2010). The Arabidopsis vacuolar anion transporter, AtCLCc, is involved in the regulation of stomatal movements and contributes to salt tolerance. *The Plant Journal, 64*, 563–576. https://doi.org/10.1111/j.1365-313X.2010.04352.x.

Kang, J., Choi, H., Im, M., & Kim, S. Y. (2002). Arabidopsis basic leucine zipper proteins that mediate stress-responsive abscisic acid signaling. *Plant Cell, 14*, 343–357.

Keller, B. U., Hedrich, R., & Raschke, K. (1989). Voltage-dependent anion channels in the plasma membrane of guard cells. *Nature, 341*, 450–453. https://doi.org/10.1038/341450a0.

Kerkeb, L., Venema, K., Donaire, J. P., & Rodríguez-Rosales, M. P. (2002). Enhanced H+/ATP coupling ratio of H+-ATPase and increased 14-3-3 protein content in plasma membrane of tomato cells upon osmotic shock. *Physiologia Plantarum*, *116*, 37–41.

Kim, T.-H., Böhmer, M., Hu, H., Nishimura, N., & Schroeder, J. I. (2010). Guard cell signal transduction network: Advances in understanding abscisic acid, CO2, and Ca2+ signaling. *Annual Review of Plant Biology*, *61*, 561–591. https://doi.org/10.1146/annurev-arplant-042809-112226.

Kinoshita, T., Doi, M., Suetsugu, N., Kagawa, T., Wada, M., & Shimazaki, K. (2001). Phot1 and phot2 mediate blue light regulation of stomatal opening. *Nature*, *414*, 656–660. https://doi.org/10.1038/414656a.

Kinoshita, T., Emi, T., Tominaga, M., Sakamoto, K., Shigenaga, A., Doi, M., et al. (2003). Blue-light- and phosphorylation-dependent binding of a 14-3-3 protein to phototropins in stomatal guard cells of broad bean. *Plant Physiology*, *133*, 1453–1463. https://doi.org/10.1104/pp.103.029629.

Kinoshita, T., & Shimazaki, K. i. (1999). Blue light activates the plasma membrane H(+)-ATPase by phosphorylation of the C-terminus in stomatal guard cells. *The EMBO Journal*, *18*, 5548–5558. https://doi.org/10.1093/emboj/18.20.5548.

Kinoshita, T., & Shimazaki, K. (2001). Analysis of the phosphorylation level in guard-cell plasma membrane H+-ATPase in response to fusicoccin. *Plant & Cell Physiology*, *42*, 424–432.

Kinoshita, T., & Shimazaki, K. (2002). Biochemical evidence for the requirement of 14-3-3 protein binding in activation of the guard-cell plasma membrane H+-ATPase by blue light. *Plant & Cell Physiology*, *43*, 1359–1365.

Kluge, C., Lahr, J., Hanitzsch, M., Bolte, S., Golldack, D., & Dietz, K.-J. (2003). New insight into the structure and regulation of the plant vacuolar H+-ATPase. *Journal of Bioenergetics and Biomembranes*, *35*, 377–388.

Köhler, B., & Blatt, M. R. (2002). Protein phosphorylation activates the guard cell Ca2+ channel and is a prerequisite for gating by abscisic acid. *The Plant Journal*, *32*, 185–194.

Kollist, H., Nuhkat, M., & Roelfsema, M. R. G. (2014). Closing gaps: Linking elements that control stomatal movement. *The New Phytologist*, *203*, 44–62. https://doi.org/10.1111/nph.12832.

Kong, D., Hu, H.-C., Okuma, E., Lee, Y., Lee, H. S., Munemasa, S., et al. (2016). L-met activates arabidopsis GLR Ca2+ channels upstream of ROS production and regulates stomatal movement. *Cell Reports*, *17*, 2553–2561. https://doi.org/10.1016/j.celrep.2016.11.015.

Koornneef, M., Reuling, G., & Karssen, C. M. (1984). The isolation and characterization of abscisic acid-insensitive mutants of Arabidopsis thaliana. *Physiologia Plantarum*, *61*, 377–383. https://doi.org/10.1111/j.1399-3054.1984.tb06343.x.

Kovermann, P., Meyer, S., Hörtensteiner, S., Picco, C., Scholz-Starke, J., Ravera, S., et al. (2007). The Arabidopsis vacuolar malate channel is a member of the ALMT family. *The Plant Journal*, *52*, 1169–1180. https://doi.org/10.1111/j.1365-313X.2007.03367.x.

Krebs, M., Beyhl, D., Görlich, E., Al-Rasheid, K. A. S., Marten, I., Stierhof, Y.-D., et al. (2010). Arabidopsis V-ATPase activity at the tonoplast is required for efficient nutrient storage but not for sodium accumulation. *Proceedings of the National Academy of Sciences of the United States of America*, *107*, 3251–3256. https://doi.org/10.1073/pnas.0913035107.

Kwak, J. M., Murata, Y., Baizabal-Aguirre, V. M., Merrill, J., Wang, M., Kemper, A., et al. (2001). Dominant negative guard cell K+ channel mutants reduce inward-rectifying K+ currents and light-induced stomatal opening in arabidopsis. *Plant Physiology*, *127*, 473–485.

Lacombe, B., Pilot, G., Gaymard, F., Sentenac, H., & Thibaud, J. B. (2000). pH control of the plant outwardly-rectifying potassium channel SKOR. *FEBS Letters*, *466*, 351–354.

Latz, A., Becker, D., Hekman, M., Müller, T., Beyhl, D., Marten, I., et al. (2007). TPK1, a Ca(2+)-regulated Arabidopsis vacuole two-pore K(+) channel is activated by 14-3-3 proteins. *The Plant Journal, 52*, 449–459. https://doi.org/10.1111/j.1365-313X.2007.03255.x.

Latz, A., Ivashikina, N., Fischer, S., Ache, P., Sano, T., Becker, D., et al. (2007). In planta AKT2 subunits constitute a pH- and Ca2+-sensitive inward rectifying K+ channel. *Planta, 225*, 1179–1191. https://doi.org/10.1007/s00425-006-0428-4.

Latz, A., Mehlmer, N., Zapf, S., Mueller, T. D., Wurzinger, B., Pfister, B., et al. (2013). Salt stress triggers phosphorylation of the Arabidopsis vacuolar K+ channel TPK1 by calcium-dependent protein kinases (CDPKs). *Molecular Plant, 6*, 1274–1289. https://doi.org/10.1093/mp/sss158.

Lawson, T., & Blatt, M. R. (2014). Stomatal size, speed, and responsiveness impact on photosynthesis and water use efficiency. *Plant Physiology, 164*, 1556–1570. https://doi.org/10.1104/pp.114.237107.

Lebaudy, A., Hosy, E., Simonneau, T., Sentenac, H., Thibaud, J.-B., & Dreyer, I. (2008). Heteromeric K+ channels in plants. *The Plant Journal, 54*, 1076–1082. https://doi.org/10.1111/j.1365-313X.2008.03479.x.

Lebaudy, A., Pascaud, F., Véry, A. A., Alcon, C., Dreyer, I., Thibaud, J. B., et al. (2010). Preferential KAT1-KAT2 heteromerization determines inward K^+ current properties in Arabidopsis guard cells. *The Journal of Biological Chemistry, 285*, 6265–6274. https://doi.org/10.1074/jbc.M109.068445.

Lebaudy, A., Vavasseur, A., Hosy, E., Dreyer, I., Leonhardt, N., Thibaud, J.-B., et al. (2008). Plant adaptation to fluctuating environment and biomass production are strongly dependent on guard cell potassium channels. *Proceedings of the National Academy of Sciences of the United States of America, 105*, 5271–5276. https://doi.org/10.1073/pnas.0709732105.

Lebaudy, A., Véry, A.-A., & Sentenac, H. (2007). K+ channel activity in plants: Genes, regulations and functions. *FEBS Letters, 581*, 2357–2366. https://doi.org/10.1016/j.febslet.2007.03.058.

Lee, S. C., Lan, W., Buchanan, B. B., & Luan, S. (2009). A protein kinase-phosphatase pair interacts with an ion channel to regulate ABA signaling in plant guard cells. *Proceedings of the National Academy of Sciences of the United States of America, 106*, 21419–21424. https://doi.org/10.1073/pnas.0910601106.

Lee, S. C., & Luan, S. (2012). ABA signal transduction at the crossroad of biotic and abiotic stress responses. *Plant, Cell & Environment, 35*, 53–60. https://doi.org/10.1111/j.1365-3040.2011.02426.x.

Lefoulon, C., Boeglin, M., Moreau, B., Véry, A.-A., Szponarski, W., Dauzat, M., et al. (2016). The Arabidopsis AtPP2CA protein phosphatase inhibits the GORK K+ efflux channel and exerts a dominant suppressive effect on phosphomimetic-activating mutations. *The Journal of Biological Chemistry, 291*, 6521–6533. https://doi.org/10.1074/jbc.M115.711309.

Leonhardt, N., Kwak, J. M., Robert, N., Waner, D., Leonhardt, G., & Schroeder, J. I. (2004). Microarray expression analyses of Arabidopsis guard cells and isolation of a recessive abscisic acid hypersensitive protein phosphatase 2C mutant. *Plant Cell, 16*, 596–615. https://doi.org/10.1105/tpc.019000.

Leung, J., Bouvier-Durand, M., Morris, P. C., Guerrier, D., Chefdor, F., & Giraudat, J. (1994). Arabidopsis ABA response gene ABI1: Features of a calcium-modulated protein phosphatase. *Science, 264*, 1448–1452.

Levchenko, V., Konrad, K. R., Dietrich, P., Roelfsema, M. R. G., & Hedrich, R. (2005). Cytosolic abscisic acid activates guard cell anion channels without preceding Ca2+ signals. *Proceedings of the National Academy of Sciences of the United States of America, 102*, 4203–4208. https://doi.org/10.1073/pnas.0500146102.

Li, J., & Assmann, S. M. (1996). An abscisic acid-activated and calcium-independent protein kinase from guard cells of fava bean. *Plant Cell, 8*, 2359–2368. https://doi.org/10.1105/tpc.8.12.2359.

Li, J., Lee, Y. R., & Assmann, S. M. (1998). Guard cells possess a calcium-dependent protein kinase that phosphorylates the KAT1 potassium channel. *Plant Physiology, 116*, 785–795.

Lim, C., Baek, W., Jung, J., Kim, J.-H., & Lee, S. (2015). Function of ABA in stomatal defense against biotic and drought stresses. *International Journal of Molecular Sciences, 16*, 15251–15270. https://doi.org/10.3390/ijms160715251.

Lin, S.-M., Tsai, J.-Y., Hsiao, C.-D., Huang, Y.-T., Chiu, C.-L., Liu, M.-H., et al. (2012). Crystal structure of a membrane-embedded H+-translocating pyrophosphatase. *Nature, 484*, 399–403. https://doi.org/10.1038/nature10963.

Linder, B., & Raschke, K. (1992). A slow anion channel in guard cells, activating at large hyperpolarization, may be principal for stomatal closing. *FEBS Letters, 313*, 27–30.

Liu, J., Elmore, J. M., & Coaker, G. (2009). Investigating the functions of the RIN4 protein complex during plant innate immune responses. *Plant Signaling & Behavior, 4*, 1107–1110.

Lohse, G., & Hedrich, R. (1992). Characterization of the plasma-membrane H(+)-ATPase from vicia faba guard cells: Modulation by extracellular factors and seasonal changes. *Planta, 188*, 206–214. https://doi.org/10.1007/BF00216815.

Mackey, D., Holt, B. F., Wiig, A., & Dangl, J. L. (2002). RIN4 interacts with Pseudomonas syringae type III effector molecules and is required for RPM1-mediated resistance in Arabidopsis. *Cell, 108*, 743–754.

MacRobbie, E. A. (1998). Signal transduction and ion channels in guard cells. *Philosophical Transactions of the Royal Society of London. Series B, Biological Sciences, 353*, 1475–1488. https://doi.org/10.1098/rstb.1998.0303.

Maierhofer, T., Diekmann, M., Offenborn, J. N., Lind, C., Bauer, H., Hashimoto, K., et al. (2014). Site- and kinase-specific phosphorylation-mediated activation of SLAC1, a guard cell anion channel stimulated by abscisic acid. *Science Signaling, 7*, ra86. https://doi.org/10.1126/scisignal.2005703.

Marten, H., Konrad, K. R., Dietrich, P., Roelfsema, M. R. G., & Hedrich, R. (2007). Ca2+-dependent and -independent abscisic acid activation of plasma membrane anion channels in guard cells of *Nicotiana tabacum*. *Plant Physiology, 143*, 28–37. https://doi.org/10.1104/pp.106.092643.

McAinsh, M. R., Gray, J. E., Hetherington, A. M., Leckie, C. P., & Ng, C. (2000). Ca2+ signalling in stomatal guard cells. *Biochemical Society Transactions, 28*, 476–481.

McLachlan, D. H., Kopischke, M., & Robatzek, S. (2014). Gate control: Guard cell regulation by microbial stress. *New Phytologist, 203*, 1049–1063. https://doi.org/10.1111/nph.12916.

Meckel, T., Hurst, A. C., Thiel, G., & Homann, U. (2005). Guard cells undergo constitutive and pressure-driven membrane turnover. *Protoplasma, 226*, 23–29. https://doi.org/10.1007/s00709-005-0106-6.

Melotto, M., Underwood, W., & He, S. Y. (2008). Role of stomata in plant innate immunity and foliar bacterial diseases. *Annual Review of Phytopathology, 46*, 101–122. https://doi.org/10.1146/annurev.phyto.121107.104959.

Melotto, M., Underwood, W., Koczan, J., Nomura, K., & He, S. Y. (2006). Plant stomata function in innate immunity against bacterial invasion. *Cell, 126*, 969–980. https://doi.org/10.1016/j.cell.2006.06.054.

Merilo, E., Laanemets, K., Hu, H., Xue, S., Jakobson, L., Tulva, I., et al. (2013). PYR/RCAR receptors contribute to ozone-, reduced air humidity-, darkness-, and CO2-induced stomatal regulation. *Plant Physiology, 162*, 1652–1668. https://doi.org/10.1104/pp.113.220608.

Merlot, S., Leonhardt, N., Fenzi, F., Valon, C., Costa, M., Piette, L., et al. (2007). Constitutive activation of a plasma membrane H(+)-ATPase prevents abscisic acid-mediated stomatal closure. *The EMBO Journal, 26*, 3216–3226. https://doi.org/10.1038/sj.emboj.7601750.

Meyer, K., Leube, M. P., & Grill, E. (1994). A protein phosphatase 2C involved in ABA signal transduction in Arabidopsis thaliana. *Science, 264*, 1452–1455.

Meyer, S., Mumm, P., Imes, D., Endler, A., Weder, B., Al-Rasheid, K. A. S., et al. (2010). AtALMT12 represents an R-type anion channel required for stomatal movement in Arabidopsis guard cells. *The Plant Journal, 63*, 1054–1062. https://doi.org/10.1111/j.1365-313X.2010.04302.x.

Meyer, S., Scholz-Starke, J., De Angeli, A., Kovermann, P., Burla, B., Gambale, F., et al. (2011). Malate transport by the vacuolar AtALMT6 channel in guard cells is subject to multiple regulation. *The Plant Journal, 67*, 247–257. https://doi.org/10.1111/j.1365-313X.2011.04587.x.

Meyerhoff, O., Müller, K., Roelfsema, M. R. G., Latz, A., Lacombe, B., Hedrich, R., et al. (2005). AtGLR3.4, a glutamate receptor channel-like gene is sensitive to touch and cold. *Planta, 222*, 418–427. https://doi.org/10.1007/s00425-005-1551-3.

Miao, Y., Lv, D., Wang, P., Wang, X.-C., Chen, J., Miao, C., et al. (2006). An Arabidopsis glutathione peroxidase functions as both a redox transducer and a scavenger in abscisic acid and drought stress responses. *Plant Cell, 18*, 2749–2766. https://doi.org/10.1105/tpc.106.044230.

Monaghan, J., & Zipfel, C. (2012). Plant pattern recognition receptor complexes at the plasma membrane. *Current Opinion in Plant Biology, 15*, 349–357. https://doi.org/10.1016/j.pbi.2012.05.006.

Mori, I. C., Murata, Y., Yang, Y., Munemasa, S., Wang, Y.-F., Andreoli, S., et al. (2006). CDPKs CPK6 and CPK3 function in ABA regulation of guard cell S-type anion- and Ca(2+)-permeable channels and stomatal closure. *PLoS Biology, 4*, e327. https://doi.org/10.1371/journal.pbio.0040327.

Mori, I. C., Uozumi, N., & Muto, S. (2000). Phosphorylation of the inward-rectifying potassium channel KAT1 by ABR kinase in vicia guard cells. *Plant & Cell Physiology, 41*, 850–856.

Morsomme, P., de Kerchove d'Exaerde, A., De Meester, S., Thinès, D., Goffeau, A., & Boutry, M. (1996). Single point mutations in various domains of a plant plasma membrane H(+)-ATPase expressed in *Saccharomyces cerevisiae* increase H(+)-pumping and permit yeast growth at low pH. *The EMBO Journal, 15*, 5513–5526.

Müller-Röber, B., Ellenberg, J., Provart, N., Willmitzer, L., Busch, H., Becker, D., et al. (1995). Cloning and electrophysiological analysis of KST1, an inward rectifying K+ channel expressed in potato guard cells. *The EMBO Journal, 14*, 2409–2416.

Murata, Y., Pei, Z. M., Mori, I. C., & Schroeder, J. (2001). Abscisic acid activation of plasma membrane Ca(2+) channels in guard cells requires cytosolic NAD(P)H and is differentially disrupted upstream and downstream of reactive oxygen species production in abi1-1 and abi2-1 protein phosphatase 2C mutants. *Plant Cell, 13*, 2513–2523.

Negi, J., Matsuda, O., Nagasawa, T., Oba, Y., Takahashi, H., Kawai-Yamada, M., et al. (2008). CO2 regulator SLAC1 and its homologues are essential for anion homeostasis in plant cells. *Nature, 452*, 483–486. https://doi.org/10.1038/nature06720.

Niittylä, T., Fuglsang, A. T., Palmgren, M. G., Frommer, W. B., & Schulze, W. X. (2007). Temporal analysis of sucrose-induced phosphorylation changes in plasma membrane proteins of Arabidopsis. *Molecular & Cellular Proteomics, 6*, 1711–1726. https://doi.org/10.1074/mcp.M700164-MCP200.

Okumura, M., Inoue, S.-I., Kuwata, K., & Kinoshita, T. (2016). Photosynthesis activates plasma membrane H+-ATPase via sugar accumulation. *Plant Physiology, 171*, 580–589. https://doi.org/10.1104/pp.16.00355.

Osakabe, Y., Arinaga, N., Umezawa, T., Katsura, S., Nagamachi, K., Tanaka, H., et al. (2013). Osmotic stress responses and plant growth controlled by potassium transporters in Arabidopsis. *Plant Cell, 25*, 609–624. https://doi.org/10.1105/tpc.112.105700.

Padmanaban, S., Lin, X., Perera, I., Kawamura, Y., & Sze, H. (2004). Differential expression of vacuolar H+-ATPase subunit c genes in tissues active in membrane trafficking and their roles in plant growth as revealed by RNAi. *Plant Physiology, 134*, 1514–1526. https://doi.org/10.1104/pp.103.034025.

Pallucca, R., Visconti, S., Camoni, L., Cesareni, G., Melino, S., Panni, S., et al. (2014). Specificity of ε and non-ε isoforms of arabidopsis 14-3-3 proteins towards the H+-ATPase and other targets. *PLoS One, 9*, e90764. https://doi.org/10.1371/journal.pone.0090764.

Palmgren, M. G. (2001). Plant plasma membrane H^+-ATPases: Powerhouses for nutrient uptake. *Annual Review of Plant Physiology and Plant Molecular Biology, 52*, 817–845. https://doi.org/10.1146/annurev.arplant.52.1.817.

Palmgren, M. G., Larsson, C., & Sommarin, M. (1990). Proteolytic activation of the plant plasma membrane H(+)-ATPase by removal of a terminal segment. *The Journal of Biological Chemistry, 265*, 13423–13426.

Palmgren, M. G., Sommarin, M., Serrano, R., & Larsson, C. (1991). Identification of an autoinhibitory domain in the C-terminal region of the plant plasma membrane H(+)-ATPase. *The Journal of Biological Chemistry, 266*, 20470–20475.

Pandey, S., Zhang, W., & Assmann, S. M. (2007). Roles of ion channels and transporters in guard cell signal transduction. *FEBS Letters, 581*, 2325–2336. https://doi.org/10.1016/j.febslet.2007.04.008.

Pei, Z. M., Kuchitsu, K., Ward, J. M., Schwarz, M., & Schroeder, J. I. (1997). Differential abscisic acid regulation of guard cell slow anion channels in Arabidopsis wild-type and abi1 and abi2 mutants. *Plant Cell, 9*, 409–423. https://doi.org/10.1105/tpc.9.3.409.

Pei, Z. M., Murata, Y., Benning, G., Thomine, S., Klüsener, B., Allen, G. J., et al. (2000). Calcium channels activated by hydrogen peroxide mediate abscisic acid signalling in guard cells. *Nature, 406*, 731–734. https://doi.org/10.1038/35021067.

Pei, Z. M., Ward, J. M., Harper, J. F., & Schroeder, J. I. (1996). A novel chloride channel in vicia faba guard cell vacuoles activated by the serine/threonine kinase, CDPK. *The EMBO Journal, 15*, 6564–6574.

Pei, n., Ward, n., & Schroeder, n. (1999). Magnesium sensitizes slow vacuolar channels to physiological cytosolic calcium and inhibits fast vacuolar channels in fava bean guard cell vacuoles. *Plant Physiology, 121*, 977–986.

Peiter, E., Maathuis, F. J. M., Mills, L. N., Knight, H., Pelloux, J., Hetherington, A. M., et al. (2005). The vacuolar Ca2+-activated channel TPC1 regulates germination and stomatal movement. *Nature, 434*, 404–408. https://doi.org/10.1038/nature03381.

Pertl-Obermeyer, H., Schulze, W. X., & Obermeyer, G. (2014). In vivo cross-linking combined with mass spectrometry analysis reveals receptor-like kinases and Ca(2+) signalling proteins as putative interaction partners of pollen plasma membrane H(+) ATPases. *Journal of Proteomics, 108*, 17–29. https://doi.org/10.1016/j.jprot.2014.05.001.

Pilot, G., Lacombe, B., Gaymard, F., Cherel, I., Boucherez, J., Thibaud, J. B., et al. (2001). Guard cell inward K+ channel activity in arabidopsis involves expression of the twin channel subunits KAT1 and KAT2. *The Journal of Biological Chemistry, 276*, 3215–3221. https://doi.org/10.1074/jbc.M007303200.

Pittman, J. K., & Hirschi, K. D. (2016). CAX-ing a wide net: Cation/H(+) transporters in metal remediation and abiotic stress signalling. *Plant Biology (Stuttgart, Germany), 18*, 741–749. https://doi.org/10.1111/plb.12460.

Popko, J., Hänsch, R., Mendel, R.-R., Polle, A., & Teichmann, T. (2010). The role of abscisic acid and auxin in the response of poplar to abiotic stress. *Plant Biology (Stuttgart, Germany), 12*, 242–258. https://doi.org/10.1111/j.1438-8677.2009.00305.x.

Pottosin, I. I., & Martínez-Estévez, M. (2003). Regulation of the fast vacuolar channel by cytosolic and vacuolar potassium. *Biophysical Journal, 84*, 977–986. https://doi.org/10.1016/S0006-3495(03)74914-5.

Raschke, K. (2003). Alternation of the slow with the quick anion conductance in whole guard cells effected by external malate. *Planta, 217*, 651–657. https://doi.org/10.1007/s00425-003-1034-3.

Rienmüller, F., Beyhl, D., Lautner, S., Fromm, J., Al-Rasheid, K. A. S., Ache, P., et al. (2010). Guard cell-specific calcium sensitivity of high density and activity SV/TPC1 channels. *Plant & Cell Physiology, 51*, 1548–1554. https://doi.org/10.1093/pcp/pcq102.

Rienmüller, F., Dreyer, I., Schönknecht, G., Schulz, A., Schumacher, K., Nagy, R., et al. (2012). Luminal and cytosolic pH feedback on proton pump activity and ATP affinity of V-type ATPase from Arabidopsis. *The Journal of Biological Chemistry, 287*, 8986–8993. https://doi.org/10.1074/jbc.M111.310367.

Roelfsema, M. R. G., & Hedrich, R. (2005). In the light of stomatal opening: New insights into "The Watergate" *The New Phytologist, 167*, 665–691. https://doi.org/10.1111/j.1469-8137.2005.01460.x.

Roelfsema, M. R. G., Levchenko, V., & Hedrich, R. (2004). ABA depolarizes guard cells in intact plants, through a transient activation of R- and S-type anion channels. *The Plant Journal, 37*, 578–588.

Roelfsema, M. R., & Prins, H. B. (1997). Ion channels in guard cells of Arabidopsis thaliana (L.) Heynh. *Planta, 202*, 18–27. https://doi.org/10.1007/s004250050098.

Roelfsema, M. R. G., Staal, M., & Prins, H. B. A. (1998). Blue light-induced apoplastic acidification of Arabidopsis thaliana guard cells: Inhibition by ABA is mediated through protein phosphatases. *Physiologia Plantarum, 103*, 466–474. https://doi.org/10.1034/j.1399-3054.1998.1030404.x.

Roelfsema, M. R., Steinmeyer, R., Staal, M., & Hedrich, R. (2001). Single guard cell recordings in intact plants: Light-induced hyperpolarization of the plasma membrane. *The Plant Journal, 26*, 1–13.

Ronzier, E., Corratgé-Faillie, C., Sanchez, F., Prado, K., Brière, C., Leonhardt, N., et al. (2014). CPK13, a noncanonical Ca2+-dependent protein kinase, specifically inhibits KAT2 and KAT1 shaker K+ channels and reduces stomatal opening. *Plant Physiology, 166*, 314–326. https://doi.org/10.1104/pp.114.240226.

Saito, S., Hamamoto, S., Moriya, K., Matsuura, A., Sato, Y., Muto, J., et al. (2018). N -myristoylation and S -acylation are common modifications of Ca 2+-regulated Arabidopsis kinases and are required for activation of the SLAC1 anion channel. *New Phytologist, 218*, 1504–1521. https://doi.org/10.1111/nph.15053.

Saponaro, A., Porro, A., Chaves-Sanjuan, A., Nardini, M., Rauh, O., Thiel, G., et al. (2017). Fusicoccin activates KAT1 channels by stabilizing their interaction with 14-3-3 proteins. *Plant Cell, 29*, 2570–2580. https://doi.org/10.1105/tpc.17.00375.

Sasaki, T., Mori, I. C., Furuichi, T., Munemasa, S., Toyooka, K., Matsuoka, K., et al. (2010). Closing plant stomata requires a homolog of an aluminum-activated malate transporter. *Plant & Cell Physiology, 51*, 354–365. https://doi.org/10.1093/pcp/pcq016.

Sato, A., Gambale, F., Dreyer, I., & Uozumi, N. (2010). Modulation of the Arabidopsis KAT1 channel by an activator of protein kinase C in xenopus laevis oocytes. *The FEBS Journal, 277*, 2318–2328. https://doi.org/10.1111/j.1742-4658.2010.07647.x.

Sato, A., Sato, Y., Fukao, Y., Fujiwara, M., Umezawa, T., Shinozaki, K., et al. (2009). Threonine at position 306 of the KAT1 potassium channel is essential for channel activity and is a target site for ABA-activated SnRK2/OST1/SnRK2.6 protein kinase. *The Biochemical Journal, 424*, 439–448. https://doi.org/10.1042/BJ20091221.

Scherzer, S., Maierhofer, T., Al-Rasheid, K. A. S., Geiger, D., & Hedrich, R. (2012). Multiple calcium-dependent kinases modulate ABA-activated guard cell anion channels. *Molecular Plant, 5*, 1409–1412. https://doi.org/10.1093/mp/sss084.

Schmidt, C., Schelle, I., Liao, Y. J., & Schroeder, J. I. (1995). Strong regulation of slow anion channels and abscisic acid signaling in guard cells by phosphorylation and dephosphorylation events. *Proceedings of the National Academy of Sciences of the United States of America, 92*, 9535–9539.

Schmidt, C., & Schroeder, J. I. (1994). Anion selectivity of slow anion channels in the plasma membrane of guard cells (large nitrate permeability). *Plant Physiology, 106*, 383–391.

Schroeder, J. I. (1988). K+ transport properties of K+ channels in the plasma membrane of vicia faba guard cells. *The Journal of General Physiology, 92*, 667–683.

Schroeder, J. I., Allen, G. J., Hugouvieux, V., Kwak, J. M., & Waner, D. (2001). GUARD CELL SIGNAL TRANSDUCTION. *Annual Review of Plant Physiology and Plant Molecular Biology, 52*, 627–658. https://doi.org/10.1146/annurev.arplant.52.1.627.

Schroeder, J. I., & Hagiwara, S. (1990). Repetitive increases in cytosolic Ca2+ of guard cells by abscisic acid activation of nonselective Ca2+ permeable channels. *Proceedings of the National Academy of Sciences of the United States of America, 87*, 9305–9309.

Schroeder, J. I., & Keller, B. U. (1992). Two types of anion channel currents in guard cells with distinct voltage regulation. *Proceedings of the National Academy of Sciences of the United States of America, 89*, 5025–5029.

Schroeder, J. I., Raschke, K., & Neher, E. (1987). Voltage dependence of K channels in guard-cell protoplasts. *Proceedings of the National Academy of Sciences of the United States of America, 84*, 4108–4112.

Schroeder, J. I., Schmidt, C., & Sheaffer, J. (1993). Identification of high-affinity slow anion channel blockers and evidence for stomatal regulation by slow anion channels in guard cells. *Plant Cell, 5*, 1831–1841. https://doi.org/10.1105/tpc.5.12.1831.

Schumacher, K., & Krebs, M. (2010). The V-ATPase: Small cargo, large effects. *Current Opinion in Plant Biology, 13*, 724–730. https://doi.org/10.1016/j.pbi.2010.07.003.

Schumacher, K., Vafeados, D., McCarthy, M., Sze, H., Wilkins, T., & Chory, J. (1999). The Arabidopsis det3 mutant reveals a central role for the vacuolar H(+)-ATPase in plant growth and development. *Genes & Development, 13*, 3259–3270.

Schwessinger, B., & Zipfel, C. (2008). News from the frontline: Recent insights into PAMP-triggered immunity in plants. *Current Opinion in Plant Biology, 11*, 389–395. https://doi.org/10.1016/j.pbi.2008.06.001.

Shigaki, T., Cheng, N. H., Pittman, J. K., & Hirschi, K. (2001). Structural determinants of Ca2+ transport in the Arabidopsis H+/Ca2+ antiporter CAX1. *The Journal of Biological Chemistry, 276*, 43152–43159. https://doi.org/10.1074/jbc.M106637200.

Shimazaki, K., Doi, M., Assmann, S. M., & Kinoshita, T. (2007). Light regulation of stomatal movement. *Annual Review of Plant Biology, 58*, 219–247. https://doi.org/10.1146/annurev.arplant.57.032905.105434.

Shimazaki, K., Tominaga, M., & Shigenaga, A. (1997). Inhibition of the stomatal blue light response by verapamil at high concentration. *Plant & Cell Physiology, 38*, 747–750.

Shope, J. C., DeWald, D. B., & Mott, K. A. (2003). Changes in surface area of intact guard cells are correlated with membrane internalization. *Plant Physiology, 133*, 1314–1321. https://doi.org/10.1104/pp.103.027698.

Sirichandra, C., Wasilewska, A., Vlad, F., Valon, C., & Leung, J. (2009). The guard cell as a single-cell model towards understanding drought tolerance and abscisic acid action. *Journal of Experimental Botany, 60*, 1439–1463. https://doi.org/10.1093/jxb/ern340.

Sottocornola, B., Gazzarrini, S., Olivari, C., Romani, G., Valbuzzi, P., Thiel, G., et al. (2008). 14-3-3 proteins regulate the potassium channel KAT1 by dual modes. *Plant Biology (Stuttgart, Germany), 10*, 231–236. https://doi.org/10.1111/j.1438-8677.2007.00028.x.

Sottocornola, B., Visconti, S., Orsi, S., Gazzarrini, S., Giacometti, S., Olivari, C., et al. (2006). The potassium channel KAT1 is activated by plant and animal 14-3-3 proteins. *The Journal of Biological Chemistry, 281*, 35735–35741. https://doi.org/10.1074/jbc.M603361200.

Stange, A., Hedrich, R., & Roelfsema, M. R. G. (2010). Ca(2+)-dependent activation of guard cell anion channels, triggered by hyperpolarization, is promoted by prolonged depolarization. *The Plant Journal*, *62*, 265–276. https://doi.org/10.1111/j.1365-313X.2010.04141.x.

Suh, S. J., Wang, Y.-F., Frelet, A., Leonhardt, N., Klein, M., Forestier, C., et al. (2007). The ATP binding cassette transporter AtMRP5 modulates anion and calcium channel activities in Arabidopsis guard cells. *The Journal of Biological Chemistry*, *282*, 1916–1924. https://doi.org/10.1074/jbc.M607926200.

Suhita, D., Raghavendra, A. S., Kwak, J. M., & Vavasseur, A. (2004). Cytoplasmic alkalization precedes reactive oxygen species production during methyl jasmonate- and abscisic acid-induced stomatal closure. *Plant Physiology*, *134*, 1536–1545. https://doi.org/10.1104/pp.103.032250.

Sutter, J.-U., Sieben, C., Hartel, A., Eisenach, C., Thiel, G., & Blatt, M. R. (2007). Abscisic acid triggers the endocytosis of the arabidopsis KAT1 K+ channel and its recycling to the plasma membrane. *Current Biology*, *17*, 1396–1402. https://doi.org/10.1016/j.cub.2007.07.020.

Svennelid, F., Olsson, A., Piotrowski, M., Rosenquist, M., Ottman, C., Larsson, C., et al. (1999). Phosphorylation of Thr-948 at the C terminus of the plasma membrane H(+)-ATPase creates a binding site for the regulatory 14-3-3 protein. *Plant Cell*, *11*, 2379–2391.

Sze, H., Schumacher, K., Müller, M. L., Padmanaban, S., & Taiz, L. (2002). A simple nomenclature for a complex proton pump: VHA genes encode the vacuolar H(+)-ATPase. *Trends in Plant Science*, 7, 157–161.

Szyroki, A., Ivashikina, N., Dietrich, P., Roelfsema, M. R., Ache, P., Reintanz, B., et al. (2001). KAT1 is not essential for stomatal opening. *Proceedings of the National Academy of Sciences of the United States of America*, *98*, 2917–2921. https://doi.org/10.1073/pnas.051616698.

Takahashi, K., Hayashi, K., & Kinoshita, T. (2012). Auxin activates the plasma membrane H+-ATPase by phosphorylation during hypocotyl elongation in Arabidopsis. *Plant Physiology*, *159*, 632–641. https://doi.org/10.1104/pp.112.196428.

Thiel, G., MacRobbie, E. A., & Blatt, M. R. (1992). Membrane transport in stomatal guard cells: The importance of voltage control. *The Journal of Membrane Biology*, *126*, 1–18.

Tian, W., Hou, C., Ren, Z., Pan, Y., Jia, J., Zhang, H., et al. (2015). A molecular pathway for CO2 response in Arabidopsis guard cells. *Nature Communications*, *6*. https://doi.org/10.1038/ncomms7057.

Ueno, K., Kinoshita, T., Inoue, S.-I., Emi, T., & Shimazaki, K.-I. (2005). Biochemical characterization of plasma membrane H+-ATPase activation in guard cell protoplasts of Arabidopsis thaliana in response to blue light. *Plant & Cell Physiology*, *46*, 955–963. https://doi.org/10.1093/pcp/pci104.

Urbach, S., Chérel, I., Sentenac, H., & Gaymard, F. (2000). Biochemical characterization of the Arabidopsis K+ channels KAT1 and AKT1 expressed or co-expressed in insect cells. *The Plant Journal*, *23*, 527–538.

Vahisalu, T., Kollist, H., Wang, Y.-F., Nishimura, N., Chan, W.-Y., Valerio, G., et al. (2008). SLAC1 is required for plant guard cell S-type anion channel function in stomatal signalling. *Nature*, *452*, 487–491. https://doi.org/10.1038/nature06608.

Vahisalu, T., Puzõrjova, I., Brosché, M., Valk, E., Lepiku, M., Moldau, H., et al. (2010). Ozone-triggered rapid stomatal response involves the production of reactive oxygen species, and is controlled by SLAC1 and OST1. *The Plant Journal*, *62*, 442–453. https://doi.org/10.1111/j.1365-313X.2010.04159.x.

Véry, A. A., Gaymard, F., Bosseux, C., Sentenac, H., & Thibaud, J. B. (1995). Expression of a cloned plant K+ channel in xenopus oocytes: Analysis of macroscopic currents. *The Plant Journal*, 7, 321–332.

Walker, D. J., Leigh, R. A., & Miller, A. J. (1996). Potassium homeostasis in vacuolate plant cells. *Proceedings of the National Academy of Sciences of the United States of America, 93*, 10510–10514.

Wang, X.-P., Chen, L.-M., Liu, W.-X., Shen, L.-K., Wang, F.-L., Zhou, Y., et al. (2016). AtKC1 and CIPK23 synergistically modulate AKT1-mediated low-potassium stress responses in Arabidopsis. *Plant Physiology, 170*, 2264–2277. https://doi.org/10.1104/pp.15.01493.

Wang, Y., He, L., Li, H.-D., Xu, J., & Wu, W.-H. (2010). Potassium channel alpha-subunit AtKC1 negatively regulates AKT1-mediated K(+) uptake in Arabidopsis roots under low-K(+) stress. *Cell Research, 20*, 826–837. https://doi.org/10.1038/cr.2010.74.

Wang, C., Hu, H., Qin, X., Zeise, B., Xu, D., Rappel, W.-J., et al. (2016). Reconstitution of CO2 regulation of SLAC1 anion channel and function of CO2-permeable PIP2;1 aquaporin as CARBONIC ANHYDRASE4 interactor. *Plant Cell, 28*, 568–582. https://doi.org/10.1105/tpc.15.00637.

Wang, Y.-F., Munemasa, S., Nishimura, N., Ren, H.-M., Robert, N., Han, M., et al. (2013). Identification of cyclic GMP-activated nonselective Ca2+-permeable cation channels and associated CNGC5 and CNGC6 genes in Arabidopsis guard cells. *Plant Physiology, 163*, 578–590. https://doi.org/10.1104/pp.113.225045.

Wang, Y., Noguchi, K., Ono, N., Inoue, S., Terashima, I., & Kinoshita, T. (2014). Overexpression of plasma membrane H+-ATPase in guard cells promotes light-induced stomatal opening and enhances plant growth. *Proceedings of the National Academy of Sciences of the United States of America, 111*, 533–538. https://doi.org/10.1073/pnas.1305438111.

Ward, J. M., & Schroeder, J. I. (1994). Calcium-activated K+ channels and calcium-induced calcium release by slow vacuolar ion channels in guard cell vacuoles implicated in the control of stomatal closure. *Plant Cell, 6*, 669–683. https://doi.org/10.1105/tpc.6.5.669.

Wasilewska, A., Vlad, F., Sirichandra, C., Redko, Y., Jammes, F., Valon, C., et al. (2008). An update on abscisic acid signaling in plants and more. *Molecular Plant, 1*, 198–217. https://doi.org/10.1093/mp/ssm022.

Wege, S., De Angeli, A., Droillard, M.-J., Kroniewicz, L., Merlot, S., Cornu, D., et al. (2014). Phosphorylation of the vacuolar anion exchanger AtCLCa is required for the stomatal response to abscisic acid. *Science Signaling, 7*, ra65. https://doi.org/10.1126/scisignal.2005140.

Wilkinson, S., & Davies, W. J. (2010). Drought, ozone, ABA and ethylene: New insights from cell to plant to community. *Plant, Cell & Environment, 33*, 510–525. https://doi.org/10.1111/j.1365-3040.2009.02052.x.

Wille, A. C., & Lucas, W. J. (1984). Ultrastructural and histochemical studies on guard cells. *Planta, 160*, 129–142. https://doi.org/10.1007/BF00392861.

Xicluna, J., Lacombe, B., Dreyer, I., Alcon, C., Jeanguenin, L., Sentenac, H., et al. (2007). Increased functional diversity of plant K+ channels by preferential heteromerization of the shaker-like subunits AKT2 and KAT2. *The Journal of Biological Chemistry, 282*, 486–494. https://doi.org/10.1074/jbc.M607607200.

Xue, S., Hu, H., Ries, A., Merilo, E., Kollist, H., & Schroeder, J. I. (2011). Central functions of bicarbonate in S-type anion channel activation and OST1 protein kinase in CO2 signal transduction in guard cell. *The EMBO Journal, 30*, 1645–1658. https://doi.org/10.1038/emboj.2011.68.

Yamauchi, S., Takemiya, A., Sakamoto, T., Kurata, T., Tsutsumi, T., Kinoshita, T., et al. (2016). The plasma membrane H+-ATPase AHA1 plays a major role in stomatal opening in response to blue light. *Plant Physiology, 171*, 2731–2743. https://doi.org/10.1104/pp.16.01581.

Zhang, T., Chen, S., & Harmon, A. C. (2016). Protein-protein interactions in plant mitogen-activated protein kinase cascades. *Journal of Experimental Botany*, *67*, 607–618. https://doi.org/10.1093/jxb/erv508.

Zhang, X., Takemiya, A., Kinoshita, T., & Shimazaki, K. (2007). Nitric oxide inhibits blue light-specific stomatal opening via abscisic acid signaling pathways in vicia guard cells. *Plant & Cell Physiology*, *48*, 715–723. https://doi.org/10.1093/pcp/pcm039.

Zhao, R., Dielen, V., Kinet, J. M., & Boutry, M. (2000). Cosuppression of a plasma membrane H (+)-ATPase isoform impairs sucrose translocation, stomatal opening, plant growth, and male fertility. *Plant Cell*, *12*, 535–546.

CHAPTER EIGHT

The Pollen Plasma Membrane Permeome Converts Transmembrane Ion Transport Into Speed

Heidi Pertl-Obermeyer*, Peter Lackner†, John W.C. Dunlop‡, Gerhard Obermeyer*,[1]

*Molecular Plant Biophysics and Biochemistry, Department of Biosciences, University of Salzburg, Salzburg, Austria
†Immunoinformatics, Department of Biosciences, University of Salzburg, Salzburg, Austria
‡Morphophysics, Department of Chemistry and Physics of Materials, University of Salzburg, Salzburg, Austria
[1]Corresponding author: e-mail address: gerhard.obermeyer@sbg.ac.at

Contents

Abstract

Pollen tubes are the fastest growing cells in nature with elongation rates of >20 μm min^{-1}. The pollen tubes grow as fast as possible through the stigma tissue towards the egg cells where they release two sperm cells for fertilization. In addition to speed, the growth process has to be precise and well balanced to find the targets and to avoid bursting of the tube, respectively. Transport of ions across the plasma membrane generates a typical polar pattern of endogenous currents that accompanies pollen tube growth, and which is thought to function as a pacemaker to synchronize cellular processes. The endogenous currents are generated by the activity of various ion transporters in the plasma membrane, which were investigated by biophysical and molecular biology approaches.

Advances in Botanical Research, Volume 87
ISSN 0065-2296
https://doi.org/10.1016/bs.abr.2018.09.008

Using an unbiased database search for pollen-expressed transporter genes in Arabidopsis and *Lilium*, a large number of putative transporters was identified that might contribute to the current pattern formation. Furthermore, we hypothesize that the distribution of the transporters to specific regions in the pollen tube is the basis for the current pattern and is achieved by a self-organization process based on the electrophoretic mobility of charged membrane proteins in an electrical field. Thus, the activity of an ion transporter causes a local change in the transmembrane potential, which in turn would attract similar transporters and repel others leading to a heterogeneous distribution of transporters in the plasma membrane and the observed current pattern.

1. INTRODUCTION

Pollen grains are the male gametophytes of higher plants. They contain the generative cell or the two sperm cells imbedded in the vegetative cell's cytoplasm, which is surrounded by a tough, protective shell (cell wall and outer exine layer) to ensure the survival of the cells during their transport by wind, insects or mammals to the stigma. During this transfer, mature pollen grains exist in a quiescent state, which is also characterized by their low water content, similarly to seeds. Immediately after landing on the stigma surface pollen grains swell by taking up water and thus, change to the active state that leads to a local protuberance at specific pores of the exine and finally, start to grow a pollen tube that penetrates the stigma tissue and grows through the entire style towards the egg cells for fertilization. The pollen tubes grow extremely fast with approximately $10\,\mu m\,min^{-1}$ with the growth activity being restricted to the tip region of the tube (apical or tip growth). The specific morphology and highly polarized growth of the pollen tube require a highly synchronized and fine-tuned progression of cellular processes. A clearly visible feature of the polar morphology is the well-identified zone of the tube tip that excludes larger organelles from the growth zone, the successive organization of organelles behind this zone, a reversed fountain-like cytoplasmic streaming pattern, tip-localized ion gradients as well as tip-localized exo- and endocytosis of membrane vesicles (Cheung & Wu, 2007; Cole & Fowler, 2006; Hepler & Winship, 2015; Michard, Alves, & Feijo, 2009; Rakusova & Geitmann, 2017; Wang, Huang, & Jauh, 2010; Wang & Jiang, 2017). Furthermore, the growing pollen tube is accompanied by fluxes of several ion species (H^+, K^+, Ca^{2+}, Cl^-), which all together generate a typical ion current pattern, e.g. large H^+ effluxes at the pollen grain, H^+ influx at the tube tip and small in- and effluxes along the tube shank (Holdaway-Clarke & Hepler, 2003;

Michard, Simon, Tavares, Wudick, & Feijo, 2017; Weisenseel, Nuccitelli, & Jaffe, 1975). The current pattern itself is not constant but changes in time corresponding with oscillating growth rates, thus leading to a chronological sequence of molecular events that determine the tube increments (Feijó et al., 2001; Messerli, Danuser, & Robinson, 1999; van Hemelryck, Bernal, Rojas, Dumais, & Kroeger, 2017) and are assumed to function as a pacemaker for tube growth (Damineli, Portes, & Feijo, 2017).

In principle, the activity of ion transporters in the plasma membrane is responsible for the generation of these ion currents and a spatial distribution of specific ion transporter activities along the pollen tube plasma membrane causes the typical ion current pattern accompanying tube growth (Feijó, Malhó, & Obermeyer, 1995; Michard et al., 2009, 2017). Therefore, a hypothesis can be proposed on how the dynamic ion transport across the plasma membrane generates the endogenous currents and drives pollen tube elongation in a possibly self-organized process. In brief, we propose that local ion transport activity is causing local changes in V_M along the PT and thus, establishes and maintains the spatial distribution of membrane proteins including ion transporters. Knowledge of the transporter proteins in pollen, understanding their biophysical transport characteristics, their physiological regulation and their localization in the plasma membrane of a growing pollen tube may serve as a proof of this hypothesis.

2. THE POLLEN PERMEOME—ION TRANSPORTER CLASSES EXPRESSED IN POLLEN

About 25 years ago, a simple model of pollen tube growth suggested the action of three ion transporters in the plasma membrane, a K^+ inward-rectifying channel, a Ca^{2+} inward channel and a H^+ ATPase, to account for the endogenous electrical current pattern and the polarity of growing pollen tubes (Feijó et al., 1995). Our view on ion transporters in pollen is still far from being complete but astonishingly many transporters that are highly expressed in pollen have been identified in the last 10 years by transcriptome studies in *Arabidopsis thaliana* (Arabidopsis: Bock et al., 2006; Loraine, McCormick, Estrada, Patel, & Qin, 2013; Pina, Pinto, Feijo, & Becker, 2005) and even in organisms with non-sequenced genome, like *Lilium longiflorum* (lily: Lang, Usadel, & Obermeyer, 2015). Additionally, the membrane proteome has been investigated in various pollen species including Arabidopsis (Grobei et al., 2009), rice (Dai et al., 2007), tomato (Paul et al., 2016) and lily (Han, Chen, Dai, Yang, & Wang, 2010; Pertl, Schulze, & Obermeyer, 2009). The characteristics and physiological roles of some key players have been reviewed recently (Michard et al., 2017). Using the large information content of these data sets, we simply excerpted all relevant data for two model pollen species, Arabidopsis and lily, by searching for all plasma membrane proteins with a high expression level in pollen without any further specification.

2.1 The Pollen Plasma Membrane Permeome

To obtain a comprehensive overview of all Arabidopsis pollen transporters, we extracted relevant data from the databases Genevestigator (Hruz et al., 2008), Aramemnon (Schwacke et al., 2003) and SUBA3/SUBA4 (Hooper et al., 2014; Hooper, Castleden, Tanz, Aryamanesh, & Millar, 2017; Tanz et al., 2013). In addition to databases, transcriptome studies that are not included in Genevestigator (e.g. Becker, Boavida, Carneiro, Haury, & Feijo, 2003; Loraine et al., 2013) and pollen proteome studies (Grobei et al., 2009; Holmes-Davies, Tanaka, Vensel, Hurkman, & McCormick, 2005; Noir, Bräutigam, Colby, Schmidt, & Panstruga, 2005; Zhou et al., 2009) were taken into account. All studies that are included in this search investigated the pollen transcriptome and proteome, respectively, and a transporter protein was assumed to be expressed in pollen with a log2 value >11 (Genevestigator, GV) corresponding roughly to an expression value with an at least twice as high expression level than average expression of all transcripts detected in Arabidopsis. The Genevestigator data search contains only experimental data from wildtype pollen (array IDs AT-00178 (Pina et al., 2005), AT-00323, AT-00465 (Wang et al., 2008) and AT-00466 (Qin et al., 2016)). In the RNAseq study, a raw read count (number of overlapping reads for each gene) of >150 designates a gene expressed in pollen (roughly corresponding to 5 RPM, reads per million (Loraine et al., 2013)). In proteome studies either the appropriate Mascot score (Perkins, Pappin, Creasy, & Cottrell, 1999) or the presence of a protein (YES) is given. A gene is also considered to be present in pollen even with a low log2 value if its presence has been published. Furthermore, a putative pollen transporter was classified as a plasma membrane (PM) protein when published experiments reported its localization in the PM or when the SUBAcon value of the SUBA3 or SUBA4 databases predicted its PM localization. However, experimental evidence overruled the SUBAcon value!

Application of these search parameters resulted in a large number of transporters expressed in Arabidopsis pollen, but only few showed higher expression levels than the average of all transporters identified in microarray studies (Fig. 1A). Similar results were obtained when comparing the microarray studies from Genevestigator with the Arabidopsis pollen RNAseq experiment (Fig. 1B). In a first simple view, these 'high expression' transporters are assumed to be important for pollen physiology although we cannot exclude that a gene showing a low expression value may not have a contribution to the pollen's life. Nevertheless, the transporters with higher gene expression levels were selected for PM location (red dots in Fig. 1) and listed in Table 1. In total, 148 plasma membrane transporters fulfilled the specification set for the database search, resulting in 23 channels, 25 pumps and 100 carriers contributing mainly to the transport of cations (H^+, K^+, Ca^{2+}), anions (Cl^-, phosphate), sugars and amino acids/peptides as well as, most surprisingly, heavy metal ions. Furthermore, a few transporter genes showed different expression levels in pollen grains (PG) than in pollen tubes (PT), which could be distinguished by selecting only PG or PT microarray experiments in Genevestigator (see green or blue gene IDs

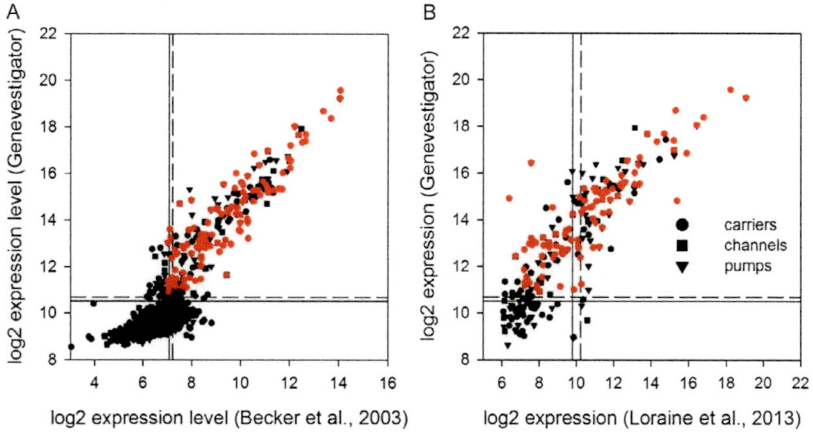

Fig. 1 Comparison between pollen transcriptome studies. (A) The expression levels (log2 values) of putative pollen transporters extracted from Genevestigator (ATH1 Affymetrix chip) were compared with the microarray experiment of Becker et al. (2003) using the 8K Affymetrix chip or (B) compared with the pollen RNAseq study of Loraine et al. (2013). The log2 values of identical genes found in the respective two databases were plotted (black dots). Red dots indicate putative plasma membrane-located transporters with log2 values >11 in Genevestigator. Solid lines give the average expression value with 99% confidence interval (dotted line) for each experiment. Details of genes represented by red dots are listed in Table 1.

Table 1 Plasma Membrane Transporters in *Arabidopsis thaliana* Pollen

	Transporter Family	Gene ID	Description	GV Expr Pollen, log2 ± S.D.	PT/PG Ratio	IEP	MW (kDa)	BECK Array	LORA RNAseq	GROB Protein	MICH Review
CHANNELS	NIP	**At5g37810**	Pollen-specific aquaporin (AtNIP4.1/AtNLM4)	12.44 ± 2.83	**1.54**	8.72	30.22	Yes	Yes	Yes	Yes
		At5g37820	Pollen-specific aquaporin (AtNIP4.2/AtNLM5)	12.44 ± 2.83	**1.54**	8.20	30.28				Yes
	TPC-type voltage-gated channel	At4g32500	Inward-rectifying K^+ channel, AtAKT5	13.38 ± 0.64		7.69	98.51	Yes	Yes		
	AKT/KAT-like K^+ channel	At2g25600	Inward-rectifying K^+ channel, AtAKT6/AtSPIK	16.97 ± 0.99		6.77	99.22	Yes	Yes	Yes	Yes
		At3g02850	Outward-rectifying K^+ channel, AtSKOR	14.23 ± 0.63		6.68	93.90	Yes	Yes		Yes
	TPK/KCO-type channel	At1g02510	Tandem-pore potassium channel, AtTPK4/AtKCO4	n.d.							Yes
	CNGC-type cation channel	**At1g15990**	Cyclic nucleotide/calmodulin-regulated channel, AtCNGC7	12.44 ± 1.30	**1.18**	9.56	81.39	Yes	Yes		Yes
		At1g19780	Cyclic nucleotide/calmodulin-regulated channel, AtCNGC8	14.87 ± 0.48		9.31	85.99	Yes	Yes		Yes
		At4g30560	Cyclic nucleotide/calmodulin-regulated channel, AtCNGC9	10.92 ± 0.62		9.85	83.63	Yes	Yes		

		At1g01340	Cyclic nucleotide/ calmodulin-regulated channel, AtCNGC10	13.02 ± 0.54		9.10	81.35	Yes	Yes		
		At3g48010	Cyclic nucleotide/ calmodulin-regulated channel, AtCNGC16	13.13 ± 0.61		8.65	81.96	Yes	Yes		Yes
		At5g14870	Ca^{2+} channel involved in PT growth, AtCNGC18	15.02 ± 0.40		8.26	80.37	Yes	Yes		Yes
	GLR-type ligand-gated cation channel	At5g48400	GLR-type calcium cation-permeable channel, AtGLR1.2	9.42 ± 0.22		6.62	84.29				Yes
		At5g27100	GLR-type calcium cation-permeable channel, AtGLR2.1	11.18 ± 0.41		6.03	100.91	Yes	Yes		Yes
		At2g32400	GLR-type calcium cation-permeable channel, AtGLR3.7	9.17 ± 0.21		7.88	103.52	Yes			
	SLAC-type guard cell anion channel	**At5g24030**	Anion channel complex (AtSLAH3)	11.65 ± 1.52	**1.20**	8.52	72.34	YES			Yes
	Mechanosensitive ion channel	At2g17010	Mechanosensitive ion channel (AtMSL8)	n.d.		8.44	89.80		Yes		Yes
	COPT-type copper cation transporter	At5g59030	Copper cation transporter (AtCOPT1)	17.67 ± 0.99		8.13	18.40	Yes	Yes		

Continued

Table 1 Plasma Membrane Transporters in *Arabidopsis thaliana* Pollen—cont'd

	Transporter Family	Gene ID	Description	GV Expr Pollen, log2 ± S.D.	PT/PG Ratio	IEP	MW (kDa)	BECK Array	LORA RNAseq	GROB Protein	MICH Review
		At3g46900	Copper cation transporter (AtCOPT2)	13.24 ± 1.89		7.19	17.06	Yes	Yes		
	AMT, ammonium transporter	At4g13510	High-affinity ammonium transporter (AtAMT1.1)	11.39 ± 0.26		7.47	53.58	Yes			
	Ca^{2+} permeable hyper-osmolality-gated channel	At1g62320	Hyperosmolality-gated calcium-permeable channel (AtOSCA1.4)	11.14 ± 1.19		9.36	88.24				
		At4g02900	Hyperosmolality-gated calcium-permeable channel (AtOSCA1.7)	13.39 ± 0.56		8.52	89.78	Yes		Yes	Yes
		At1g58520	Hyperosmolality-gated calcium-permeable channel (AtOSCA2.1)	11.32 ± 3.4 7	**1.78**	8.28	84.85				
		At1g10090	Hyperosmolality-gated calcium-permeable channel (AtOSCA2.2)	14.70 ± 0.33		8.37	86.77	Yes			
PUMPS	ABC subfamily A transporter	At3g47730	ABC subfamily A (AtABCA2/AtATH1)	15.42 ± 0.27		7.92	108.69	Yes	Yes		
		At3g47780	ABC subfamily A (AtABCA7/AtATH6)	14.31 ± 0.48		9.08	105.05	Yes	Yes		
		At5g61730	ABC fatty acid transporter (AtABCA9/AtATH11)	12.37 ± 0.64		7.85	104.50	Yes			

		At5g61690	ABC subfamily A (AtABCA11/AtATH15)	12.37 ± 0.64		8.16	101.17		Yes	Yes
		At5g61700	ABC subfamily A (AtABCA12/AtATH16)	13.66 ± 1.27	**0.93**	8.51	99.00	Yes	Yes	
	MDR-type ABC subfamily B transporter	At2g39480	ABC subfamily B (AtABCB6/AtMDR6)	11.63 ± 0.52		5.99	155.89	Yes		
		At4g18050	ABC subfamily B (AtABCB9/AtMDR9)	13.91 ± 2.38	**1.32**	7.52	134.37	Yes		Yes
		At1g10680	ABC subfamily B (AtABCB10/AtMDR10/AtPGP10)	14.44 ± 0.87		7.22	134.48	Yes		
	PDR-type ABC subfamily G transporter	At2g29940	ABC abscisic acid exporter (AtABCG31/AtPDR3)	13.02 ± 0.89		8.77	159.88	Yes		Yes
	NAP12-like ABC subfamily G transporter	At5g60740	ABC subfamily G (AtABCG28/AtWBC29)	15.89 ± 0.68		8.77	123.35	Yes	Yes	Yes
		At2g37010	ABC-type transporter (AtNAP12)	14.85 ± 1.36		9.00	120.59	Yes	Yes	
	P1B, heavy metal ATPase	At1g63440	Copper/silver cation-transporting P1B-type ATPase (AtHMA5)	12.78 ± 0.34		6.37	108.34	Yes	Yes	Yes
	P2B, Ca^{2+} ATPase	At2g22950	Calcium cation ATPase, AtACA7	12.57 ± 0.60		5.47	110.78	Yes	Yes	Yes

Continued

Table 1 Plasma Membrane Transporters in *Arabidopsis thaliana* Pollen—cont'd

Transporter Family	Gene ID	Description	GV Expr Pollen, log2 ± S.D.	PT/PG Ratio	IEP	MW (kDa)	BECK Array	LORA RNAseq	GROB Protein	MICH Review
	At5g57110	Calmodulin-regulated calcium cation ATPase, AtACA8	11.28 ± 0.45		7.92	116.18	Yes	Yes		
	At3g21180	Calcium cation ATPase, AtACA9	18.04 ± 0.92		6.73	118.78	Yes	Yes	Yes	Yes
P3A, H^+ ATPase	At2g18960	Plasma membrane proton ATPase (AtAHA1/AtOST2)	12.72 ± 0.51		6.67	104.23		Yes		
	At2g07560	Plasma membrane proton ATPase (AtAHA6)	13.89 ± 1.03		6.08	105.02	Yes	Yes	Yes	Yes
	At3g60330	Plasma membrane proton ATPase (AtAHA7)	15.53 ± 0.79		6.83	105.53	Yes	Yes	Yes	Yes
	At3g42640	Plasma membrane proton ATPase (AtAHA8)	19.24 ± 0.51		5.51	104.14	Yes	Yes	Yes	Yes
	At1g80660	Plasma membrane proton ATPase (AtAHA9)	12.84 ± 1.02		6.33	105.21	Yes	Yes	Yes	Yes
P4, component of phospholipid flippase	At5g04930	Phospholipid flippase, AtALA1	14.56 ± 0.41		6.23	130.34	Yes	Yes		
	At5g44240	Phospholipid flippase, AtALA2/AtAVI1	12.57 ± 0.57		5.78	128.76	Yes	Yes	Yes	

		At1g54280	Phospholipid flippase, AtALA6	16.36 ± 0.49	7.30	140.11	Yes	Yes	Yes
		At3g13900	Phospholipid flippase, AtALA7	15.60 ± 0.53	7.35	140.31	Yes	Yes	
		At1g26130	Phospholipid flippase, AtALA12	14.79 ± 0.69	6.15	133.80	Yes	Yes	
CARRIERS	SULTR-type transporter	At3g51895	Sulphate transporter (AtSultr3.1/AtST1)	11.86 ± 0.63	9.66	74.502	Yes	Yes	
	LHT-type amino acid transporter	At4g35180	Lysine/histidine transporter (AtLHT7)	18.69 ± 0.21	9.13	52.30	Yes	Yes	
	ProT-type proline transporter	At2g39890	Proline transporter (AtProT1)	17.68 ± 0.71	9.76	48.46	Yes	Yes	
	ANT-type amino acid transporter	At3g11900	Amino acid transporter (AtANT1)	15.66 ± 0.35	5.16	46.47	Yes	Yes	Yes
		At2g42005	Amino acid transporter (AtANTL1)	14.81 ± 0.48	7.99	44.96		Yes	
		At4g38250	Amino acid transporter (AtANTL2)	14.81 ± 0.48	6.66	47.21	Yes	Yes	
	GAT-type GABA transporter	At5g41800	GAT-type GABA transporter	11.49 ± 1.33	8.95	49.86	Yes		
	SLC32-like transporter	At2g39130	Unspecified transporter	12.81 ± 0.27	5.58	60.07	Yes	Yes	Yes
		At5g02180	Unspecified transporter	15.07 ± 0.48	4.78	59.67	Yes	Yes	

Continued

Table 1 Plasma Membrane Transporters in *Arabidopsis thaliana* Pollen—cont'd

Transporter Family	Gene ID	Description	GV Expr Pollen, log2 ± S.D.	PT/PG Ratio	IEP	MW (kDa)	BECK Array	LORA RNAseq	GROB Protein	MICH Review
	At5g16740	Unspecified transporter	13.53 ± 0.40		8.99	45.81				
SLC38-like transporter	At1g80510	Unspecified transporter	13.53 ± 0.40		6.31	52.96	Yes	Yes	Yes	
	At3g30390	Unspecified transporter	15.20 ± 0.41		6.50	49.52	Yes	Yes		
	At5g38820	Unspecified transporter	13.57 ± 0.36		5.90	49.68	Yes	Yes		
LAT-type amino acid transporter	At1g31830	LAT, polyamine transporter (AtLAT4/AtPAR1/AtPQR2)	13.37 ± 0.36		6.16	54.83	Yes	Yes		
CAT-type cationic amino acid transporter	At3g03720	Cationic amino acid transporter (AtCAT4)	15.51 ± 0.55		6.37	86.38	Yes	Yes	Yes	
CCC, cation: Cl^- co-transporter	At1g30450	Cation-chloride co-transporter (AtCCC)	n.d.		6.45	106.65		Yes		Yes
VGT-type hexose transporter	At3g03090	VGT subfamily vacuolar glucose transporter (AtVGT1)	16.66 ± 0.64		7.60	53.55	Yes	Yes		
INT-type inositol/proton transporter	**At2g35740**	Hypothetical protein sequence (AtINT3ps)	11.19 ± 1.84	**1.35**	7.88	63.18	Yes			
	At4g16480	INT subfamily inositol: proton symporter (AtINT4)	14.52 ± 2.37	**0.96**	8.77	62.90	Yes	Yes	Yes	

	ERD6-type mono-saccharide transporter	At1g75220	ERD monosaccharide transporter (AtERDL6)	16.23 ± 0.63		8.65	52.90	Yes		Yes
		At3g05150	ERD6-type monosaccharide transporter	17.20 ± 0.60		6.90	51.03	Yes		Yes
		At3g05155	ERD6-type monosaccharide transporter	11.39 ± 0.81		5.87	35.72			
		At3g05160	ERD6-type monosaccharide transporter	11.39 ± 0.81		7.72	50.30			
		At3g05165	ERD6-type monosaccharide transporter	11.39 ± 0.81		8.03	51.30	Yes		
	STP-type monosaccharide transporter	**At3g19930**	STP monosaccharide: proton symporter (AtSTP4)	12.72 ± 2.97	**1.53**	8.25	57.10	Yes	Yes	Yes
		At3g05960	STP monosaccharide: proton symporter (AtSTP6)	11.23 ± 1.40		8.00	55.91		Yes	Yes
		At1g50310	STP monosaccharide: proton symporter (AtSTP9)	11.13 ± 0.37		6.75	56.29	Yes	Yes	Yes
		At3g19940	STP monosaccharide: proton symporter (AtSTP10)	11.29 ± 3.3 2	**1.71**	8.26	56.20			
		At5g23270	STP monosaccharide: proton symporter (AtSTP11)	18.38 ± 0.91		6.36	56.73	Yes	Yes	
		At4g21480	STP monosaccharide: proton symporter (AtSTP12)	12.91 ± 0.94		9.43	55.57	Yes	Yes	

Continued

Table 1 Plasma Membrane Transporters in *Arabidopsis thaliana* Pollen—cont'd

Transporter Family	Gene ID	Description	GV Expr Pollen, log2 ± S.D.	PT/PG Ratio	IEP	MW (kDa)	BECK Array	LORA RNAseq	GROB Protein	MICH Review
PLT-type polyol transporter	At2g20780	PLT polyol/monosacch.: proton symporter (AtPMT4/AtPLT4)	14.25 ± 0.37		5.59	56.80	Yes			
SUT/SUC-type sugar transporter	At1g71880	Sucrose-proton symporter (AtSUC1)	11.00 ± 0.91		9.16	54.86	Yes	Yes	Yes	Yes
	At1g66570	Hypothetical protein sequence (AtSUC7ps)	9.33 ± 1.28	**1.44**	9.29	52.88				
	At2g14670	Sucrose-proton symporter (AtSUC8)	9.72 ± 1.46	**1.51**	9.65	52.75				
	At5g06170	Ultra-high affinity sucrose-proton symporter (AtSUC9)	11.30 ± 3.08	**1.62**	9.32	52.58				
NiaP-type nicotinate transporter	At3g13050	Pyridine-3-carboxylate transporter (AtNiaP)	15.36 ± 0.46		8.13	54.58	Yes			
OCT-type organic cation transporter	**At1g79360**	Organic cation/carnitine transporter (AtOCT2)	12.93 ± 1.74	**1.25**	7.26	57.75		Yes		
	At3g20660	Organic cation/carnitine transporter (AtOCT4)	14.52 ± 0.76		6.01	57.51	Yes	Yes		
	At5g54860	BT1-type small solute transporter	12.03 ± 1.14	**1.15**	9.49	53.89	Yes		Yes	

	PHT1-type phosphate transporter	**At3g54700**	PHT1-type phosphate transporter (AtPHT1;7)	14.91 ± 1.60	**1.19**	8.32	58.34		Yes	Yes
	G3P-type organic phosphate/glycerol-3-P permease	At2g13100	Phosphate/glycerol-3-phosphate permease (AtG3Pp5)	15.29 ± 0.42		7.63	53.06	Yes		
	ZIF/TOM-type metal cation transporter	At5g13750	ZIF/TOM-type potassium cation transporter (AtZIFL1)	13.08 ± 0.46		7.65	52.30	Yes	Yes	
	TCR-type metal cation transporter	At2g16980	Tetracycline permease-like transporter (AtTCR1)	11.47 ± 0.97		7.88	44.38	Yes		
	NRT1/PTR-type oligopeptide transporter	**At5g01180**	NRT1/PTR subfamily NPF8 oligopeptide carrier (AtNPF8.2/AtPTR5)	12.82 ± 1.29	**1.18**	8.03	63.31	Yes	Yes	Yes
	NRT2-type nitrate anion transporter	At1g08090	High-affinity nitrate transporter (AtNRT2.1/AtACH1)	11.27 ± 0.39		8.67	57.71			
	Unspecific transporter	At2g22730	(Yeast AMF1)-like ammonium transporter	12.64 ± 0.51		7.92	55.67			
		At5g64500	(Yeast AMF1)-like ammonium transporter	12.64 ± 0.51		4.79	52.16	Yes		
	KVAG-type nucleotide sugar transporter	At5g05820	Phosphometabolite transporter (AtUUAT3)	11.54 ± 0.45		10.23	33.90	Yes	Yes	

Continued

Table 1 Plasma Membrane Transporters in *Arabidopsis thaliana* Pollen—cont'd

Transporter Family	Gene ID	Description	GV Expr Pollen, log2 ± S.D.	PT/PG Ratio	IEP	MW (kDa)	BECK Array	LORA RNAseq	GROB Protein	MICH Review
URGT/UXT-type nucleotide sugar transporter	At1g76670	UDP-rhamnose/-galactose transporter (AtURGT1/AtUDP-GalT2)	11.53 ± 0.73		8.05	37.70	Yes	Yes		
	At1g21070	UDP-rhamnose/-galactose transporter (AtURGT2)	12.88 ± 1.34	**0.85**	8.62	38.05	Yes	Yes		
Unspecified nucleotide sugar transporter	At5g55950	Nucleotide sugar transporter	12.35 ± 0.49		10.06	42.65	Yes	Yes	Yes	
UTR-1/5 nucleotide sugar transporter	**At1g14360**	UDP-galactose/UDP-glucose transporter (AtUTr3)	13.02 ± 1.13	**1.11**	9.82	36.72	Yes	Yes	Yes	
	At3g46180	UDP-galactose/UDP-glucose transporter (AtUTr5)	11.06 ± 0.81		9.36	38.86	Yes	Yes		
PUP-type organic cation transporter	**At1g28220**	Purine permease (AtPUP3)	11.14 ± 1.10	**1.18**	8.55	38.87	Yes			
SIAR1-like amino acid transporter	At2g37450	MtN21/SIAR/UmamiT-type transporter (AtUmamiT13)	11.18 ± 0.77		8.79	32.22	Yes	Yes		
WAT1-like auxin transporter	At3g45870	MtN21/WAT/UmamiT-type transporter (AtUmamiT03)	12.04 ± 0.45		9.38	41.88	Yes	Yes	Yes	

	Unspecified transporter	At3g28050	MtN21/UmamiT-type transporter (AtUmamiT41)	11.04 ± 1.08		8.53	40.06	Yes		
		At4g30420	MtN21/UmamiT-type transporter (AtUmamiT34)	11.46 ± 1.06		8.60	41.28	Yes		
		At5g47470	MtN21/UmamiT-type transporter (AtUmamiT07)	15.23 ± 0.78		8.19	39.56	Yes		
		At3g26670	Unspecified transporter	13.89 ± 0.66		8.05	48.41	Yes		
		At2g21120	Host factor for antiviral RNA interference (AtAVI2)	13.19 ± 0.38		6.77	35.58	Yes		
		At3g23870	ENOR3-like membrane protein of unknown function	13.57 ± 0.85		9.41	36.29	Yes		
		At4g13800	ENOR3-like membrane protein of unknown function	10.22 ± 1.46	**1.49**	8.55	36.60	Yes		
		At5g11960	Unspecified transporter	15.25 ± 0.59		9.86	37.39	Yes		
		At4g32140	Metabolite transporter	11.31 ± 0.45		6.53	43.40	Yes	Yes	
	<u>CHX-type H^+/cation antiporter</u>	At1g79400	Cation-proton exchanger (AtCHX2)	15.35 ± 0.50		6.85	88.21	Yes	Yes	Yes
		At1g08135	Cation-proton exchanger (AtCHX6b)	14.79 ± 0.67		7.41	90.12		Yes	
		At2g28180	Cation-proton exchanger (AtCHX8)	17.39 ± 0.63		7.20	93.70	Yes	Yes	

Continued

Table 1 Plasma Membrane Transporters in *Arabidopsis thaliana* Pollen—cont'd

Transporter Family	Gene ID	Description	GV Expr Pollen, log2 ± S.D.	PT/PG Ratio	IEP	MW (kDa)	BECK Array	LORA RNAseq	GROB Protein	MICH Review
	At3g44930	Cation-proton exchanger (AtCHX10)	15.33 ± 0.67		6.51	88.15	Yes	Yes		
	At3g44920	Cation-proton exchanger (AtCHX11)	15.33 ± 0.67		6.60	88.44		Yes		
	At3g44910	Cation-proton exchanger (AtCHX12)	10.77 ± 1.97	**1.38**	6.38	85.82				
	At1g06970	Potassium cation-proton exchanger (AtCHX14)	14.28 ± 1.13	**0.90**	7.02	92.17	Yes	Yes		Yes
	At2g13620	Cation-proton exchanger (AtCHX15)	15.19 ± 0.53		6.00	89.86	Yes	Yes	Yes	
	At2g31910	Cation-proton exchanger (AtCHX21)	10.70 ± 1.92	**1.36**	5.31	91.99	Yes			Yes
	At5g01680	Cation-proton exchanger (AtCHX26)	10.40 ± 2.03	**1.42**	7.38	87.04		Yes		
	At5g01690	Cation-proton exchanger (AtCHX27)	14.33 ± 1.05		8.35	84.46	Yes	Yes		
KEA-type H^+/K^+(cation) antiporter	At5g51710	Potassium cation efflux antiporter (AtKEA5)	13.60 ± 0.81		6.12	61.60	Yes	Yes		

	CCX-type cation/Ca^{2+} exchanger	**At5g17850**	Cation calcium exchanger (AtCCX2/AtCAX8)	13.22 ± 1.10	**1.12**	6.82	61.18	Yes	Yes	
		At3g14070	Cation calcium exchanger (AtCCX3/AtCAX9)	13.15 ± 0.34		5.99	70.16	Yes	Yes	
		At1g54115	Cation calcium exchanger (AtCCX4/AtCAX10)	12.52 ± 0.53		5.49	70.80		Yes	
	MHX-type H^{+}/Mg^{2+} exchanger	At2g47600	Magnesium/proton cation exchanger (AtMHX1)	15.27 ± 0.84		6.19	59.80	Yes	Yes	
	NCL/EF-CAX-type cation exchanger	**At1g53210**	Sodium:calcium cation exchanger, AtNCL/AtEFCAX1	12.97 ± 1.43	**1.16**	4.97	63.42	Yes	Yes	Yes
	Mn-CDF-type Mn/cation transporter	At1g16310	Mn-CDF-type metal cation transporter (AtMTP10)	11.43 ± 0.72		7.01	49.11	Yes	Yes	
	DTX-type metabolite transporter	At5g52450	MATE-related efflux carrier (AtDTX16)	14.05 ± 1.36		8.11	52.48	Yes	Yes	
		At4g00350	MATE-related efflux carrier (AtDTX34)	16.84 ± 1.91		4.66	59.05	Yes	Yes	
	PILS-type auxin transporter	At2g17500	Auxin transporter (AtPILS5)	17.34 ± 0.56		6.89	43.40	Yes	Yes	

Continued

Table 1 Plasma Membrane Transporters in *Arabidopsis thaliana* Pollen—cont'd

Transporter Family	Gene ID	Description	GV Expr Pollen, log2 ± S.D.	PT/PG Ratio	IEP	MW (kDa)	BECK Array	LORA RNAseq	GROB Protein	MICH Review
BOR-type borate transporter	At5g25430	Boron transporter (AtBOR6)	15.51 ± 0.40		7.82	75.56	Yes	Yes		
YSL-type iron chelator	At5g24380	Fe(III)-phytosiderophore uptake mediator (AtYSL2)	13.12 ± 1.05		9.29	72.99	Yes			
	At1g65730	Fe(III)-phytosiderophore uptake mediator (AtYSL7)	11.74 ± 1.40		9.32	75.71	Yes		Yes	
OPT-type oligopeptide transporter	At5g55930	OPT-type transporter (AtOPT1)	19.58 ± 0.87		8.47	85.35	Yes	Yes		
	At1g09930	OPT-type transporter (AtOPT2)	11.34 ± 0.73		8.96	82.34		Yes	Yes	
	At5g53520	OPT-type transporter (AtOPT8)	12.89 ± 0.49		9.11	82.08	Yes	Yes		
PHO-type phosphate transporter	At3g29060	Inorganic phosphate transporter, AtPHO1-H9	16.03 ± 0.47		9.86	92.45	Yes		Yes	
ZIP13-like metal cation transporter	At3g08650	Divalent transition metal cation transporter (AtZIP13)	15.33 ± 0.43		7.55	61.93	Yes	Yes		
HAK/KUP/KT-type K^+ transporter	At4g23640	K^+ cation transporter, AtKUP4/AtTRH1/AtKT3	12.78 ± 0.47		9.64	86.85	Yes	Yes	Yes	

	SWEET-type sugar efflux transporter	At5g62850	Sugar efflux transporter (AtSWEET5)	12.96 ± 0.61		8.87	27.21	Yes			Yes
		At5g40260	Sugar efflux transporter (AtSWEET8/AtRPG1)	11.17 ± 1.08	**1.13**	9.59	26.89	Yes			
	Unspecific transporter	At1g11540	Unspecified transporter	11.10 ± 0.48		9.46	40.28	Yes			
		At2g36630	Unspecified transporter	14.45 ± 0.54		8.60	50.26	Yes			
	CLC-type anion/ H^+ antiporter	At5g49890	Voltage-gated anion channel (AtCLC-c)	16.01 ± 0.69		8.46	85.24	Yes	Yes	Yes	

The information on membrane transporter proteins from different databases: Aramemnon 8.1, Genevestigator (GV) and SUBA3/SUBA4 and from transcriptome studies BECK (Becker et al., 2003) and LORA (Loraine et al., 2013) as well as proteome studies (Holmes-Davies et al., 2005; Noir et al., 2005) and GROB (Grobei et al., 2009) was extracted to identify transporters highly expressed in the plasma membrane of pollen of *Arabidopsis thaliana*. See text for details of the selection criteria. Transporters listed in the recent review of Michard et al. (2017) are also marked (MICH). Gene IDs in bold indicate transporters with a differential expression in PGs and PTs shown by the PT/PG expression ratio (bold green >1, bold blue <1). Underlined transporter families were also identified in lily pollen. n.d. = no data.

in Table 1). A PT to PG expression level ratio >1 reflects an increase in expression of the respective gene for pollen tube growth and increases up to 80% could be observed.

In addition to Arabidopsis the large-sized pollen of lily serves as a model organism that enables many physiological experiments, which have not yet been performed with the tiny Arabidopsis pollen, e.g. turgor pressure measurements (Benkert, Obermeyer, & Bentrup, 1997) or metabolomics (Obermeyer, Fragner, Lang, & Weckwerth, 2013). To reveal the lily pollen permeome, the RNAseq data from Lang et al. (2015) translated into amino acid code served as a database for a re-analysis of the PM proteome (interface 38/45 from Pertl et al., 2009). In contrast to Arabidopsis, the identified peptides could not be linked up with the corresponding gene, but to transporter categories (BIN classes (Thimm et al., 2004)), which were further divided into transporter families similar to Table 1, thus allowing a coarse comparison between the pollen permeome of Arabidopsis and lily.

In the lily pollen PM permeome, 588 unigenes could be identified that code for putative plasma membrane-localized transporter proteins. By simply assuming that a high number of identified unigenes correlates with a high abundance of the respective protein in the pollen, the transporter categories with the highest unigene numbers might be important for pollen physiology. For instance the well-known PM H^+ ATPases have been identified in both databases with values corresponding to high expression levels (Arabidopsis AHA8: GV log2 = 19.24 and lily P3A PM H^+ ATPases with 27 unigenes). Strong experimental evidence verified their crucial role in the physiological function of the pollen (reviewed by Lang, Pertl-Obermeyer, Safiarian, & Obermeyer, 2014). In consequence, other not yet characterized transporters or transport classes showing similarly high values, e.g. OPT-type oligopeptide transporters (Arabidopsis OPT1: GV log2 = 19.58, lily unigene no. = 10) or CHX-type H^+/cation exchangers (Arabidopsis CHX8: GV log2 = 17.39, lily unigene no. = 29), might be equally important. Although such comparisons are pretty crude, the above approach of unbiased screening of databases might have revealed transporters with novel and possibly essential functions in pollen grains and tubes.

2.2 Ion Transport

The above hypothesis explains the nature and the pattern of the ion current driving tube elongation by knowing the underlying membrane transporter activities. Using the overlapping results of the Arabidopsis and lily pollen database searches, a generally effectual PM permeome may be constructed that is responsible for driving PT growth. The necessary key players were identified in both database searches as well as in independent studies performing micro arrays, RNAseq or proteomics, and all show high expression values in at least two studies. In the following, the transport characteristics of the identified ion transporters are described and their specific function during pollen germination and tube growth shortly summarized. A complementary view of the specific pollen ion transporters has been published previously by Michard et al. (2017) who listed and discussed the function of all published ion transporters. Here, we used a basically unbiased approach without any precondition to identify putative candidate proteins for transport in pollen.

However, understanding ion transport in a cellular system needs knowledge of the biophysical transport properties of all involved transporters. Electrogenic transport implies that net charges are translocated across the PM. This translocation in turn changes the membrane voltage (V_M) while the transporter proteins as well as the transport process itself are directly affected by V_M. In principle, the sum of all electrogenic transporters in a membrane determines V_M and is affecting each individual transport process. In addition, all transport processes of a system/cell are also connected via the transported substrates and for the case of H^+ by the pH of the compartments/organelles. Both, the V_M and the H^+ concentration (pH), allow a direct modulation of the transporters in addition to other cellular signalling processes and, therefore, may play an important role in the composition and installation of the dynamic transport processes at the PM. In the following paragraphs, we shortly describe the respective ion transport and mainly refer to review articles for more details.

2.2.1 Primary Active Transport and the PM Proton Pump

The major component of primary active transport systems in the pollen PM is constituted by proton pumps (**PM H^+ ATPase**) belonging to the P3B-type family. By using the energy from ATP hydrolysis, the pump transports H^+ across and, thus, hyperpolarizes the PM and generates an electrochemical gradient of H^+ that fuels secondary active transport systems. The PM H^+ ATPase is a house-keeping enzyme, also controlling the pH of the cytosol and of the extracellular compartment (cell wall) and its crucial role in pollen physiology has been discussed recently (Lang et al., 2014). An

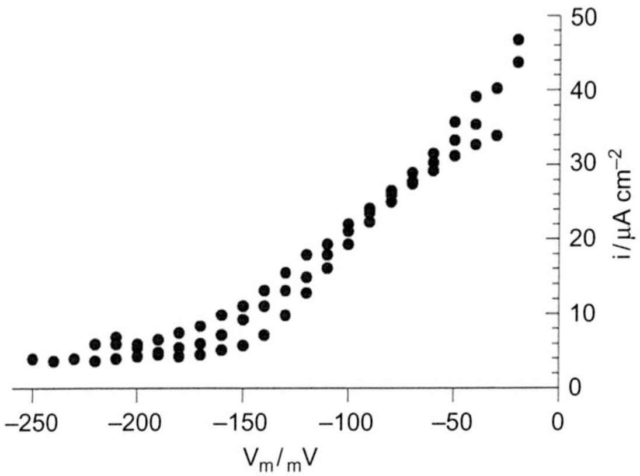

Fig. 2 Current voltage (IV)-relation of plasma membrane ATPase in lily pollen. IV curves between 250 and 0 mV were measured by two electrode voltage clamp in intact lily pollen grains under normal conditions and in the presence of 0.3 mM KCN and 0.4 mM SHAM (salicylhydroxamic acid) to inhibit ATP synthesis. The difference between both conditions is assumed to represent the ATP-dependent current mainly generated by the PM H^+ ATPase (Obermeyer & Blatt, 1995).

example of the current-voltage (IV) curve of the ATP-dependent current in pollen is shown in Fig. 2 to demonstrate the voltage dependence of the ATPase-generated H^+ current. However, this IV curves contain all electrogenic ATP-fuelled currents at the PM, including the possible contributions of the **Ca^{2+} ATPase** (ACA), the heavy metal ATPases (P1B, see Section 2.3) and probably to a minor proportion, the ABC transporters that have been detected in the pollen PM (Tables 1 and 2). The localization of AtACA9 to the plasma membrane and its important role in PT growth have been clearly demonstrated (Schiott et al., 2004) but so far, it is not known whether the Ca^{2+} transport is electrogenic because the highly endergonic transport of Ca^{2+} is probably balanced by a still unknown number of 'antiported' H^+ to account for the large energy difference (Bonza & DeMichaelis, 2010).

ABC transporters are also classified as primary active transporters and are known to be involved in pollen development and exine formation (Zhang & Liu, 2014). ABC proteins transport a large variety of substrates ranging from small inorganic and organic molecules (e.g. amino acids, sugars, nucleosides, vitamins and metal clusters) to larger organic compounds, including peptides, lipid molecules, oligonucleotides and polysaccharides (Verrier et al., 2008; Wilkens, 2015). Their putative function in transport of small solutes or ions during PT growth has not been investigated in detail. In guard cells, evidence for a sulphonylurea-inhibitable ABC

Table 2 Plasma Membrane Transporter in *Lilium longiflorum* Pollen

Functional Category (BIN name, MapMan (Thimm et al., 2004))	Transporter Family	No. Unigenes	Total No. Unigenes
transport.ABC transporters and multidrug resistance systems	ABC subfamily A transporter	5	
	ABC subfamily C transporter	33	
	ABC subfamily F transporter	2	
	ABC subfamily I transporter	6	
	MDR-type ABC subfamily B transporter	33	
	NAP12-like ABC subfamily G transporter	10	
	PDR-type ABC subfamily G transporter	15	
	WBC-type ABC subfamily G transporter	12	
	ZIF/TOM-type metal cation transporter	9	**125**
transport.amino acids	AAP-type amino acid transporter	6	
	ANT-type amino acid transporter	3	
	APC, amino acid-polyamine-organocation	2	
	AUX/LAX-type auxin transporter	1	
	CAT-type cationic amino acid transporter	7	
	GABP-type amino acid/GABA transporter	6	
	GAT-type GABA transporter	6	

Continued

Table 2 Plasma Membrane Transporter in *Lilium longiflorum* Pollen—cont'd

Functional Category (BIN name, MapMan (Thimm et al., 2004))	Transporter Family	No. Unigenes	Total No. Unigenes
	LAT-type amino acid transporter	4	
	LHT-type amino acid transporter	3	
	ProT-type proline transporter	2	
	SLC32-like transporter	12	
	SLC38-like transporter	15	
	Unspecific transport	1	**68**
transport.ammonium	AMT, ammonium transporter	8	**8**
transport.calcium	CCX-type cation:calcium cation exchanger	8	**8**
transport.cyclic nucleotide or calcium regulated channels	CNGC-type cyclic nucleotide-gated cation channel	21	**21**
transport.Major Intrinsic Proteins	NIP	11	
transport.Major Intrinsic Proteins	PIP	3	
transport.Major Intrinsic Proteins	TIP	1	**15**
transport.membrane system unknown	G3P-type organic phosphate/glycerol-3-phosphate permease	4	**4**
transport.metabolite transporters at the mitochondrial membrane	SLAC-type guard cell anion channel	4	**4**

transport.metal	CHX-type H^+/cation antiporter	29	
	COPT-type copper cation transporter	4	
	IAR-type metal cation transporter	1	
	Mn-CDF-type manganese cation transporter	2	
	ZIP13-like metal cation transporter	2	
	ZIP-type metal cation transporter	4	
	Zn-CDF-type zinc cation transporter	1	**43**
transport.misc	FBT, folate-biopterin transporter	5	
	MATE, Multidrug/Oligosaccharidyl-lipid/ Polysaccharide Flippase	22	
	Major Facilitator Symport (MFS) unspecific	4	
	NAT/NCS2, Nucleobase/Ascorbate Transporter	4	
	OCT/NiAP family	2	
	OCT-type organic cation transporter	1	
	PILS-type auxin transporter	4	
	Unknown function	2	**44**
transport.NDP-sugars at the ER	UTR-1/5-type nucleotide sugar transporter	4	**4**
transport.nucleotides	PUP-type organic cation transporter	5	**5**

Continued

Table 2 Plasma Membrane Transporter in *Lilium longiflorum* Pollen—cont'd

Functional Category (BIN name, MapMan (Thimm et al., 2004))	Transporter Family	No. Unigenes	Total No. Unigenes
transport.p- and v-ATPases	P1B, heavy metal ATPase	24	
	P2B, Ca^{2+} ATPase	2	
	P3A, H^{+} ATPase	27	
	P4, component of phospholipid flippase	21	
	P5, cation transporting ATPase	7	**81**
transport.peptides and oligopeptides	NRT1/PTR subfamily NPF3 nitrite transporter	2	
	NRT1/PTR subfamily NPF4 AIT/SP1-like transporter	2	
	NRT2-like nitrate anion transporter	3	
	NRT1/PTR-type oligopeptide transporter	1	
	OPT-type oligopeptide transporter	10	
	POT/PTR, proton-dependent oligopeptide transporter	1	
	YSL-type iron chelator transporter	6	**25**
transport.phosphate	PHT1, PHT2, PHT4-type phosphate transporter	11	**11**
transport.potassium	AKT/KAT-like K^{+} channel	5	
	Component of K^{+} channel, β subunit	1	
	KEA-type proton:potassium cation antiporter	6	
	KUP, potassium uptake permease	20	
	Unknown function	4	**36**

transport.sugars	ERD6-type monosaccharide transporter	2	
	INT-type inositol/proton transporter	7	
	OCT/NiAP family	4	
	OCT-type organic cation transporter	2	
	PLT-type polyol transporter	2	
	STP-type monosaccharide transporter	23	
	VGT-type hexose transporter	2	
	SUT/SUC-type sugar transporter	2	**44**
transport.sulphate	SULTR-type transporter	18	**18**
transport.unspecified anions	BOR-type borate transporter	4	
	CCC-type cation:chloride co-transporter	3	
	CLC-type anion/H^+ antiporter	6	**17**
transport.unspecified cations	ENT-type nucleoside transporter	2	
	MRS/MGT-type metal cation transporter	1	
	OCT-type organic cation transporter	1	
	SOS1-type proton:lithium/sodium cation antiporter	2	
	Unknown function	1	**7**
Total transporter unigenes			**588**

Data from the membrane proteome study of *Lilium longiflorum* pollen (Pertl et al., 2009) were re-analysed using the lily pollen transcriptome (Lang et al., 2015) as database. Annotations of lily pollen unigenes was performed with Mercator (Lohse et al., 2014). Unigene sequences were translated into amino acid sequences and served as a database for mass spectrometry analysis of the plasma membrane proteome using MaxQuant software. The numbers of unigenes of the pollen transcriptome database that were identified as plasma membrane transporters are given. Transporter family names also present in Arabidopsis pollen are underlined (Table 1).

transporter-mediated anion current has been observed (Forestier, Frangne, Eggmann, & Klein, 2003; Leonhardt et al., 2001).

Usually **V-type ATPases** are located at the tonoplast and other endomembranes, but for specific isoforms localization at the PM is assumed and reported in plants (Ratajcak, 2000) and pollen (Pertl et al., 2009). However, their role and localization at the plasma membrane is still to be investigated and therefore, the V-ATPase is not considered in the present pollen PM permeome.

2.2.2 K^+ Transport

Potassium ion transport in plants is mainly mediated by members of five transporter families: *Shaker*-type K^+ channels, tandem-pore K^+ channels (TPK), HAK/KUP/KT transporters, HKT and CHX (Y. Wang & Wu, 2013). In almost all plant cells studied so far by electrophysiological methods, K^+ conductance of the PM is dominated by the voltage-dependent opening of inward and outward rectifying channels of the ***Shaker*** family (Very et al., 2014), and the K^+ concentration is assumed as a major component in adjusting the osmotic potential of the cells (Sharma, Dreyer, & Riedelsberger, 2013). In PGs similar K^+ channel currents were detected (Fig. 3; lily (Griessner & Obermeyer, 2003; Obermeyer & Blatt, 1995); *Brassica* (Fan, Wang, & Wu, 2003; Fan, Wu, & Yang, 1999); *Arabidopsis* (Fan, Wang, Wang, & Wu, 2001; Mouline et al., 2002)). The responsible K^+ inward rectifying channels AtAKT6/SPIK and LilKT1 were identified in Arabidopsis and lily pollen, respectively (Mouline et al., 2002; Safiarian et al., 2015), whereas the K^+ outward rectifier AtSKOR or orthologous channels in other species are assumed to be responsible for the outward currents. These channels from the *Shaker* family are assumed to carry the bulk flow of endogenous K^+ currents accompanying tube growth and account for the K^+ uptake and release necessary for osmoregulation (Pertl, Poeckl, Blaschke, & Obermeyer, 2010). So far, no explicit localization of GFP-tagged LilKT1 channels could be detected at the plasma membrane of lily PTs although the K^+ current pattern suggests K^+ efflux at the tube shank and a large K^+ influx at the tube tip (reviewed in Holdaway-Clarke & Hepler, 2003).

In addition to the K^+ channels of the *Shaker* family, a large number of K^+:H^+ carriers from other families are associated with pollen tube growth (Li, Xu, Alli, & Yu, 2018) (Table 1). For instance, TRH1, a member of the **HAK/KUP/KT** family, is an essential component of K^+ homeostasis of root hair tip growth (Rigas et al., 2001) and might have a similar function in tip growth of PTs. In rice pollen, the interaction of high-affinity

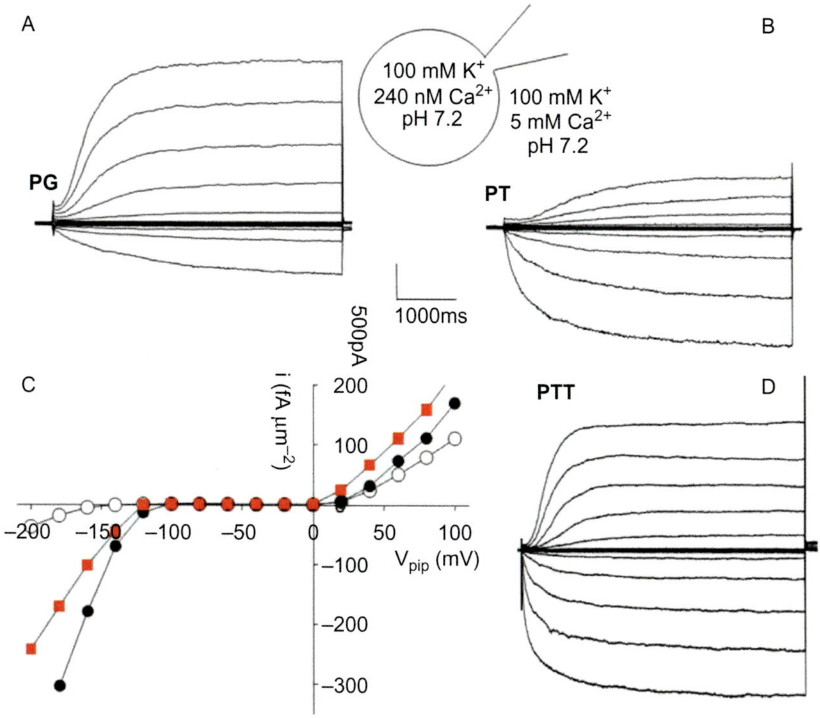

Fig. 3 K^+ channels in lily pollen protoplasts. Different types of protoplasts from *Lilium longiflorum* pollen were isolated and K^+ currents were measured by means of the patch-clamp method in whole cell configuration as described by Griessner and Obermeyer (2003) in detail. Voltage steps from −200 to +100 mV in 20 mV increments were applied and the respective currents recorded. (A) Protoplasts that were enzymatically isolated from pollen grains (PG), (B) from pollen tubes (PT) and (D) from pollen tube tips (PTT). (C) Normalized whole-cell currents of PG (○), PT (•) and PTT (□) protoplasts.

K^+ transporters (OsHAK1, 19, 20) with the receptor-like kinase RUPO was essential for K^+ homeostasis and prevented pollen tube from bursting (Liu et al., 2017).

Another set of K^+ transporters that were found in the pollen permeome data are members of the **KEA** (K^+ efflux antiporter) family, which may permeate K^+ either in a channel-like mechanism or via an H^+ antiporter mechanism (Fujisawa, Ito, & Krulwich, 2007). AtKEA5 showed an increase in expression in seedlings after osmotic shock (Zheng, Pan, Fan, & Qiu, 2013). However, its role in pollen physiology has not been studied yet.

Furthermore, high expression levels of 18 genes of the large cation/H^+ exchanger family (**CHX**) have been detected in male gametophytes and an important role in K^+ homeostasis and osmoregulation has been suggested (Sze et al., 2004). Interestingly, the expression of CHX12, CHX21 and CHX26 even increases in PTs (Table 1) indicating a necessary fine-tuning of K^+ uptake in cooperation with the PM H^+ ATPase to regulate the cytosolic osmotic potential.

2.2.3 Ca^{2+} Transport

Since the discovery of the tip-localized $[Ca^{2+}]_{cyt}$ gradient in growing PTs (Miller, Callaham, Gross, & Hepler, 1992; Nobiling & Reiss, 1987; Obermeyer & Weisenseel, 1991; Rathore, Cork, & Robinson, 1991) that works like a navigation system to guide the tube towards the ovules, many different models with putative and hypothetical calcium ion transporters that are assumed to contribute to the generation of this gradient have been proposed (see Liu & Hussey, 2014 for a recent model). However, experimental evidence for Ca^{2+} transporters that facilitate Ca^{2+} uptake in plant cells is rare and mainly based on electrophysiological characterization of Ca^{2+}-permeable channels, which also allow nonselective cation transport (e.g. Wu et al., 2011). The search for Ca^{2+} channel genes revealed only two families with animal counterparts, cyclic nucleotide-gated channels (CNGCs) and glutamate receptor-like (GLRs), and remains challenging (Swarbeck, Colaco, & Davies, 2013).

In pollen, **GLR-like channels** (e.g. GLR1.2, GLR3.7) were characterized that allow Ca^{2+} influx into the tube tip upon binding of the amino acid D-serine (Michard et al., 2011) and which are involved in Ca^{2+} homeostasis (Wudick et al., 2018).

Members of the **CNGC** family are involved in many physiological processes ranging from heat stress and thermosensing (Finka, Cuendet, Maathuis, Saidi, & Goloubinoff, 2012) to nutrient uptake (Quiles-Pando et al., 2013). An active role for CNGC 7, CNGC 8 and CNGC 16 has been attributed to pollen physiology (Tunc-Ozdemir, Rato, et al., 2013; Tunc-Ozdemir, Tang, et al., 2013; see also Table 1). Most pollen from *cngc7* and *cngc8* knock-out mutants burst immediately after germination and fluorescence-tagged CNGC 7 was localized in the PM of pollen tube flanks, not at the tip. Pollen mutants lacking CNGC16 were hypersensitive to external calcium chloride and CNGCs are involved in heat response signalling. In addition, CNGC18 was characterized as a Ca^{2+}-permeable channel responsible for tube guidance to the ovule (Gao et al., 2016).

Four isoforms of the **OSCA** (reduced hyperosmolality-induced Ca^{2+} increase) family showed a high expression level in Arabidopsis pollen with

AtOSCA2.1 expression level increasing nearly by 80% in PTs compared to PGs (Table 1). OSCA proteins form hyperosmolality-gated Ca^{2+} permeable channels that allow Ca^{2+} influxes upon hyperosmotic treatment and are involved in osmotic signalling in guard and root cells (Yuan et al., 2014), thus assigning them a role as putative osmosensors in pollen, too. However, no OSCA orthologues could be noticed in lily pollen, so far.

High expression levels in Arabidopsis pollen were also found for **CCX/CAX-type** (calcium/cation exchangers) transporters that were also detected in the lily pollen PM proteome (Table 2). Apart from Ca^{2+}/Na^{+} exchangers also belonging to this family, many transporters use a pH gradient to energize the transport of Ca^{2+} and other divalent cations across membranes of acidic organelles (e.g. vacuoles (Pittman & Hirschi, 2016)) but a localization in the PM is also predicted for isoforms listed in Table 1. A stoichiometry of Ca^{2+} and one H^{+} per transport cycle favours an electrogenic transport mechanism, which therefore may be affected by membrane voltage and the pH gradient across the membrane.

2.2.4 Anion Transport

Anion transport in pollen has been detected and characterized by ion-sensitive vibrating probe (Messerli, Smith, Lewis, & Robinson, 2004; Zonia, Cordeiro, Tupy, & Feijó, 2002) and patch-clamp measurements (Tavares et al., 2011). Patch-clamp experiments revealed a Ca^{2+}-dependent, large Cl^{-} conductance channel in lily pollen protoplasts from grains and tubes that might be responsible for the large anion currents detected with the vibrating probe technique. The presented database search shows a high expression level of **SLAC-type anion channels** in Arabidopsis and lily pollen, and recent experiments identified AtSLAH 3 in Arabidopsis pollen as a target for Ca^{2+}-dependent protein kinases (Gutermuth et al., 2013). Furthermore, other anion carriers were detected in pollen of both species. A cation/chloride cotransporter (**CCC transporter**) is expressed in Arabidopsis and lily pollen (Tables 1 and 2 (Colmenero-Flores et al., 2007)). Although **ClC transporters** (chloride/H^{+} antiporters) have been mainly associated with the tonoplast and endomembranes, the predicted PM localization for the ClC-c (At5g49890, Table 1) allocates these carriers a putative function in adjusting the membrane voltage due to coupling the influx of H^{+} with an efflux of chloride or anions, thus leading to local depolarization of the PM.

Putative transporters for nitrate and sulphate have been found, too. The **NRT2-type transporters** contribute to the high affinity NO_3^{-} uptake system and AtNRT2.1 was observed in Arabidopsis roots (Yong, Kotur, & Glass, 2010). NRT1-type transporters facilitate nitrate uptake in roots and as for NRT1.1 can also be involved in nitrate signalling (Bougyon et al., 2015). The **SULTR-type transporters** were identified in both pollen databases. Although the SUBA database predicted a PM localization for SULTR3.1, it has been localized in chloroplasts of Arabidopsis leaves (Cao et al., 2013). Its role and localization in pollen is still unknown. Phosphate transporters of the **PHT family** were also found in pollen and a high phosphate concentration could be detected in the pollen when plants were grown on phosphate-rich soil (Lau & Stephenson, 1994). Despite this, and although they could also be detected in maize pollen (Nagy et al., 2006), their role during PT growth has not been studied yet.

In patch-clamp experiments the **mechanosensitive channel MSL8** exhibited a permeability coefficient ratio $P_{Cl^-} : P_{Na^+} = 6.3$ which correlates well with an anion selective channel (Hamilton, Jensen, Maksaev, Katims, & Sherp, 2015). This mechanosensitive channel maintains cellular integrity during rehydration and an optimum osmotic potential for germination and, thus, is likely to be involved in osmoregulation and osmosensing in pollen.

2.3 Metabolite Transport

Although PGs are filled with nutrients during the last phases of pollen development, they are probably also fed by nutrients supplied by stylar tissue (Reinders, 2016). Recently, the possible function of sugars and amino acids, and their corresponding transporters during pollen development, tube growth and fertilization has been characterized (Borghi & Fernie, 2017). Boron transporters (BOR and NIP), which may provide the supply with an essential element for tube growth, were also detected in addition to different nucleotide sugar transporters, organic cation (OCT), pyridine and purine permeases (PUP) and oligopeptide transporters, whose function in pollen is still unknown.

2.3.1 Sugar Transport

On their way to the ovule, pollen tubes consume large amounts of energy supplied by aerobic respiration of the mitochondria (Linskens & Schrauwen, 1966) or by ethanolic fermentation if oxygen supply becomes limited in the stylar tissue (Obermeyer et al., 2013; Tagede, Dupuis, & Kuhlemeier, 1999). In both cases the production of energy relies on the stocks of sugars, starch and lipids in the mature PG. In addition to energy, lipids also are needed for the synthesis of new membrane phospholipids and sugars for cell wall synthesis. Furthermore, degradation of starch and catabolism of lipids add additional osmolytes to the cytosol and thus, change the osmotic pressure, which needs to be balanced by water uptake and/or efflux of osmolytes. In lily pollen, a proton-coupled co-transport of sucrose and glucose has been measured (Deshusses, Gumber, & Loewus, 1981), and several sugar transporters might be responsible for this transport (Table 2, transport.sugars). In Arabidopsis, the main sugar transporter families expressed in pollen are the monosaccharide specific **ERD6-type** and **STP-type** and the sucrose transporter family **SUT/SUC** as well as the **SWEET efflux** transporter (Table 1). The pollen grain is pre-loaded with mRNAs of many STP and ERD6 isoforms with AtSTP11 and ERDL6 showing some of the highest expression values. Furthermore, expression of AtSTP4, AtSTP10 and AtSTP11 increases during PT growth by 50–70% (Büttner, 2010). Similarly high increases in expression values were also observed for SUT/SUC-type transporter isoforms (e.g. AtSUC7, 8, and 9 in Table 1). Members of the sugar efflux transporter family (SWEET) showed only moderate expression values. Although the expression of other sugar transporters can be observed in pollen (INT: inositol/H^+, PLT: polyol monosaccharide:H^+), the main components for sugar transport are supposedly built by the four previously mentioned families, which allow H^+-coupled transport of sugars in pollen. However, it has to be kept in mind that high sugar uptake rates will cause local depolarization of the PM due to co-transport with protons and charges need to be balanced by K^+ efflux as demonstrated in phloem loading (Deeken et al., 2003), which might also contribute to balancing the osmotic pressure of the pollen tube cytosol.

2.3.2 Amino Acid/Peptide Transport

Similarly to sugars, amino acids are also loaded and stored in the mature PG and amino acid transporters of the **LHT**, **ProT**, **ANT** and **OPT families** were highly expressed with highest expression values >15. Similar to other sink organs, PGs are supplied by the tapetum tissue with all necessary nutrients. In contrast to sugars, amino acids may carry a charge across the PM depending on the isoelectric point of the amino acid species and pH value, in addition to the co-transported protons. Probably due to the large variation in amino acids, transporters of eight families are expressed in pollen. The physiological function of amino acid transport in pollen is still under investigation and the absorbed amino acids may contribute to the osmotic balance in addition to protein synthesis as has been suggested for proline

(Lehmann, Funck, Szabados, & Rentsch, 2010). Surprisingly, the oligopeptide carrier OPT1 (At5g55930) shows the highest expression level of all observed pollen PM transport proteins (Table 1). The expression of OPT1, OPT3 and OPT8 in pollen was confirmed experimentally (Stacey, Osawa, Patel, Gassmann, & Stacey, 2006) but their function is far from being clear (Lubkowitz, 2011).

2.3.3 Boron Transport

Boric acid is an essential component of almost all synthetic germination media for pollen culture (e.g. Dickinson, 1978; Rodriguez-Enriquez, Mehdi, Dickinson, & Grant-Downton, 2013) and it is assumed that boric acid cross-links and thus stabilizes the newly built pectin layer at the growing tip (Iwai et al., 2006). Furthermore, external addition of boric acid stimulated the PM H^+ ATPase in lily pollen (Obermeyer, Kriechbaumer, Strasser, Maschessnig, & Bentrup, 1996) but it is not known if boric acid acted from the external or cytosolic side. Boric acid may diffuse passively through the plasma membrane and the identified **BOR-type** transporters in pollen might be involved in the extrusion of excess of boron but so far, they were only characterized as boron efflux transporters for xylem loading (Takano, Miwa, & Fujiwara, 2008; Takano et al., 2002). Although the boron-transporting aquaporin homologues NIP5.1 and NIP6.1 are only slightly expressed in Arabidopsis pollen, the highly expressed **NIP4.1** and **NIP4.2** might contribute to boron uptake due to their ability of transporting the small solutes ammonium, urea, boric acid and H_2O_2, too (Perez di Giorgio et al., 2016).

2.4 Heavy Metal Ion Transport

Most surprisingly a large number of metal transporters were identified by the database searches in pollen of both species. One would expect these transporters to be highly active in roots to allow uptake of heavy metal micro-nutrients (e.g. copper, iron) or to be involved in their detoxification (Andres-Colas et al., 2006; Williams & Mills, 2005). Their role in pollen in which metal ions severely disturb germination and tube growth remains to be clarified (Sawidis & Reiss, 1995). In addition to the **P1B ATPase AtHMA5** several transition metal transporters with a predicted localization in the PM are likely highly expressed in pollen, e.g. **ZIP13-like** transition metal cation transporter, **Mn-CDF-type** (metal tolerance protein, MTP), **COPT-type** (Tables 1 and 2). In Arabidopsis, the high affinity copper transporter COPT1 was highly expressed in trichomes, stomata, root tips as well as in pollen and transgenic plants with reduced COPT1 expression showed malformations of mature PGs (Sancenon et al., 2004). Most of the studied MTPs were also involved in detoxification and were detected in organelle membranes of plant cells and not in the PM as predicted for the pollen (Ricachenevsky, Menguer, Sperotto, Williams, & Fett, 2014). Similarly, the **DTX-type, MATE-related efflux** carriers DTX16 and DTX34 have a physiological function in detoxification (Li, He, Pandey, Tsuchiya, & Luan, 2002). In summary, the presence and the sometimes high level of expression of heavy metal transporters are remarkable and their possible roles still need to be studied.

3. CONCERTED ACTION OF ION TRANSPORT LEADS TO SPATIAL SELF-ORGANIZATION AND DRIVES TUBE GROWTH

Our search of public databases with very simple selection criteria has yielded a list of putative PM-localized transport proteins with an at least twice as high expression levels than average expression in Arabidopsis (Table 1). It has to be noted that not all transport proteins might have been picked by this search or some of the listed transporters might not be relevant at all for PT growth, and obviously all PM proteins like receptor kinases or structural proteins without transport activities are lacking. As reasoned later, these membrane proteins may establish and maintain local distributions of transporters but ion transporters will set the pattern of the three-dimensional allocation in the plasma membrane. After listing most of the essential ion and nutrient transporters, we are confronted with the following questions: how do they all work together and more interestingly, how do they all function in the same membrane, interfering with each other due to their electrogenic transport properties, their voltage dependency and sharing of common substrates (e.g. H^+)? A well-studied system of the concerted action of ion transporters that finally increases the turgor pressure, thereby leading to cell volume increase, are the guard cells (Jezek & Blatt, 2017; Mäser, Leonhardt, & Schroeder, 2003; Pandey, Zhang, & Assmann, 2007; Roelfsema & Hedrich, 2005). Simply speaking, the temporal succession of specific ion transport processes in the guard cell leads to a proton pump-energized net influx of K^+, which increases the osmotic pressure that drives water uptake and, finally, the sharp increase in turgor pressure due to the stiff cell walls allows limited volume changes and opens the stomata. As already pointed out by Feijo and co-workers (Michard et al., 2017), the PT elongation might function similarly but with a spatial organization of the transport events that finally allow an influx of water to extend the PT. All key ion transporters necessary to open or close the stomata are also present in the pollen especially those that are localized in the PM (Jezek & Blatt, 2017; Michard et al., 2017). They may not be exactly the same gene or isoform but resemble in function. As described above, the pollen contains PM H^+ ATPases, K^+ inward and outward channels, Ca^{2+} permeable transporters as well as anion channels that carry the bulk of charge fluxes detected in the endogenous current pattern. Other transport systems driven by the PM H^+ gradient may fine-tune the current pattern along the tube.

3.1 Examples of Heterogeneous Distribution in the Plasma Membrane

The net flux of charges or ions at a specific membrane location depends on the activity and the amount of the respective transporter at this position. An example is given by the channel-mediated K^+ currents in lily pollen (Fig. 3). The K^+ inward- and outward-rectifying currents have been measured in protoplasts prepared from pollen grains or from two regions of the pollen tube, shank and tip. It can be clearly seen that the K^+ outward current is larger than the K^+ inward current at the PG whereas the K^+ inward current is higher in the PT protoplast and both the inward and outward currents have similar levels in PT tip protoplasts (Fig. 3C). The permeome database (Table 2) suggests that, in Arabidopsis, inward and outward K^+ currents are generated by the voltage-dependent opening of AtAKT6/SPIK-like and AtSKOR-like channels, respectively. In lily pollen, an AKT-like channel has been identified in lily pollen (LilKT1 (Safiarian et al., 2015)). The higher activity of LilKT1 in the pollen tube might be achieved by incorporating more K^+ inward channels into the tube PM via exocytosis or by spatially regulating the activity of the K^+ channels at their different locations. A similar situation has to be assumed for the K^+ outward channel: either its abundance or its activity is higher in the tube tip than in the tube shank. These two possibilities imply either a local incorporation of K^+ channels into the PM or spatially distributed regulatory systems for K^+ channels.

Differences in the abundance of membrane proteins between specific regions of a single cell can be achieved by localized transport to, and exocytosis of vesicles with, the plasma membrane. This process includes sorting and clustering in the ER and Golgi membranes of transporters with the same final destination and their transfer into transport vesicles targeted to a specific membrane region. In the growing pollen, the majority of transport vesicles heads towards the tube tip to deliver cell wall and membrane materials as well as transport proteins. As such, most studies have focussed on the tip region (see recent reviews by Grebnev, Ntefidou, & Kost, 2017; Rakusova & Geitmann, 2017). When looking at the distribution of the PM H^+ ATPase in lily pollen, much less ATPase proteins are observed in the PM of the tip region with no ATPases being observed in the secretory vesicles (Obermeyer, Lützelschwab, Heumann, & Weisenseel, 1992). Also, PM H^+ ATPases are lacking in the tip of tobacco PTs (Certal et al., 2008). This experimental evidence indicates that other exocytosis sites in the PT exist in addition to the tip region, which however have not yet been observed. Recently, cornichon proteins were reported to be responsible for sorting, trafficking and localization GLRs in Arabidopsis pollen (Wudick et al., 2018). Although the presence of other potential sites for specific exocytosis and membrane protein insertion is likely, this still does not resolve the problem of what determines the location of these exocytosis sites and their organization.

An almost similar problem exists in nerve cells where a heterogeneous distribution of Na^+, K^+ and Ca^{2+} channels determines the function and the cell type, with different types of voltage-gated Na^+, K^+ and Ca^{2+} channels at the presynaptic nerve terminals, at the nodes of Ranvier, and at dendritic region where they generate neurotransmitter release, action potentials and postsynaptic potentials, respectively (Lai & Jan, 2006). In the nerve cells, the differential distribution of ion channels is achieved by a combination of directed targeting, transcytosis (after being firstly incorporated into a target subcellular region, proteins are incorporated and relocated to another region) and selective retention via interaction with anchor proteins. All these trafficking steps require specific amino acid motifs in the channel sequence, e.g. for ER export, axonal targeting or ankyrin binding. In plants, the trafficking of KAT1 to the PM has been investigated in more details showing an insertion via regulated exo- and endocytosis with a di-acidic motif involved (Meckel, Hurst, Thiel, & Homann, 2004; Mikosch, Hurst, Hertel, & Homann, 2006; Sutter, Campanoni, Tyrrell, & Blatt, 2006). These phenomena finally resulted in a speckled or brindled localization to PM subdomains similar to what has been reported for the pollen LilKT1 channel (Safiarian et al., 2015). Further amino acid motif-dependent incorporation into the PM has been reported from the KAT2 K^+ channel and a PIP2 aquaporin (Chevalier, Bienert, & Chaumont, 2014; Nieves-Cordones et al., 2014).

Brindled patterns of channel localization were also reported from the synaptic membrane of motor nerve terminals, in which Ca^{2+} channels were detected with a fluorescent, specific inhibitor ω-conotoxin (Cohen, Jones, & Angelides, 1991). The patterning of Ca^{2+} channels was inversely correlated with acetylcholine receptors in the same membrane leading to zones concentrated with Ca^{2+} channels adjacent to zones with acetylcholine receptors and so on (Fig. 2).

3.2 Pattern Formation by Electrophoretic Mobility of Membrane Proteins

Proteins in a membrane can be viewed as charged particles floating in a semi-fluidic two-dimensional plane. Application of a constant external electrical field will manipulate the membrane potential locally and redistribute such particles according to their charges, thus leading to accumulation of negatively charged proteins in regions of the cell membrane close to the anode (Jaffe, 1977). If we assume active ion transporters in a biological membrane, their own activity changes the electric field across the membrane in a proximate area and may result in an accumulation of similar transporters in specific areas (Fig. 4). For instance, an active K^+ outward rectifier will depolarize whereas a H^+ pump hyperpolarizes the PM with the change in voltage potential (V_M) depending on the number of shifted charges per time (=current). Given the large number of different electrogenic transporters in the pollen PM, the activity of the transporters changes V_M. So far, the V_M was not studied systematically, but in lily PGs a value of approximately $-120\,mV$ was measured resembling the equilibrium potential for K^+ (Obermeyer & Blatt, 1995) whereas in growing lily PTs a much higher V_M of $-62 \pm 21\,mV$ has been detected (Engel et al., 1997). These differences in V_M may correlate with the activity of different transporters in the PM of pollen grains and tubes, e.g. higher K^+ inward channel activity in pollen tubes depolarizes the PM (Fig. 3A and B).

In theory, ion channels in a fluid mosaic model membrane have the potential of self-organization and spontaneous pattern formation due to the interaction of ion channel activity and membrane potential (Fromherz, 1988; Fromherz & Kaiser, 1991; Hilt & Zimmermann, 2007). A spatial distribution of ion channels can be achieved by considering four basic principles of biomembranes: (i) membrane proteins move in a viscous fluid due to lateral forces or diffusion, (ii) specific ion conductances and the associated reversal potentials regulate the membrane potential, (iii) the specific ion conductances are caused by channel proteins, and (iv) membrane proteins are electrically charged and move freely in an applied electrical field (Fromherz, 1988). With these considerations various spatial patterns of ion channel densities in differentially shaped cells could be modelled.

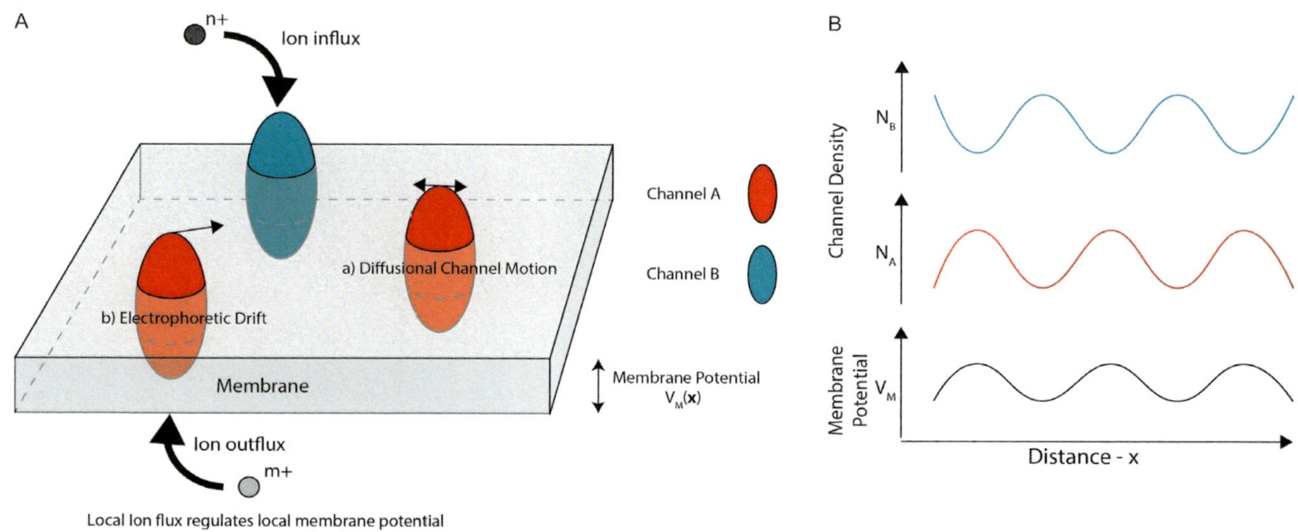

Fig. 4 Spontaneous pattern formation of mobile ion channels in the plasma membrane. (A) The local ion flux through mobile ion channel species *'A'* and *'B'* within the plasma membrane influences the local membrane potential. In turn these membrane channels can move due to (a) diffusion and (b) electrophoretic drift arising from local variations in the membrane potential. (B) Such processes can lead to accumulation of like channels with similar physical properties in certain regions in the membrane giving rise to modulations in channel density and local membrane potential.

How can we transfer these considerations to the pollen system? One might assume that in the mature PG the transporter proteins are arbitrarily distributed in the PM and during re-hydration the PM H^+ ATPases are activated, the PG swells and turgor pressure increases leading to incidental activity of K^+ and anion channels to adjust V_M, pH_{cyt} and other cytosolic ion concentrations. Due to their occasional activity, local changes in V_M cause an electrophoretic separation of H^+ pumps and channels generating an electrical dipole and the site of the major influx of positive charges (=efflux of negative charges) determines the future germination site or pore (Weisenseel & Jaffe, 1976). During tube growth the hypothesis becomes more and more complicated as one does not know at which site new transporters are incorporated. If all transporter proteins are inserted in the tube tip together with the new membrane material, they have to travel long distances in the membrane to find their specific location. This may take a long time for an individual membrane protein by just diffusing along the tube's membrane. But considering a constant supply of new transporter proteins at the tip, which are only active until they reach their final destination at the tube surface, and taking into account the fast growth speed of circa $10\,\mu m\,min^{-1}$, a membrane protein inserted into the tube tip PM will be able to reach a position $50\,\mu m$ behind the tip, for instance, in just 5 min. In addition, a K^+ inward channel incorporated into the tip PM will become active only at a tube position where $V_M < -100\,mV$ (Fig. 3), which occurs at a tube position with a high activity of PM H^+ ATPases. The most critical point in this hypothesis is that V_M is not an intrinsic parameter but changes along the pollen tube! When considering a PT as an electric cable with a high cytosolic resistance, currents dissipate mainly across the membrane and only a minor component along the cytosol according to the principles of Ohm's law, meaning stable local differences in V_M along the PT. So far, we do not know much about the electrical properties of the pollen tube and only indirect evidence of the cable characteristics from tubular cells was obtained (Meharg, Maurousset, & Blatt, 1994) whereas other electrophysiology studies in plant tubular cells simply neglected the cable characteristics (Gutermuth et al., 2018; Lew, 1996). However, in the pollen system one may consider an additional direct link between transporters, namely the electrochemical potential difference of H^+ across the plasma membrane. The transporters using H^+-coupled transport of other ions or uncharged molecules, e.g. amino acids or sugars, simply interfere with other transport processes via the chemical potential gradient (pH difference) across the PM without exhibiting electrogenic transport properties.

So far, all other signalling cascades and biological regulation pathways of proteins including transporters have been neglected in this view. Our point of view here is to emphasize the idea that the self-organization of the transport proteins in the membrane is the major step to determine the spatial organization, which are then fine-tuned by other regulation mechanisms, e.g. phosphorylation, Ca^{2+}/calmodulin binding, 14-3-3 proteins or other factors that influence the activity and as a consequence the localization of the respective transporter. It also might be necessary to consider membrane protein complexes as functional transporter units (Pertl-Obermeyer, Schulze, & Obermeyer, 2014) that float within the membrane in response to their activity, and furthermore the mobility of single transporters or transporter units might be restricted to specific membrane areas, e.g. lipid rafts (Wu, Sanchez Rodriguez, Pertl-Obermeyer, Obermeyer, & Schulze, 2013).

Nevertheless, the spatial changes in V_M along the PT have to be measured to support and verify this hypothesis. Modern microscopy techniques (single plane illumination microscopy) and fluorescence labelling of specific membrane transporters in combination with electrophysiological techniques (two-electrode voltage clamp) can detect V_M and the localization of the transporter while measuring its activity. Specific blockade of the activity of the labelled transporter is expected to alter V_M in the proximity of the transporter and finally, lead to a re-distribution of the transporter, provided that this protein is not held in position by cytoskeletal elements, as discussed above.

4. CONCLUSION

Our knowledge about the role of pollen ion transporters is constantly increasing and an ever increasing number of transport proteins with important functions during tube growth are described, resulting in a large level of information that can be extracted from published databases. However, to evaluate the contribution of each transporter to the complex and fine-regulated system of tip growth, we need experimental information on its biophysical transport properties, its trafficking route towards the PM and its specific spatial location in the PM. An additional level of complexity is brought by the regulation of the membrane transporters by signalling pathways and their interaction with other membrane proteins to build functional complexes in specific membrane compartments.

ACKNOWLEDGEMENTS

Research in GOs lab is partially financed by the Austrian Research Fund (FWF, P29626). We thank Ana Gimeno for helping with the data search. Thanks to Prof. Dr. José Feijó (Univ. Maryland, USA) for providing the raw array data of Becker et al. (2003).

REFERENCES

Andres-Colas, N., Sancenon, V., Rodriguez-Navarro, S., Mayo, S., Thiele, D. J., Ecker, J. R., et al. (2006). The Arabidopsis heavy metal P-type ATPase HMA5 interacts with metallochaperones and functions in copper detoxification of roots. *The Plant Journal*, *45*, 225–236.

Becker, J. D., Boavida, L. C., Carneiro, J., Haury, M., & Feijo, J. (2003). Transcriptional profiling of Arabidopsis tissues reveals the unique characteristics of the pollen transcriptome. *Plant Physiology*, *133*, 713–725.

Benkert, R., Obermeyer, G., & Bentrup, F.-W. (1997). The turgor pressure of growing lily pollen tubes. *Protoplasma*, *198*, 1–8.

Bock, K. W., Honys, D., Ward, J. M., Padmanaban, S., Nawrocki, E. P., Hirschi, K. D., et al. (2006). Integrating membrane transport with male gametophyte development and function through transcriptomics. *Plant Physiology*, *140*, 1151–1168.

Bonza, M. C., & DeMichaelis, M. I. (2010). The plant Ca^{2+}-ATPase repertoire: Biochemical features and physiological functions. *Plant Biology*, *13*, 421–430.

Borghi, M., & Fernie, A. R. (2017). Floral metabolism of sugars and amino acids: Implications for pollinators' preferences and seed and fruit set. *Plant Physiology*, *175*, 1510–1524.

Bougyon, E., Brun, F., Meynard, D., Kubes, M., Pervent, M., Leran, S., et al. (2015). Multiple mechanisms of nitrate sensing by Arabidopsis nitrate transceptor NRT1.1. *Nature Plants*, *1*, 15015. https://doi.org/10.1038/nplants.2015.15.

Büttner, M. (2010). The Arabidopsis sugar transporter (AtSTP) family: An update. *Plant Biology*, *12*, 35–41.

Cao, M.-J., Wang, Z., Wirtz, M., Hell, R., Oliver, D. J., & Xiang, C.-B. (2013). SUTR3.1 is a chloroplast-localised sulfate transporter in Arabidopsis thaliana. *The Plant Journal*, *73*, 607–616.

Certal, A. C., Almeida, R. B., Carvalho, L. M., Wong, E., Moreno, N., Michard, E., et al. (2008). Exclusion of a proton ATPase from the apical membrane is associated with cell polarity and tip growth in *Nicotiana tabacum* pollen tubes. *The Plant Cell*, *20*, 614–634.

Cheung, A. Y., & Wu, H.-M. (2007). Structural and functional compartmentalization in pollen tubes. *Journal of Experimental Botany*, *58*, 75–82.

Chevalier, A. S., Bienert, G. P., & Chaumont, F. (2014). A new LxxxA motif in the transmembrane helix 3 of maize aquaporins belonging to the plasma membrane intrinsic protein PIP2 group is required for their trafficking to the plasma membrane. *Plant Physiology*, *166*, 125–138.

Cohen, M. W., Jones, O. T., & Angelides, K. J. (1991). Distribution of Ca^{2+} channels on frog motor nerve terminals revealed by fluorescent ω-conotoxin. *Journal of Neuroscience*, *11*, 1032–1039.

Cole, R. A., & Fowler, J. E. (2006). Polarized growth: Maintaining focus on the tip. *Current Opinion in Plant Biology*, *9*, 579–588.

Colmenero-Flores, J. M., Martınez, G., Gamba, G., Vazquez, N., Iglesias, D. J., Brumos, X., et al. (2007). Identification and functional characterization of cation-chloride cotransporters in plants. *The Plant Journal*, *50*, 278–292.

Dai, S., Chen, T., Chong, K., Xue, Y., Liu, S., & Wang, T. (2007). Proteomics identification of differentially expressed proteins associated with pollen germination and tube growth reveals characteristics of germinated *Oryza sativa* pollen. *Molecular & Cellular Proteomics*, *6*, 207–230.

Damineli, D. S. C., Portes, M. T., & Feijo, J. (2017). One thousand and one oscillator at the pollen tube tip: The quest for a central pacemaker revisited. In G. Obermeyer & J. Feijo (Eds.), *Pollen tip growth: From biophysical aspects to systems biology* (pp. 391–411). Cham, Switzerland: Springer Nature.

Deeken, R., Geiger, D., Fromm, J., Korroleva, O., Ache, P., Langenfeld-Heyser, R., et al. (2003). Loss of the AKT2/3 potassium channel affects sugar loading into the phloem of Arabidopsis. *Planta*, *216*, 334–344.

Deshusses, J., Gumber, S. C., & Loewus, F. A. (1981). Sugar uptake in lily pollen. A proton symport. *Plant Physiology*, *67*, 793–796.

Dickinson, D. B. (1978). Influence of borate and penta-erythritol concentrations on germination and tube growth of *Lilium longiflorum* pollen. *Journal of the American Society for Horticultural Science*, *103*, 413–416.

Engel, E., Richter, K., Obermeyer, G., Briza, P., Kungl, A. J., Simon, B., et al. (1997). Immunological and biological properties of Bet v 4, a novel birch pollen allergen with two EF-hand calcium-binding domains. *Journal of Biological Chemistry*, *272*, 28630–28637.

Fan, L.-M., Wang, Y.-F., Wang, H., & Wu, W.-H. (2001). *In vitro* pollen germination and characterization of the inward potassium currents in *Arabidopsis* pollen grain protoplasts. *Journal of Experimental Botany*, *52*, 1603–1614.

Fan, L.-M., Wang, Y.-F., & Wu, W.-H. (2003). Outward K^+ channels in *Brassica chinensis* pollen protoplasts are regulated by external and internal pH. *Protoplasma*, *220*, 143–152.

Fan, L.-M., Wu, W.-W., & Yang, H.-Y. (1999). Identification and characterization of the inward K^+ channel in the plasma membrane of *Brassica* pollen protoplasts. *Plant Cell Physiology*, *40*(8), 859–865.

Feijó, J. A., Malhó, R., & Obermeyer, G. (1995). Ion dynamics and its possible role during in vitro pollen germination and tube growth. *Protoplasma*, *187*, 155–167.

Feijó, J. A., Sainhas, J., Holdaway-Clarke, T., Cordeiro, S., Kunkel, J. G., & Hepler, P. K. (2001). Cellular oscillations and the regulation of growth: The pollen tube paradigm. *BioEssays*, *23*(1), 86–94.

Finka, A., Cuendet, A. F., Maathuis, F. J., Saidi, Y., & Goloubinoff, P. (2012). Plasma membrane cyclic nucleotide-gated calcium channels control land plant thermal sensing and acquired thermotolerance. *The Plant Cell*, *24*, 3333–3348.

Forestier, C., Frangne, N., Eggmann, T., & Klein, M. (2003). Differential sensitivity of plant and yeast MRP (ABCC)-mediated organic anion transport processes towards sulfonylureas. *FEBS Letters*, *554*, 23–29.

Fromherz, P. (1988). Self-organization of the fluid mosaic of charged channel proteins in membranes. *Proceedings of the National Academy of Sciences of the United States of America*, *85*, 6353–6357.

Fromherz, P., & Kaiser, B. (1991). Stationary patterns in membranes by nonlinear diffusion of ion channels. *Europhysics Letters*, *15*, 313–318.

Fujisawa, M., Ito, M., & Krulwich, T. A. (2007). Three two-component transporters with channel-like properties have monovalent cation/proton antiport activity. *Proceedings of the National Academy of Sciences of the United States of America*, *104*, 13289–13294.

Gao, Q. F., Gu, L. L., Wang, H. Q., Fei, C. F., Fang, X., Hussain, J., et al. (2016). Cyclic nucleotide-gated channel 18 is an essential Ca^{2+} channel in pollen tube tips for pollen tube guidance to ovules in Arabidopsis. *Proceedings of the National Academy of Sciences of the United States of America*, *113*, 3096–3101.

Grebnev, G., Ntefidou, M., & Kost, B. (2017). Secretion and endocytosis in pollen tubes: Models of tip growth in the spot light. *Frontiers in Plant Science*, *8*, 154. https://doi.org/10.3389/fpls.2017.00154.

Griessner, M., & Obermeyer, G. (2003). Characterization of whole-cell K^+ currents across the plasma membrane of pollen grain and pollen tube protoplasts of *Lilium longiflorum*. *Journal of Membrane Biology*, *193*, 99–108.

Grobei, M. A., Qeli, E., Brunner, E., Rehrauer, H., Zhang, R., Roschitzki, B., et al. (2009). Deterministic protein inference for shotgun proteomics data provides new insights into *Arabidopsis* pollen development and function. *Genome Research, 19*, 1786–1800.

Gutermuth, T., Herbell, S., Lassig, R., Brosche, M., Romeis, T., Feijo, J., et al. (2018). Tip-localized Ca^{2+}-permeable channels control pollen tube growth via kinase-dependent R- and S-type anion channel regulation. *New Phytologist, 218*(3), 1089–1105. https://doi.org/10.1111/nph.15067.

Gutermuth, T., Lassig, R., Portes, M. T., Maierhofer, T., Romeis, T., Borst, J. W., et al. (2013). Pollen tube growth regulation by free anions depends on the interaction between the anion channel SLAH3 and calcium-dependent protein kinases CPK2 and CPK20. *The Plant Cell, 25*, 4525–4543.

Hamilton, E. S., Jensen, G. S., Maksaev, G., Katims, A., Sherp, A. M., & Haswell, E. S. (2015). Mechanosensitive channel MSL8 regulates osmotic forces during pollen hydration and germination. *Science, 350*, 438–441.

Han, B., Chen, S., Dai, S., Yang, N., & Wang, T. (2010). Isobaric tags for relative and absolute quantification-based comparative proteomics reveals the features of plasma membrane-associated proteomes of pollen grains and pollen tubes from *Lilium davidii*. *Journal of Integrative Plant Biology, 52*, 1043–1058.

Hepler, P. K., & Winship, L. J. (2015). The pollen tube clear zone: Clues to the mechanism of polarized growth. *Journal of Integrative Plant Biology, 57*, 79–92.

Hilt, M., & Zimmermann, W. (2007). Hexagonal, square and stripe patterns of the ion channel density in biomembranes. *Physical Review, E75*, 1–16. https://doi.org/10.1103/PhysRevE.75.016202.

Holdaway-Clarke, T., & Hepler, P. K. (2003). Control of pollen tube growth: Role of ion gradients and fluxes. *New Phytologist, 159*, 539–563.

Holmes-Davies, R., Tanaka, C. K., Vensel, W. H., Hurkman, W. J., & McCormick, S. (2005). Proteome mapping of mature pollen of *Arabidopsis thaliana*. *Proteomics, 5*, 4864–4884.

Hooper, C., Castleden, I., Tanz, S. K., Aryamanesh, N., & Millar, A. H. (2017). SUBA4: The interactive data analysis centre for Arabidopsis subcellular protein locations. *Nucleic Acids Research, 45*, D1064–D1074.

Hooper, C., Tanz, S. K., Castleden, I., Vacher, M., Small, I., & Millar, A. H. (2014). SUBAcon: A consensus algorithm for unifying the subcellular localization data of the Arabidopsis proteome. *Bioinformatics, 30*, 3356–3364.

Hruz, T., Laule, O., Wessendorp, F., Bleuler, S., Oertle, L., Widmayer, P., et al. (2008). Genevestigator V3: A reference expression database for the meta-analysis of transcriptomes. *Advances in Bioinformatics, 2008*, 420747. https://doi.org/10.1155/2008/420747.

Iwai, H., Hokura, A., Oishi, M., Chida, H., Ishii, T., Sakai, S., et al. (2006). The gene responsible for borate cross-linking of pectin Rhamnogalacturonan-II is required for plant reproductive tissue development and fertilization. *Proceedings of the National Academy of Sciences of the United States of America, 103*, 16592–16597.

Jaffe, L. F. (1977). Electrophoresis along cell membranes. *Nature, 265*, 600–602.

Jezek, M., & Blatt, M. R. (2017). The membrane transport system of the guard cell and its integration for stomatal dynamics. *Plant Physiology, 174*, 487–519.

Lai, H. C., & Jan, L. Y. (2006). The distribution and targeting of neuronal voltage-gated ion channels. *Nature Reviews. Neuroscience*, 7, 548–562.

Lang, V., Pertl-Obermeyer, H., Safiarian, M. J., & Obermeyer, G. (2014). Pump up the volume—A central role for the plasma membrane H^+ pump in pollen germination and tube growth. *Protoplasma, 251*, 477–488. https://doi.org/10.1007/s007909-013-0555-2.

Lang, V., Usadel, B., & Obermeyer, G. (2015). *De novo* sequencing and analysis of the lily pollen transcriptome: An open access data source for an orphan plant species. *Plant Molecular Biology, 87*, 69–80. https://doi.org/10.1007/s11103-014-0261-2.

Lau, T. C., & Stephenson, A. G. (1994). Effects of soil phosphorus on pollen production, pollen size, pollen phosphorus content, and ability to sire seeds in *Cucurbita pepo* (Cucurbitaceae). *Sexual Plant Reproduction*, *7*, 215–220.

Lehmann, S., Funck, D., Szabados, L., & Rentsch, D. (2010). Proline metabolism and transport in plant development. *Amino Acids*, *39*, 949–962.

Leonhardt, N., Bazin, I., Richard, P., Marin, E., Vavasseur, A., & Forestier, C. (2001). Antibodies to the CTFR modulate the turgor pressure of guard cell protoplasts via slow anion channels. *FEBS Letters*, *494*, 15–18.

Lew, R. R. (1996). Pressure regulation of the electrical properties of growing *Arabidopsis thaliana* L. root hairs. *Plant Physiology*, *112*, 1089–1100.

Li, L., He, Z., Pandey, G. K., Tsuchiya, T., & Luan, S. (2002). Functional cloning and characterization of a plant efflux carrier for multidrug and heavy metal detoxification. *Journal of Biological Chemistry*, *277*, 5360–5368.

Li, W., Xu, G., Alli, A., & Yu, L. (2018). Plant HUK/KUP/KT K^+ transporters: Function and regulation. *Seminars in Cell and Developmental Biology*, *74*, 133–141.

Linskens, H. F., & Schrauwen, J. (1966). Measurement of oxygen tension changes in the style during pollen tube growth. *Planta*, *71*, 98–106.

Liu, J., & Hussey, P. (2014). Dissecting the regulation of pollen tube growth by modeling the interplay of hydrodynamics, cell wall and ion dynamics. *Frontiers in Plant Science*, *5*, 392. https://doi.org/10.3389/fpls.2014.00392.

Liu, L., Zheng, C., Kuang, B., Wei, L., Yan, L., & Wang, T. (2017). Receptor-like kinase RUPO interacts with potassium transporters to regulate pollen tube growth and integrity in rice. *PLoS Genetics*, *12*, e1006085. https://doi.org/10.1371/journal.pgen.1006085.

Lohse, M., Nagel, A., Herter, T., May, P., Schroda, M., Zrenner, R., et al. (2014). Mercator: A fast and simple web server for genome scale functional annotation of plant sequence data. *Plant, Cell & Environment*, *37*(5), 1250–1258. https://doi.org/10.1111/pce.12231.

Loraine, A. E., McCormick, S., Estrada, A., Patel, K., & Qin, P. (2013). RNA-Seq of Arabidopsis pollen uncovers novel transcription and alternative splicing. *Plant Physiology*, *162*, 1092–1109.

Lubkowitz, M. (2011). The oligopeptide transporters: A small gene family with a diverse group of substrates and functions? *Molecular Plant*, *4*, 407–415.

Mäser, P., Leonhardt, N., & Schroeder, J. I. (2003). The clickable guard cell: Electronically linked model of guard cell signal transduction pathways. In K. Torii (Ed.), *The Arabidopsis book*. American Society of Plant Biologists.

Meckel, T., Hurst, A. C., Thiel, G., & Homann, U. (2004). Endocytosis against high turgor: Intact guard cells of *Vicia faba* constitutively endocytose fluorescently labelled plasma membrane and GFP-tagged K^+-channel KAT1. *The Plant Journal*, *39*, 182–193.

Meharg, A. A., Maurousset, L., & Blatt, M. R. (1994). Cable correction of the membrane currents recorded from root hairs of *Arabidopsis thaliana* L. *Journal of Experimental Botany*, *45*, 1–6.

Messerli, M., Danuser; G., & Robinson, K. R. (1999). Pulsatile influxes of H^+, K^+ and Ca^{2+} lag growth pulses of *Lilium longiflorum* pollen tubes. *Journal of Cell Science*, *112*, 1497–1509.

Messerli, M. A., Smith, P. J. S., Lewis, R. C., & Robinson, K. R. (2004). Chloride fluxes in lily pollen tubes: A critical reevaluation. *The Plant Journal*, *40*, 799–812.

Michard, E., Alves, F., & Feijo, J. A. (2009). The role of ion fluxes in polarized cell growth and morphogenesis: The pollen tube as an experimental paradigm. *International Journal of Developmental Biology*, *53*, 1609–1622. https://doi.org/10.1387/ijdb.072296em.

Michard, E., Lima, P. T., Borges, F., Silva, A. C., Portes, M. T., Carvalho, J. E., et al. (2011). Glutamate receptor-like genes form Ca^{2+} channels in pollen tubes and are regulated by pistil D-serine. *Science*, *332*, 434–437.

Michard, E., Simon, A. A., Tavares, B., Wudick, M. M., & Feijo, J. A. (2017). Signaling with ions: The keystone for apical cell growth and morphogenesis in pollen tubes. *Plant Physiology*, *173*, 91–111.

Mikosch, M., Hurst, A. C., Hertel, B., & Homann, U. (2006). Diacidic motif is required for efficient transport of the K^+ channel KAT1 to the plasma membrane. *Plant Physiology*, *142*, 923–930.

Miller, D. D., Callaham, D. A., Gross, D. J., & Hepler, P. K. (1992). Free Ca^{2+} gradient in growing pollen tubes. *Journal of Cell Science*, *101*, 7–12.

Mouline, K., Very, A.-A., Gaymard, F., Boucherez, J., Pilot, G., Devic, M., et al. (2002). Pollen tube development and competitive ability are impaired by disruption of a Shaker K^+ channel in *Arabidopsis*. *Genes & Development*, *16*, 339–350.

Nagy, R., Vasconcelos, M. J., Zhao, S., McElver, J., Bruce, W., Amrhein, N., et al. (2006). Differential regulation of five Pht1 phosphate transporters from maize (*Zea mays* L.). *Plant Biology*, *8*, 186–197.

Nieves-Cordones, M., Chavanieu, A., Jeanguenin, L., Alcon, C., Szoponarski, W., Estaran, S., et al. (2014). Distinct amino acids in the C-linker domain of the Arabidopsis K^+ channel KAT2 determine its subcellular localization and activity at the plasma membrane. *Plant Physiology*, *164*, 1415–1429.

Nobiling, R., & Reiss, H.-D. (1987). Quantitative analysis of calcium gradients and activity in growing pollen tubes of *Lilium longiflorum*. *Protoplasma*, *139*, 20–24.

Noir, S., Bräutigam, A., Colby, T., Schmidt, J., & Panstruga, R. (2005). A reference map of the *Arabidopsis thaliana* mature pollen proteome. *Biochemical and Biophysical Research Communications*, *337*, 1257–1266.

Obermeyer, G., & Blatt, M. R. (1995). Electrical properties of intact pollen grains of *Lilium longiflorum*: Characteristics of the non-germinating pollen grain. *Journal of Experimental Botany*, *46*, 803–813.

Obermeyer, G., Fragner, L., Lang, V., & Weckwerth, W. (2013). Dynamic adaption of metabolic pathways during germination and growth of lily pollen tubes after inhibition of the electron transport chain. *Plant Physiology*, *162*, 1822–1833.

Obermeyer, G., Kriechbaumer, R., Strasser, D., Maschessnig, A., & Bentrup, F.-W. (1996). Boric acid stimulates the plasma membrane H^+ ATPase of ungerminated lily pollen grains. *Physiologia Plantarum*, *98*, 281–290.

Obermeyer, G., Lützelschwab, M., Heumann, H.-G., & Weisenseel, M. H. (1992). Immunolocalisation of H^+ ATPases in the plasma membrane of pollen grains and pollen tubes of *Lilium longiflorum*. *Protoplasma*, *171*, 55–63.

Obermeyer, G., & Weisenseel, M. H. (1991). Calcium channel blocker and calmodulin antagonists affect the gradient of free calcium ions in lily pollen tubes. *European Journal of Cell Biology*, *56*, 319–327.

Pandey, S., Zhang, W., & Assmann, S. M. (2007). Roles of ion channels and transporters in guard cell signal transduction. *FEBS Letters*, *581*, 2325–2336.

Paul, P., Chaturvedi, P., Selymesi, M., Ghatak, A., Mesihovic, A., Scharf, K.-D., et al. (2016). The membrane proteome of male gametophyte in *Solanum lycopersicum*. *Journal of Proteomics*, *131*, 48–60.

Perez di Giorgio, J. A., Bienert, G. P., Ayub, N. D., Yaneff, A., Barberini, M. L., Mecchia, M. A., et al. (2016). Pollen-specific aquaporins NIP4;1 and NIP4;2 are required for pollen development and pollination in *Arabidopsis thaliana*. *The Plant Cell*, *28*, 1053–1077.

Perkins, D. N., Pappin, D. J., Creasy, D. M., & Cottrell, J. S. (1999). Probability-based protein identification by searching sequence databases using mass spectrometry data. *Electrophoresis*, *20*(18), 3551–3567.

Pertl, H., Poeckl, M., Blaschke, C., & Obermeyer, G. (2010). Osmoregulation in Lilium pollen grains occurs via modulation of the plasma membrane H^+ ATPase activity by 14-3-3 proteins. *Plant Physiology*, *154*, 1921–1928.

Pertl, H., Schulze, W. X., & Obermeyer, G. (2009). The pollen organelle membrane proteome reveals highly spatial-temporal dynamics during germination and tube growth of lily pollen. *Journal of Proteome Research*, *8*, 5142–5152.

Pertl-Obermeyer, H., Schulze, W. X., & Obermeyer, G. (2014). *In vivo* cross-linking combined with mass spectrometry analysis reveals receptor-like kinases and Ca^{2+} signalling proteins as putative interaction partners of pollen plasma membrane H^{+} ATPases. *Journal of Proteomics*, *108*, 17–29.

Pina, C., Pinto, F., Feijo, J. A., & Becker, J. D. (2005). Gene family analysis of the *Arabidopsis* pollen transcriptome reveals biological implications for cell growth, division control and gene expression regulation. *Plant Physiology*, *138*, 744–756.

Pittman, J. K., & Hirschi, K. (2016). CAX-ing a wide net: Cation/H^{+} transporters in metal remediation and abiotic stress signalling. *Plant Biology*, *18*, 741–749.

Qin, Y., Leydon, A. R., Manziello, A., Pandey, R., Mount, D., Denic, S., et al. (2016). Penetration of the stigma and style elicits a novel transcriptome in pollen tubes, pointing to genes critical for growth in a pistil. *PLoS Genetics*, *12*(7), e1006210. https://doi.org/10.1371/journal.pgen.1006210.

Quiles-Pando, C., Rexach, J., Navarro-Gochicoa, M. T., Camacho-Cristobal, J. J., Herrera-Rodriguez, M. B., & Gonzales-Fontes, A. (2013). Boron deficiency increases the levels of cytosolic Ca^{2+} and expression of Ca^{2+}-related genes in *Arabidopsis thaliana* roots. *Plant Physiology and Biochemistry*, *65*, 55–60.

Rakusova, H., & Geitmann, A. (2017). Pollen tip growth: Control of cellular morphogenesis through intracellular trafficking. In G. Obermeyer & J. Feijo (Eds.), *Pollen tip growth: From biophysical aspects to systems biology* (pp. 129–148). Cham, Switzerland: Springer Nature.

Ratajcak, R. (2000). Structure, function and regulation of the plant vacuolar H^{+} translocating ATPase. *Biochimica et Biophysica Acta*, *1465*, 17–36.

Rathore, K. S., Cork, R. J., & Robinson, K. R. (1991). A cytoplasmic gradient of Ca^{2+} is correlated with the growth of lily pollen tubes. *Developmental Biology*, *148*, 612–619.

Reinders, A. (2016). Fuel for the road—Sugar transport and pollen tube growth. *Journal of Experimental Botany*, *67*, 2121–2123.

Ricachenevsky, F. K., Menguer, P. K., Sperotto, R. A., Williams, L. E., & Fett, J. P. (2014). Roles of plant metal tolerance proteins (MTP) in metal storage and potential use in biofortification strategies. *Frontiers in Plant Science*, *4*, 144. https://doi.org/10.3389/fpls.2013.00144.

Rigas, S., Debrosses, G., Haralampidis, K., Vicente-Agullo, F., Feldmann, K. A., Garbov, A., et al. (2001). *THR1* encodes a potassium transporter required for tip growth in Arabidopsis root hairs. *The Plant Cell*, *13*, 139–151.

Rodriguez-Enriquez, M. J., Mehdi, S., Dickinson, H. G., & Grant-Downton, R. T. (2013). A novel method for efficient *in vitro* germination and tube growth of Arabidopsis thaliana pollen. *New Phytologist*, *197*(2), 668–679. https://doi.org/10.1111/nph.12037.

Roelfsema, M. R. G., & Hedrich, R. (2005). In the light of stomatal opening: New insights into 'the Watergate'. *New Phytologist*, *167*, 665–691.

Safiarian, M. J., Pertl-Obermeyer, H., Lughofer, P., Hude, R., Bertl, A., & Obermeyer, G. (2015). Lost in traffic? The K^{+} channel of lily pollen, LilKT1, is detected at the endomembranes inside yeast cells, tobacco leaves and lily pollen. *Frontiers in Plant Science*, *6*, 47. https://doi.org/10.3389/fpls.2015.00047.

Sancenon, V., Puig, S., Mateu-Andres, I., Dorcey, E., Thiele, D. J., & Penarrubia, L. (2004). The Arabidopsis copper transporter COPT1 functions in root elongation and pollen development. *Journal of Biological Chemistry*, *279*, 15348–15355.

Sawidis, T., & Reiss, H.-D. (1995). Effects of heavy metals on pollen tube growth and ultrastructure. *Protoplasma*, *185*, 113–122.

Schiott, M., Romanowski, S., Baekgaard, L., Jakobsen, M. K., Palmgren, M. G., & Harper, J. F. (2004). A plant plasma membrane Ca^{2+} pump is required for normal pollen tube growth and fertilization. *Proceedings of the National Academy of Sciences of the United States of America*, *101*(25), 9502–9507.

Schwacke, R., Schneider, C., van der Graaff, E., Fischer, K., Catoni, E., Desimone, M., et al. (2003). ARAMEMNON, a novel database for Arabidopsis integral membrane proteins. *Plant Physiology*, *131*, 16–26.

Sharma, T., Dreyer, I., & Riedelsberger, J. (2013). The role of K^+ channels in uptake and redistribution of potassium in the model plant *Arabidopsis thaliana*. *Frontiers in Plant Science*, *4*, 224. https://doi.org/10.3389/fpls.2013.00224.

Stacey, M. G., Osawa, H., Patel, A., Gassmann, W., & Stacey, G. (2006). Expression analyses of Arabidopsis oligopeptide transporters during seed germination, vegetative growth and reproduction. *Planta*, *223*, 291–305.

Sutter, J. U., Campanoni, P., Tyrrell, M., & Blatt, M. R. (2006). Selective mobility and sensitivity to SNAREs is exhibited by the Arabidopsis KAT1 K^+ channel at the plasma membrane. *The Plant Cell*, *18*, 935–954.

Swarbeck, S. M., Colaco, R., & Davies, J. M. (2013). Plant calcium-permeable channels. *Plant Physiology*, *163*, 514–522.

Sze, H., Padmanaban, S., Cellier, F., Honys, D., Cheng, N.-H., Bock, K. W., et al. (2004). Expression patterns of a novel AtCHX gene family highlight potential roles in osmotic adjustment and K^+ homeostasis in pollen development. *Plant Physiology*, *136*, 2532–2547.

Tagede, M., Dupuis, I., & Kuhlemeier, C. (1999). Ethanolic fermentation: New functions for an old pathway. *Trends in Plant Science*, *4*, 320–325.

Takano, J., Miwa, A., & Fujiwara, T. (2008). Boron transport mechanisms: Collaboration of channels and transporters. *Trends in Plant Science*, *13*, 451–457.

Takano, J., Noguchi, K., Yasumori, M., Kobayashi, M., Gajdos, Z., Miwa, K., et al. (2002). Arabidopsis boron transporter for xylem loading. *Nature*, *420*, 337–340.

Tanz, S. K., Castleden, I., Hooper, C., Vacher, M., Small, I., & Millar, A. H. (2013). SUBA3: A database for integrating experimentation and prediction to define the SUBcellular location of proteins in Arabidopsis. *Nucleic Acids Research*, *41*, D1185–D1191.

Tavares, B., Dias, P. N., Domingos, P., Moura, T. F., Feijo, J. A., & Bicho, A. (2011). Calcium-regulated anion channels in the plasma membrane of *Lilium longiflorum* pollen protoplasts. *New Phytologist*, *192*, 45–60.

Thimm, O., Blasing, O., Gibon, Y., Nagel, A., Meyer, S., Kruger, P., et al. (2004). MAPMAN: A user-driven tool to display genomics data sets onto diagrams of metabolic pathways and other biological processes. *The Plant Journal*, *37*, 914–939.

Tunc-Ozdemir, M., Rato, C., Brown, E., Rogers, S., Mooneyham, A., Frietsch, S., et al. (2013). Cyclic nucleotide-gated channels 7 and 8 are essential for male reproductive fertility. *PLoS One*, *8*, e55277.

Tunc-Ozdemir, M., Tang, C., Ishka, M. R., Brown, E., Groves, N. R., Myers, C. T., et al. (2013). A cyclic nucleotide-gated channel (CNGC16) in pollen is critical for stress tolerance in pollen reproductive development. *Plant Physiology*, *161*, 1010–1020.

van Hemelryck, M., Bernal, R., Rojas, E., Dumais, J., & Kroeger, J. (2017). A fresh look at growth oscillations in pollen tubes: Kinematic and mechanistic descriptions. In G. Obermeyer & J. Feijo (Eds.), *Pollen tip growth: From biophysical aspects to systems biology* (pp. 369–390). Cham, Switzerland: Springer Nature.

Verrier, P. J., Bird, D., Burla, B., Dassa, E., Forestier, C., Geisler, M., et al. (2008). Plant ABC proteins—A unified nomenclature and updated inventory. *Trends in Plant Science*, *13*, 151–159.

Very, A.-A., Nieves-Cordones, M., Daly, M., Khan, I., Fizames, C., & Sentenac, H. (2014). Molecular biology of K^+ transport across the plant cell membrane: What do we learn from comparison between plant species? *Journal of Plant Physiology*, *171*, 748–769.

Wang, H.-J., Huang, J.-C., & Jauh, G.-Y. (2010). Pollen germination and tube growth. *Advances in Botanical Research, 54*, 1–52.

Wang, H., & Jiang, L. (2017). Polar protein exocytosis: Lessons from plant pollen tube. In G. Obermeyer & J. Feijo (Eds.), *Pollen tip growth: From biophysical aspects to systems biology* (pp. 107–127). Cham, Switzerland: Springer Nature.

Wang, Y., & Wu, W.-H. (2013). Potassium transport and signalling in higher plants. *Annual Review of Plant Biology, 64*, 451–476.

Wang, Y., Zhang, W.-Z., Song, L.-F., Zou, J.-J., Su, Z., & Wu, W.-H. (2008). Transcriptome analyses show changes in gene expression to accompany pollen germination and tube growth in *Arabidopsis*. *Plant Physiology, 148*, 1201–1211.

Weisenseel, M. H., & Jaffe, L. A. (1976). The major growth current through lily pollen tubes enters as K^+ and leaves as H^+. *Planta, 133*, 1–7.

Weisenseel, M. H., Nuccitelli, R., & Jaffe, L. A. (1975). Large electrical currents traverse growing pollen tubes. *Journal of Cell Biology, 66*, 556–567.

Wilkens, S. (2015). Structure and mechanism of ABC transporters. *F1000Prime Reports, 7*, 1–9. https://doi.org/10.12703/P7-14.

Williams, L. E., & Mills, R. F. (2005). P_{1B}-ATPases—An ancient family of transition metal pumps with diverse functions in plants. *Trends in Plant Science, 10*, 491–502.

Wu, X. N., Sanchez Rodriguez, C., Pertl-Obermeyer, H., Obermeyer, G., & Schulze, W. X. (2013). Sucrose-induced receptor kinase SIRK1 regulates a plasma membrane aquaporin in Arabidopsis. *Molecular & Cellular Proteomics, 12*, 2856–2873.

Wu, J., Wang, S., Gu, Y., Zhang, S., Publicover, S. J., & Franklin-Tong, V. E. (2011). Self-incompatibility in *Papaver rhoeas* activates nonspecific cation conductance permeable to Ca^{2+} and K^+. *Plant Physiology, 155*, 963–973.

Wudick, M. M., Portes, M. T., Michard, E., Rosas-Santiago, P., Lizzio, M. A., Nunes, C. O., et al. (2018). Cornichon sorting and regulation of GLR channels underlie pollen tube homeostasis. *Science, 360*, 533–536.

Yong, Z., Kotur, Z., & Glass, D. M. (2010). Characterisation of an intact two-component high-affinity nitrate transporter from Arabidopsis roots. *The Plant Journal, 63*, 739–748.

Yuan, F., Yang, H., Xue, Y., Kong, D., Ye, R., Li, C., et al. (2014). OSCA1 mediates osmotic-stress-evoked Ca^{2+} increases vital for osmosensing in Arabidopsis. *Nature, 514*, 367–371.

Zhang, D., & Liu, H. (2014). Exine export in pollen. In M. Geisler (Ed.), *Plant ABC transporters* (pp. 49–62). Cham, Switzerland: Springer.

Zheng, S., Pan, T., Fan, L., & Qiu, Q.-S. (2013). A novel AtKEA gene family, homolog of bacterial K^+/H^+ antiporters, plays potential roles in K^+ homeostasis and osmotic adjustment in Arabidopsis. *PLoS One, 8*(11), e81463. https://doi.org/10.1371/journal.pone.0081463.

Zhou, J., Song, L.-F., Zhang, W., Wang, Y., Ruan, S., & Wu, W.-H. (2009). Comparative proteomic analysis of *Arabidopsis* mature pollen and germinated pollen. *Journal of Integrative Plant Biology, 51*, 438–455.

Zonia, L., Cordeiro, S., Tupy, J., & Feijó, J. A. (2002). Oscillatory chloride efflux at the pollen tube apex has a role in growth and cell volume regulation and is targeted by inositol 3,4,5,6-tetrakisphosphate. *The Plant Cell, 14*, 2233–2249.

CHAPTER NINE

Xylem Ion Loading and Its Implications for Plant Abiotic Stress Tolerance

Tetsuya Ishikawa*, Tracey Ann Cuin*, Nadia Bazihizina*,†, Sergey Shabala*,‡,1

*Tasmanian Institute for Agriculture, College of Science and Engineering, University of Tasmania, Hobart, TAS, Australia
†Department of Agrifood Production and Environmental Sciences, University of Florence, Florence, Italy
‡Department of Horticulture, Foshan University, Foshan, China
[1]Corresponding author. e-mail address: sergey.shabala@utas.edu.au

Contents

Abstract

Plant adaptive potential is critically dependent on efficient communication and coordination of resource allocation and signalling between above and below-ground plant parts. Control of xylem ion loading plays an important role in this process. This review focuses on the molecular identity, tissue-specific expression patterns, and transcriptional and post-translational regulation of transporters mediating xylem loading of Na^+, K^+, and Cl^- in plants grown under abiotic stress conditions such as drought, salinity, and soil flooding. The data are discussed in the context of breeding crops for stress resilience, which remains one of the highest priorities for dealing with the global food security challenge. This resilience was present in wild ancestors, but has been lost during

Advances in Botanical Research, Volume 87
ISSN 0065-2296
https://doi.org/10.1016/bs.abr.2018.09.006

domestication of crop species and exacerbated by the selection for higher yielding cultivars over the last 100 years. Thus, progress in the field requires a major rethink of current paradigms in crop breeding and the targeting of previously unexplored genes and traits. We argue that control of xylem ion loading is one of these traits and represents an unexplored opportunity for genetic improvement of plant germplasms.

1. INTRODUCTION

Abiotic stresses such as drought, salinity, temperature extremes and flooding severely limit crop production worldwide, with estimated losses of over US$120 billion p.a. (FAO, 2005). With the current trends in climate change, these costs will increase. At the same time, global food security requires food production to increase by ~70% by 2050 to match population growth (Tester & Langridge, 2010). This goal is not achievable by simply maintaining current agronomical and breeding practices. There is a growing recognition of the limitations of the first Green Revolution, primarily because the technologies developed did not consider the constraints to production in more marginal environments such as those subjected to drought, salinity or flooding (Pingali, 2012). Tolerance to these stresses was present in wild ancestors, but has been lost during domestication of crop species and exacerbated by the selection for higher yielding cultivars over the last 100 years (Huang et al., 2016). Nonetheless, future agricultural production will inevitably need to utilise marginal land due to increasing urbanisation, global climate trends and the need to match food production to population growth. In their latest concept paper, the Global Plant Council, a peak body comprising a coalition of national and international societies, and affiliates representing plant, crop and agricultural and environmental sciences across the globe, has identified breeding for stress resilience as one of the highest priorities for dealing with the global food security challenge (Davies & Ribaut, 2017). This requires a major rethink of the current paradigms in crop breeding and the targeting of previously unexplored genes and traits. One such trait is the control of xylem ion loading.

2. ESSENTIALITY OF XYLEM ION LOADING FOR ABIOTIC STRESS TOLERANCE

Plant adaptive potential is critically dependent on efficient communication and coordination of resource allocation and signalling between above and below-ground plant parts (Shabala, Bose, Fuglsang, & Pottosin, 2016).

This resource allocation becomes especially critical under stress conditions. To maintain cell turgor, so enabling leaf expansion growth under the conditions of reduced soil water potential that are present in saline conditions, plants need to undergo osmotic adjustment. From an energy cost-benefit point of view, this is best achieved by increased delivery of the major inorganic osmolytes (K^+, Na^+, Cl^-) to the shoot. However, because some of these nutrients are cytotoxic, their uptake must be restricted to essential and precisely regulated quantities. In this context, the optimal scenario for a plant would be to quickly send the amount of Na^+ and Cl^- to the shoot that is required to rapidly achieve full osmotic adjustment and maintain the normal growth rate (i.e. no yield penalties) (Shabala, 2013). Once this is achieved, plants need to reduce the rate of xylem Na^+ and Cl^- loading to the absolute minimum required for driving cell turgor in newly growing tissues. Such control can be achieved at the xylem-parenchyma boundary. Not surprisingly, the ability of plants to control xylem ion loading has been repeatedly named as central to salt tolerance (Berthomieu et al., 2003; Bose, Shabala, et al., 2014; Mäser, Gierth, & Schroeder, 2002; Munns & Tester, 2008; Shabala, 2013; Zhu, Zhou, Shabala, & Shabala, 2017).

Control of xylem ion loading is also essential for plant adaptation to drought, which also reduces the soil water potential. Transcription levels of SKOR (an outward-rectifying K^+ efflux channel in the stele) in roots of the xerophytic *Z. xanthoxylum* plants are several fold higher in plants exposed to osmotic stress (Hu, Ma, Kumar, Duan, & Zhang, 2016). The reported increase in K^+ delivery to the shoot could contribute to improved drought tolerance via two concurrent mechanisms: directly by contributing to shoot cell sap osmolarity, and indirectly, through regulation of the plant hydraulic conductance. Supporting evidence for the latter mechanism comes from work by Oddo et al. (2011), who showed that increases in xylem sap K^+ in *Laurus nobilis* resulted in a marked increase in plant hydraulic conductance, most likely the result of the interaction of potassium ions with the pectin matrix of intervessel pits.

Equally essential is the control of xylem ion loading in plant adaptive responses to flooding. In aerated roots, the respiration rate in the stele is sixfold higher than in the cortex (Aguilar, Turner, Gibbs, Armstrong, & Sivasithamparam, 2003). The stele is the first to sense oxygen deprivation caused by soil waterlogging or flooding. Hypoxia in the stele inhibits H^+-ATPase activity in the xylem parenchyma, thus diminishes the H^+ gradients, inhibiting secondary energy-dependent ion transport and channel conductances (Colmer & Greenway, 2011). This has major implications for the delivery of essential nutrients to the shoot. Oxygen deficiency in the stele also reduces the concentrations of K^+, Na^+ and Cl^- in the xylem parenchyma cells at the subapical region (Kotula et al., 2015), impeding ion translocation.

Root stellar tissues harbour a large number of membrane transporter proteins that mediate ion loading into the xylem. These are regulated (gated in case of channels) by various signalling molecules and metabolites whose level, in turn, is strongly affected by stress. For example, soil mineral deficiencies result in a substantial accumulation of γ-aminobutyric acid (GABA) in tomato (Sung et al., 2015). GABA is a non-protein amino acid, which modulates anion fluxes across the plasma membrane (Ramesh et al., 2015). A dramatic (several orders of magnitude) increase in the GABA content is one of the most prominent biochemical alterations under flood conditions (Kinnersley & Turano, 2000). In addition, evidence for a role of GABA in sequestering salt load into epidermal bladder cells in a halophyte species (*C. quinoa*) has been recently provided (Kiani-Pouya et al., 2017). Xylem loading of cytokinins and ABA were shown to be involved in the regulation of ion fluxes through xylem parenchyma cells (Lips, 1997), and the levels of both these hormones are altered dramatically under stress conditions. Thus, understanding the molecular identity and control modes of key ion transporters at the xylem-parenchyma interface may offer plant breeders new and previously unexplored opportunities for creating stress-tolerant germplasms.

3. THE MOLECULAR IDENTITY OF THE KEY TRANSPORT SYSTEMS MEDIATING XYLEM ION LOADING

3.1 Sodium

Despite its critical importance for salinity tolerance, it is still a matter of debate whether the process of Na^+ loading is thermodynamically active or passive (Tester & Davenport, 2003). Most likely, both active and passive transport systems are involved, but their respective roles may differ depending on the length of time since salinity onset (Shabala, 2013). Hence, a number of candidate genes encoding different transport systems are likely to be involved.

3.1.1 SOS1

Given that the reported membrane potential values for parenchyma cells are in the range of −110 to −140 mV (Wegner et al., 2011) and the xylem Na^+ content typically exceeds 10 mM under saline conditions (Shabala et al.,

2010), a thermodynamically active xylem loading mechanism is likely to be required. One of the most likely candidates for active loading is the *SOS1* (salt over sensitive 1)-encoded Na^+/H^+ antiporter present at the xylem/parenchyma interface. SOS1 belongs to the cation proton antiporter (CPA) subfamily of proteins and consists of 10–12 transmembrane domains with a 700 amino acid-long tail, which faces the cytoplasm (reviewed by Zhu, 2003).

Bacterial Na^+/H^+ antiporters: NhaA (from *Escherichia coli*) and NhaP1 (from *Methanococcus jannaschii*) are dimers in the membrane containing 12 and 13 transmembrane helices, respectively (reviewed by Núñez-Ramírez et al., 2012). In plants, the SOS1 protein is a homodimer, which contains a membrane domain similar to that found in other antiporters of the family and an elongated, large and structured cytosolic domain. Both the transmembrane and cytosolic moieties contribute to the dimerisation of the antiporter (reviewed by Núñez-Ramírez et al., 2012).

SOS1 mRNA is unstable under normal, unstressed conditions, but its stability is substantially increased under salt stress and other ionic and dehydration stresses (Chung, Zhu, Bressan, Hasegawa, & Shi, 2008). The driving force for its antiport activity is a H^+ gradient, created by H^+-translocating ATPases. SOS1 also operates at the plasma membrane of epidermal root cells where it expels Na^+ out of the root, back to the rhizosphere. It has also been suggested to operate as a Na^+ sensor (Sánchez-Barrena, Martínez-Ripoll, & Albert, 2013; Shabala, Wu, & Bose, 2015).

In response to salt stress, the activity of SOS1 is controlled via phosphorylation by SOS2 (a member of calcium-induced protein kinase; CIPK24) (Chinnusamy, Schumaker, & Zhu, 2004), with calcium binding proteins, SOS3 (CBL4) or SCaBP8/CBL10, being required for the activity of this kinase (Quan et al., 2007).

3.1.2 HKT

The high-affinity HKT transporters are present in the xylem parenchyma cells and, despite their name (high affinity K^+ transporters), they show high permeability for Na^+ (Rubio, Gassmann, & Schroeder, 1995; Rubio, Schwarz, Gassmann, & Schroeder, 1999; Schachtman & Schroeder, 1994). Biophysical transport and phylogenetic analysis have classified the HKT family into two subgroups, Class 1 and Class 2 (Platten et al., 2006). Class 2 HKTs generally show co-transport of Na^+-K^+, while Class 1 shows preferential transport of Na^+ over other cations (reviewed by Horie, Hauser, & Schroeder, 2009). HKT1 is related to the bacterial K^+

transporters such as bacterial Trk and fungal Ktr (Corratgé-Faillie et al., 2010), and an evolutionary correlation is proposed between Trk/Ktr/HKT-type K^+ transporters, K^+ channels and Kdp-type K^+ pumps (present in prokaryotic cells) (reviewed by Hamamoto et al., 2015). AtHKT1;1 consists of a single polypeptide with four repeated domains (reviewed by Hamamoto et al., 2015; Shabala, White, Djordjevic, Ruan, & Mathesius, 2016), and its major role is assigned to Na^+ retrieval from the xylem (Byrt et al., 2007; Davenport et al., 2007; Horie et al., 2009; Mäser, Eckelman, et al., 2002; Mäser, Gierth, et al., 2002; Ren et al., 2005; Sunarpi et al., 2005). Far less is known about the role of Class 2 HKT transporters. From a biophysical point of view, these transporters are highly suited for active xylem Na^+ loading under conditions that favour passive outward K^+ movement across the parenchyma cell plasma membrane (e.g. the depolarised membrane potential that occurs under saline conditions; Wegner et al., 2011). Supporting this notion, strong OsHKT2;1 expression was reported in the stellar root tissues of rice (Jabnoune et al., 2009).

3.1.3 NSCC

Xylem Na^+ loading could also be mediated by a passive mechanism that involves non-selective cation channels (NSCCs). NSCCs show diverse characteristics in terms of regulation and functional expression (Demidchik & Maathuis, 2007). NSCCs do not strongly differentiate between cations, particularly between K^+ and Na^+; K^+/Na^+ permeability ratios range between 0.3 and 3 (see Pottosin & Dobrovinskaya, 2014 for review). The two major known families of NSCCs are the cyclic nucleotide gated channels (CNGC; 20 genes in Arabidopsis) and the ionotropic glutamate receptors (GLR; 20 genes in Arabidopsis), some of which show high expression in root tissues (Demidchik, Davenport, & Tester, 2002; Gobert, Park, Amtmann, Sanders, & Maathuis, 2006; Lam et al., 1998). The potential importance of CNGC and GLR in generating cytosolic Ca^{2+} signals in response to stress conditions has been proposed (Swarbreck, Colaço, & Davies, 2013) and their engagement in Na^+ transport at the xylem/symplast boundary is also highly plausible.

3.1.4 CNGC

CNGC are considered likely candidates for the voltage-independent NSCCs (Demidchik et al., 2002; Kaplan, Sherman, & Fromm, 2007).

Heterologously expressed CNGCs act as cation transporters (Gobert et al., 2006; Li, Borsics, Harrington, & Christopher, 2005) and electrophysiological studies have shown that both AtCNGC1 and AtCNGC4 can transport K^+ and Na^+ (Balagué et al., 2003; Leng, Mercier, Hua, Fromm, & Berkowitz, 2002). Furthermore, CNGC10 is suggested to mediate K^+ efflux and Na^+ influx at the root-soil interface (Guo, Babourina, Christopher, Borsics, & Rengel, 2008) and based on altered shoot cation (K^+, Ca^{2+} and Mg^{2+}) content in *atcngc10* mutants, it has also been suggested that this GNGC mediates xylem ion loading (Guo, Babourina, Christopher, Borsic, & Rengel, 2010). However, direct evidence is needed to validate this.

3.1.5 GLR

GLRs are an important class of NSCCs that transport Na^+, K^+ and Ca^{2+} across the plasma membrane (Demidchik, 2014). At least three GLRs (GLR3.2, GLR3.4 or GLR3.3) are expressed in Arabidopsis root tissues (Vincill, Clarin, Molenda, & Spalding, 2013). Although a direct role for GLR-mediated xylem Na^+ loading remains to be shown, such a scenario is plausible (Vincill et al., 2013).

3.1.6 Aquaporins

Aquaporins are abundant in the parenchyma cells of woody angiosperms (Almeida-Rodriguez & Hacke, 2012; Sakr et al., 2003; Secchi, Pagliarani, & Zwieniecki, 2017). Aquaporins are small proteins (21–34 kDa), which consist of six membrane-spanning alpha-helices integrated by five loops. Initially, aquaporins were regarded as merely water channels (Maurel, Kado, Guern, & Chrispeels, 1995), but recently, Byrt et al. (2017) revealed a new activity for one of most highly expressed aquaporins, AtPIP2;1 (plasma membrane intrinsic protein). AtPIP2:1 was found to induce a Na^+ conductance that is inhibited by low external pH and Ca^{2+} when expressed in heterologous oocyte systems. Given that AtPIP2;1 ion conductance exhibits characteristics typical of NSCCs (reviewed by Wegner, 2017), it was suggested as one of these (Byrt et al., 2017).

3.1.7 CCC

Cation chloride cotransporters (CCC) mediate symport of Cl^-, Na^+ and K^+ (Colmenero-Flores et al., 2007). Although there are uncertainties regarding their plasma membrane location (Chen, Yamaji, Fujii-Kashino, & Ma, 2016; Geilfus, 2018; Henderson, Wege, & Gilliham, 2018), their preferential expression at the xylem/symplast boundary of *Arabidopsis* has been shown (Colmenero-Flores et al., 2007). This suggests a role in long-distance ion transport. Functional support for this role came from pharmacological studies by Zhu et al. (2017), who demonstrated a significant reduction in the magnitude of Na^+ efflux from barley root stelar tissue in the presence of bumetanide, a known inhibitor of mammalian CCC. Because Cl^- transport into the xylem is thermodynamically passive (see next section), it may provide a driving force for Na^+ (secondary) active loading into the xylem.

3.2 Chloride

Xylem Cl^- loading is considered a thermodynamically passive process, the result of a negative plasma membrane potential in the xylem parenchyma cells and outwardly-directed chloride concentration gradients (Li, Tester, & Gilliham, 2017; Li, Yu, et al., 2017; Teakle & Tyerman, 2010). Two key anion conductances have been identified in xylem parenchyma cells by patch-clamp studies. One is a quickly activating anion conductance (X-QUAC), the dominant conductance for xylem Cl^- loading, and the other, a more rarely observed, slowly activating anion conductance (X-SLAC) (Gilliham & Tester, 2005; Köhler & Raschke, 2000). More recently, genes encoding the candidate anion channels previously identified by the patch-clamp studies have been discovered. These include the NPF2.4 protein that belongs to the Nitrate Transporter 1/Peptide Transporter (NRT1/PTR) family and two anion channels, SLAH1 and SLAH3, from the SLAC/SLAH (slow-type) family (Cubero-Font et al., 2016; Li et al., 2016).

The NPF2.4 protein in Arabidopsis was the first plasma membrane protein shown to directly catalyse Cl^- transport into the root xylem (Li et al., 2016; Li, Tester, et al., 2017; Li, Yu, et al., 2017). NPF2.4 overexpression resulted in a 23% increase in shoot Cl^- compared with null segregant lines in seedlings exposed to 75 mM NaCl. When expressed in *Xenopus laevis* oocytes with negative membrane potentials equivalent to those in the stele (−100 mV or lower), AtNPF2.4 generated a pH-independent Cl^- efflux, with a four-times greater permeability for Cl^- than NO_3^-. However, the properties of this transporter (lack of rectification, Na^+ and K^+-dependence of the currents and the greater permeability to Cl^- over NO_3^-) indicate that it does not correspond to any of the anion channels detected in patch-clamp studies on xylem parenchyma cells (Li et al., 2016). Two subsequent studies identified SLAH1 and SLAH3 as key players in xylem Cl^- loading. Both co-localised to the xylem pole pericycle cells adjacent to the xylem vessels (Cubero-Font et al., 2016), with the properties for SLAH3 conforming to the X-SLAC based currents found in xylem parenchyma cells. SLAH1, though, was found to be electrically silent. Nonetheless, co-expression studies showed that SLAH1 interacts with SLAH3 forming SLAH1/SLAH3 heteromers. When expressed in *Xenopus* oocytes, these heteromers elicit macroscopic Cl^- currents that override the kinase and nitrate-dependent activation of SLAH3. Further oocyte experiments revealed that the interaction with SLAH1 alters the electrical properties of SLAH3, enhancing the Cl^- conductance and the Cl^-/NO_3^- selectivity of the SLAH3/SLAH1 heteromers compared with the kinase-activated SLAH3 homomers, which are characterised by a 10-fold greater permeability to NO_3^- compared with Cl^- (Cubero-Font et al., 2016). This indicates that SLAH1 and SLAH3 act in tandem to regulate Cl^- and NO_3^- loading in the xylem. Thus, SLAH1 abundance is a key player that determines the degree of Cl^- exclusion from the shoot by activating and influencing the anion selectivity of SLAH3.

However, we are still lacking the candidate gene(s) encoding the X-QUAC type channel, the dominant conductance for Cl^- loading into the xylem (Gilliham & Tester, 2005; Köhler & Raschke, 2000). Hence, in addition to NPF2.4 and SLAH3/SLAH1, other transport pathways are in place in roots for xylem loading of Cl^-. Transcriptional comparisons have been used in the search for additional candidate transport proteins (reviewed in Li, Tester, et al., 2017). Two members of the NRT1/PTR family, AtNPF7.2 and AtNPF7.3, have been identified and given their electrophysiological feature could, respectively, catalyse xylem Cl^- retrieval and loading. Another candidate is the Aluminium-Activated Malate Transporter AtALM12, which is permeable to Cl^- and expressed in the stele of roots (for details see Li, Tester, et al., 2017; Li, Yu, et al., 2017). A further potential contender is the Cation Chloride (Cl) Cotransporters, discussed above.

3.3 Potassium

K^+ contributes about 50% to the xylem sap ion content (Siebrecht, Herdel, Schurr, & Tischner, 2003), with concentrations ranging from 3 to 5 mM (Jiang, Belfield, Cao, Smith, & Harberd, 2013; Watson, Pritchard, & Malone, 2001). Control of K^+ xylem loading is vital for supplying adequate K^+ to the photosynthesising shoot tissue. However, far less is known about xylem loading of K^+ and its regulation compared to knowledge of root uptake and transport processes in guard cells—surface cells are much easier to access than cells located deep in the root.

Because membrane potential values for xylem parenchyma cells reportedly range from −80 mV (De Boer & Volkov, 2003) to −130 mV (Shabala et al., 2010), channel-mediated xylem loading of K^+ is feasible. It was shown in the 1990s that two xylem parenchyma-located channel types contribute to loading, K^+ selective Outward Rectifying Channels (KORC) and Non-selective Outward Rectifying Cation (NORC) channels (Wegner & De Boer, 1997; Wegner & Raschke, 1994). While this early work was on barley, the KORC channel was soon after identified in Arabidopsis as the Shaker-like channel Stellar K^+ Outward-Rectifier, SKOR (Gaymard et al., 1998).

Although SKOR is frequently cited as the main transport system involved in K^+ xylem loading in Arabidopsis, *skor* knockout lines only have a 40% reduction in xylem sap K^+ and a 50% reduction in shoot K^+ (Gaymard et al., 1998). Thus a large proportion of K^+ xylem loading is unaccounted for. In these respects, NORCs were proposed early on as being involved in xylem K^+ loading but their molecular identity is still unknown.

Also localised to the plasma membrane of xylem parenchyma cells is the Arabidopsis AtHKT1 (Berthomieu et al., 2003; Mäser, Eckelman, et al., 2002; Mäser, Gierth, et al., 2002; Sunarpi et al., 2005). Despite its name (High Affinity Potassium Transport) and being initially proposed as a K^+ transporter (Schachtman & Schroeder, 1994), it is now apparent that its primary role is in Na^+ transport (Horie et al., 2009; Ren et al., 2005; Sunarpi et al., 2005). Thus, there is no evidence of a significant direct role in xylem K^+ loading in the literature for any of the HKTs in any plant species, although an indirect effect cannot be ruled out (see below). Also, by virtue of its triple function, the Cation Chloride Co-transporter (CCC) may mediate xylem K^+ loading, using Cl^- efflux as a driving force. It was shown that in rice, OsCCC1 RNAi lines accumulated less K^+ than wildtype plants, indicating a role of CCC in K^+ homeostasis (Kong et al., 2011). So despite a paucity of information in the literature, CCCs could contribute towards that 'missing' 50% of K^+ xylem loading in Arabidopsis.

Recently, Han, Wu, Wu, and Wang (2016) proposed that the KT/HAK/KUP transporter KUP7 mediates K^+ release into the xylem. Xylem sap K^+ content was reduced in *kup7* loss-of-function mutants. Localisation analysis indicated targeting of KUP7 to the plasma membrane, with high expression in the stellar tissues, and KUP7 mediation of K^+ transport was shown in yeast complementation assays. However, the evidence is not unequivocal because *kup7* lines also had lower root K^+ uptake rates, which could be the cause of the reduced xylem content. Furthermore, as a member of the high affinity KT/HAK/KUP transporter family, KUP7 may rather be involved in K^+ uptake into cells than release. For example, it may function in uptake into xylem parenchyma cells for later release to the xylem? However, some HAK transporters have been shown to mediate K^+ efflux in addition to K^+ uptake under some conditions (Bañuelos, Garciadeblas, Cubero, & Rodriguez-Navarro, 2002; Garciadeblás, Benito, & Rodriguez-Navarro, 2002; Osakabe et al., 2013).

Finally, and very recently, Li, Tester, et al. (2017) and Li, Yu, et al. (2017) reported that NRT1.5/NPF7.3 from the Nitrate Transporter 1/Peptide Transporter (NRT1/PTR) family functions as a H^+/K^+ antiporter mediating K^+ release from root parenchyma cells into the xylem. Traditionally, this NRT1/PTR family has been cited as nitrate/peptide transporters, but this work shows that NPF7.3/NRT1.5 is important for NO_3^--dependent K^+ translocation in Arabidopsis (Li, Yu, et al., 2017), possibly in association with SKOR (Drechsler et al., 2015).

Most of our current knowledge regarding xylem loading of K^+ is based on the model dicot Arabidopsis and may not apply to all species. Indeed, the literature suggests different systems of xylem loading of K^+ in monocot species. Rice (*Oryza sativa*), for example, possesses two putative outward Shaker channel subunits (Pilot, Gaymard, Mouline, Cherel, & Sentenac, 2003; Véry et al., 2014): OsK5.1 (or OsSKOR) and OsK5.2 (or OsGORK) (Kim et al., 2015; Pilot et al., 2003). While OsK5.1 is mainly expressed in root vasculature (Kim et al., 2015), like AtSKOR, OsK5.2 is expressed in both roots and shoots (Kim et al., 2015). Similar to AtSKOR, OsK5.2 mediates K^+ translocation into the xylem sap in the root, and like the guard cell-localised K^+ outward rectifying channel AtGORK (Ache et al., 2000; Ivashikina et al., 2001), it is involved in K^+ release from guard cells and stomatal movements (Nguyen et al., 2017). Thus, one channel in rice, AsK5.2, combines the functions of two Arabidopsis homologues, AtSKOR and AtGORK.

It is not yet known whether the expression patterns of OsK5.1 and OsK5.2 are shared by homologous genes in other monocots. In maize, the expression pattern of only one gene from the OsK5.2/OsGORK K^+-selective, outward-rectifying subgroup has been partially characterised (Büchsenschütz et al., 2005). Like OsK5.2, the maize channel, *ZORK*, is expressed in both guard cells and roots, and therefore may have a dual function.

In Arabidopsis, AtAKT1 mediates low-affinity K^+ uptake in roots (Lagarde et al., 1996), while AtHAK5 enables high-affinity uptake (Rubio, Aleman, Nieves-Cordones, & Martinez, 2010) and there is no evidence of their expression in stellar tissue. In contrast, strong expression of the rice OsAKT1 (Li et al., 2014), OsHAK1 (Bañuelos et al., 2002) and OsHAK5 (Yang et al., 2014) is detected in the vascular bundles as well as in root epidermal cells. This suggests that they could have a direct or indirect role in K^+ translocation, in addition to their roles in root K^+ uptake. Although these transporters could mediate K^+ accumulation in xylem parenchyma cells to enable efficient K^+ release into the xylem, similar to the Arabidopsis AtKUP7, it has been proposed that they themselves mediate K^+ transfer to the xylem (Nieves-Cordones, Martinez, Benito, & Rubio, 2016). Support for this is that some HAK transporters can allow K^+ efflux, as noted above (Bañuelos et al., 2002; Garciadeblás et al., 2002; Osakabe et al., 2013).

4. STRESS-INDUCED REGULATION OF XYLEM ION LOADING AND ITS IMPLICATIONS

4.1 Sodium

4.1.1 Transcriptional Changes

In response to salt stress, both *SOS1* and *HKT* transcript levels are regulated (Tounsi, Feki, Hmidi, Masmoudi, & Brini, 2017), a process far better optimised in salt-tolerant halophyte species. For example, the halophyte *Salicornia dolichostachya* substantially increases its expression of *SdSOS1* in the root under salinity compared with the taxonomically related glycophyte *Spinacia oleracea* (Katschnig, Bliek, Rozema, & Schat, 2015). This process was accompanied by the down-regulation of xylem Na^+ unloading by HKT1;1, leading to increased salt accumulation in the shoot. It should be kept in mind that, when accumulated in shoots, Na^+ can serve as a 'cheap' osmoticum for energetically efficient osmotic adjustment, and this rapid elevation of xylem Na^+ loading may only be transient (Bose, Shabala, et al., 2014; Shabala, 2013).

After completion of osmotic adjustment, the maintenance of non-toxic shoot Na^+ levels and a favourable Na^+/K^+ ratio is crucial for plant survival under prolonged salinity. The expression levels of *PtSOS1* and *PtHKT1;5* in *Puccinellia tenuiflora* (halophytic grass) depend on the capacity of the vacuole to sequester Na^+, mediated by PtNHX1 (Zhang et al., 2017). *PtSOS1* expression is enhanced under mild salinity, a condition under which

PtNHX would efficiently sequester Na^+ to the vacuole. Under more severe conditions, when the vacuolar Na^+ capability is exceeded, strong expression of *PtHKT1;5* (mediating xylem Na^+ unloading) is observed (Zhang et al., 2017).

Compared with Class 1 *HKTs*, regulation of *HKT2* expression in the root stele and its role in abiotic stress tolerance are less well understood. When *OsHKT2;1* is heterologously expressed with *OsHKT1* in oocytes, it shows the characteristics of a typical Na^+ transporter (Horie et al., 2009, 2001). A compensatory role of OsHKT2;1 in K^+-starved rice plants was also suggested (Horie et al., 2007). The same upregulation of Class 2 *HKT* genes by K^+ starvation was found in species such as wheat, barley and rice (Garciadeblás, Senn, Banuelos, & Rodriguez-Navarro, 2003; Horie et al., 2001; Wang, Gassmann, Rubio, Schroeder, & Glass, 1998). Taken together, it can be hypothesised that *HKT2* expression in the root stele mediates loading of Na^+ into the xylem as a substitution for K^+ under saline conditions, when the availability of the latter is limited.

4.1.2 Post-translational Regulation and Signalling

SOS1 is proposed as the most likely candidate for xylem Na^+ loading, and cytosolic Ca^{2+} plays a crucial role in its activation. An increase in cytosolic Ca^{2+} is referred to as a 'master switch' in a wide range of cell responses, which includes stress responses (Knight, Brandt, & Knight, 1998). The activity of SOS1 is controlled by SOS2 (CIPK24), which is itself activated by the SOS3 protein (CBL4). In brief, CBL4 dimerises upon calcium binding, and interacts with the NAF (asparagine, alanine, phenylalanine) domain of CIPK24. This induces the release of the C-terminal autoinhibition domain of CIPK24 (Quintero et al., 2011). Activated CIPK24 phosphorylates the SOS1 C-terminus, which removes the SOS1 autoinhibitory domain (Quintero et al., 2011), and SOS1 is activated. Although CIPK24 may also affect the activity of HKT1 (Laurie et al., 2002; Rus et al., 2004), its role in Ca^{2+} signalling and control of HKTs is unknown. Most studies proposed that Ca^{2+} signalling occurs as a result of osmotic rather than ionic stress, and that salt/osmotic induced Ca^{2+} elevation is only short-lived (Ca^{2+} is restored within 1–2 min) (reviewed by Maathuis, 2014). This phenomenon can support a role for SOS1 in osmotic adjustment by means of Na^+ accumulation.

In addition to Ca^{2+}, changes in cytosolic pH are also reported under stressing environments, including flooding (Felle, 2005) and salinity (D'Onofrio & Lindberg, 2009). This may also induce SOS1 activation (reviewed by Núñez-Ramírez et al., 2012). Indeed, activation of bacterial Na^+/H^+ antiporters can be triggered by changes in pH and in the nonvascular plant *Physcomitrella patens*, SOS1 activity is coupled with cytosolic pH changes (Fraile-Escanciano, Kamisugi, Cuming, Rodríguez-Navarro, & Benito, 2010). Consequently, in higher plants, the mechanism of SOS1 regulation could be a concerted activation by pH and by SOS2-SOS3 (reviewed by Núñez-Ramírez et al., 2012).

While abscisic acid (ABA) plays a key role in signalling under abiotic stress conditions (e.g. drought and salinity), its accumulation also leads to changes in H^+ flux into the xylem. Stimulation by ABA of H^+ extrusion into the xylem followed by a pH decrease was observed in onion roots (Clarkson & Hanson, 1986). In addition, increases in net H^+ uptake into xylem parenchyma cells (H^+ efflux from the xylem) by direct application of ABA were observed in pea and quinoa (Sun et al., 2017). Although the mechanism of ABA-induced H^+ flux into the xylem is still unknown, this process could be essential for fuelling the activity of SOS1 at the xylem/symplast boundary. In addition to activation of SOS1, ABA application to the barley root stele induced a (direct or indirect) down-regulation of Na^+ retrieval from the xylem (Zhu et al., 2017), likely due to the repression of HKT1 activity. The transcriptional factor abscisic acid insensitive (ABI4), which is known to down-regulate AtHKT1;1 in Arabidopsis (Shkolnik-Inbar, Adler, & Bar-Zvi, 2013), could be involved in this process. Taken together, ABA accumulation and pH changes in the xylem could lead to increased xylem Na^+ concentrations.

Stress conditions frequently result in the increased production of reactive oxygen species (ROS) (e.g. H_2O_2, $O_2^{\bullet}$, $^{\bullet}OH$ and $^1O_2^{\bullet}$) (Miller, Shulaev, & Mittler, 2008; Mittler, 2002), which are likely to influence xylem Na^+ loading and, in turn, the xylem Na^+ content. Apoplastic H_2O_2 production, mediated by NADPH oxidases and causally related ROS-induced Ca^{2+} uptake systems in the root stele, operates upstream of xylem Na^+ loading (Zhu et al., 2017). H_2O_2 has been shown to activate Na^+-permeable NSCCs (Demidchik, Shabala, & Davies, 2007; Zepeda-Jazo et al., 2011), which would result in high xylem Na^+ concentrations. Also possible is an increase in SOS1 activity due to a ROS-induced increase in cytosolic Ca^{2+} (Quintero et al., 2011). ROS also contributes to up-regulation of H^+-ATPases (reviewed by Assaha, Ueda, Saneoka, Al-Yahyai, & Yaish, 2017), which would increase the driving force for SOS1. However, the detailed mechanisms of ROS-induced xylem Na^+ loading are unclear.

NSCC-mediated Na^+ currents have been observed in root stelar tissue (Zhao, Song, He, & Zhu, 2007). Given their voltage-dependence, any change in membrane potential, including those induced by abiotic stress, is likely to affect NSCC-mediated xylem Na^+ loading. One of such factors is reduced ATP availability under stress conditions (e.g. hypoxia; Gibbs & Greenway, 2003). Under oxygen deficiency, xylem parenchyma cells are significantly depolarised. Salinity also causes depolarisation of the root stele (Wegner et al., 2011). This stress-induced depolarisation of the plasma membrane at the xylem/symplast boundary would induce passive xylem loading of Na^+ through NORC or other voltage-dependent NSCCs.

Cyclic nucleotides are second messengers in plant adaptive response to the environment and their concentration in the cell increases within seconds after salinity and osmotic stress onset (reviewed by Julkowska & Testerink, 2015; Shabala et al., 2015). Cyclic guanosine monophosphate (cGMP) is one such cyclic nucleotide and its concentration rapidly increases under salinity (reviewed by Donaldson, Ludidi, Knight, Gehring, & Denby, 2004). Cyclic nucleotide-gated channels (CNGCs) are ligand-gated cation channels, which can be activated by direct binding with cyclic nucleotides (Chin, Moeder, & Yoshioka, 2009). Representing a sub-class of NSCCs, they potentially modulate xylem Na^+ loading when expressed in the stellar tissue. In addition to CNGCs, cGMP activated-aquaporins could also mediate Na^+ transport into the xylem. Mammalian Aquaporin-1 (AQP1) is activated by intracellular cGMP, and the phosphorylation of tyrosine 253 located at the carboxyl terminal domain was found to govern the responsiveness of the AQP1 ionic conductance to cGMP (Campbell, Birdsell, & Yool, 2012). Water permeability through AtPIP2;1 is also regulated by phosphorylation (Prado et al., 2013), but the regulation of its ionic conductance (including Na^+) is not yet clear. Due to similar cyclic nucleotide signalling cascades present in animal and plant cells (Isner, Nühse, & Maathuis, 2012; Maurel et al., 1995), the mechanism known to control ion and water transport in mammalian AQP1 (including phosphorylation) should also be explored in AtPIP2;1 (reviewed by Kourghi et al., 2018). In plants, the level of cGMP is increased by ABA perception (Isner et al., 2012), and ABA modulates aquaporin conductance in response to water stress (Parent et al., 2009). Thus, it is plausible that plant aquaporins (e.g. PIP2;1) mediate xylem Na^+ loading in response to osmotic stress caused by drought or salinity (Fig. 1).

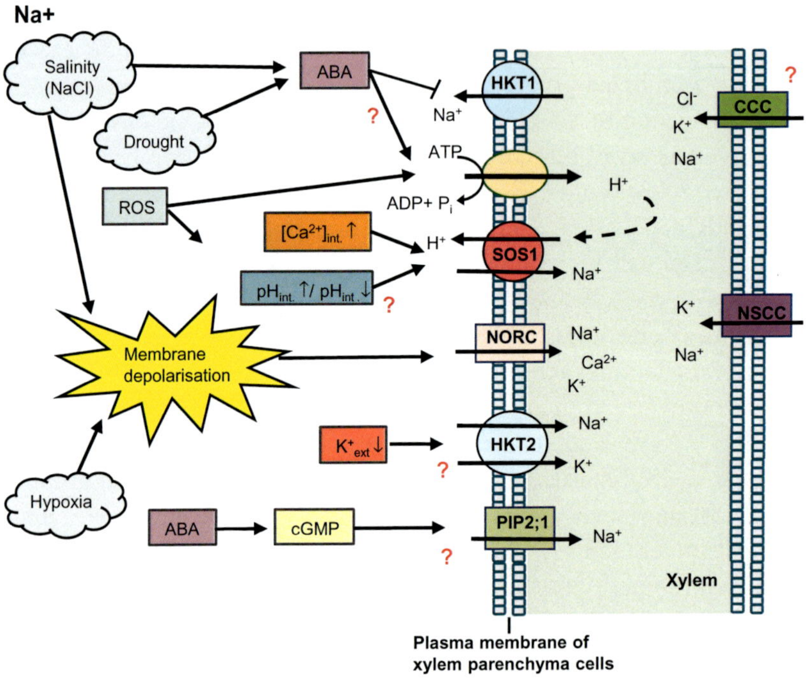

Fig. 1 Membrane transporters mediating xylem Na^+ loading under abiotic stress condition and signalling pathways involved in their control. *CCC*, chloride-cation cotransporter; *HKT*, High Affinity Potassium Transporter; *NORC*, non-selective outward-rectifying channel; *NSCC*, non-selective cation channel; *PIP*, plasma membrane intrinsic protein; *SOS*, Salt Overly Sensitive.

4.2 Chloride

4.2.1 Transcriptional Changes

Chloride is the predominant anion in saline soils. Elevated Cl^- concentrations are metabolically toxic to plants and associated with crop yield reduction in salinised plants (Li, Tester, et al., 2017; Li, Yu, et al., 2017; White & Broadley, 2001). Therefore, the dominant view on Cl^- regulation in non-halophytes is that the rate of Cl^- transport and exclusion from shoots correlates with salt tolerance. The key gatekeepers in controlling this process are the transporters within the root stelar cells, which are responsible for xylem loading of Cl^- (Li, Tester, et al., 2017; Li, Yu, et al., 2017). Consistent with this, the expression of known Cl^- transport systems (e.g. NPF2.4, SLAH1) has been found to decrease after exposure to salt stress. In Arabidopsis for instance, expression profiling by qRT-PCR showed that a 5 days treatment with 75 mM NaCl reduced the root abundance of NPF2.4 mRNA by almost 90% compared to plants grown at 2 mM Cl^- (Li et al., 2016).

Salt stress was also found to decrease SLAH1 expression in roots in a dose dependent manner, underpinning the view that SLAH1 fine-tunes the degree of Cl^- exclusion from the transpirational stream via the SLAH3 pathway. In contrast, SLAH3 transcript abundance was affected in the short (24 h) but not in the long-term (Cubero-Font et al., 2016; Qiu, Henderson, Tester, Roy, & Gilliham, 2016). This transcriptional regulation of SLAH1 might therefore represent a molecular valve to ensure NO_3^- loading while limiting Cl^- entry into the xylem during salinity stress.

Very little is known about transcriptional regulation of Cl^- transporters in plants following drought stress. As for saline plants, drought stress significantly reduces SLAH1 and SHLA3 transcript abundance in Arabidopsis (Cubero-Font et al., 2016). Given that ABA is involved in both salt and drought stress responses, this similar regulation suggests that in both stresses, ABA acts in the stele and transcriptionally regulates Cl^- transport into the xylem (Cubero-Font et al., 2016; Gilliham & Tester, 2005).

4.2.2 Post-translational Regulation and Signalling

Anion channels and transporters play a prominent role in a wide spectrum of physiological functions in plants and are under tight post-translational regulation (Barbier-Brygoo et al., 2011; Roberts, 2006). Classic examples are the ABA-mediated regulation of SLAC1 and SLAH3 in guard cells (Hedrich & Geiger, 2017) and the CIPK23-dependent phosphorylation of the NO_3^- transporter NRT1.1, which fine-tunes NO_3^- uptake at the root plasma membrane (Jacquot, Li, Gojon, Schulze, & Lejay, 2017; Liu & Tsay, 2003). Although our knowledge of the post-translational modifications and the signalling pathways involved in the regulation of these key anion channels in response to environmental stimuli is fairly advanced, the translation of that knowledge into transport regulation in xylem parenchyma cells may prove difficult. Indeed, given their differential role in a wide range of physiological processes throughout the plant, similar anion channel homologues appear to have pronounced differences in their mode of regulation by cytosolic messengers in different cells (e.g. Colcombet, Lelievre, Thomine, Barbier-Brygoo, & Frachisse, 2005; Frachisse, Colcombet, Guern, & Barbier-Brygoo, 2008). For example, while ABA activates Cl^- efflux through OST1-activated SLAC1 in guard cells (Hedrich & Geiger, 2017; Lind et al., 2015), ABA inhibits Cl^- efflux to the xylem apoplast in xylem parenchyma cells via an ABA-stimulated rise in cytosolic Ca^{2+} (Gilliham & Tester, 2005). This raises the question as to whether other regulatory proteins could act as intermediaries and alter channel regulation by cytosolic messengers, thus explaining this differential regulation in different cells. Possible candidates are aquaporins, whose intermediary role in ABA- and pathogen-triggered stomatal closure has recently been uncovered in Arabidopsis. Epidermal assays showed that AtPIP2;1 is required for ABA-dependent ROS signalling in guard cells and ABA-triggered stomatal closure (Rodrigues et al., 2017). This, coupled with the fact that AtPIP2;1 was found to interact with OST1–SLAC1 and CPK6/23–SLAC1 complexes and indirectly regulated SLAC1 activity in Arabidopsis guard cells (Wang et al., 2016), invites speculation that aquaporins could play a similar intermediary role in the ABA mediated regulation of Cl^- transport in xylem parenchyma cells.

Other candidate cellular signals that deserve further attention are GABA and ATP, which are known to post-translationally affect anion channel activity (e.g. ALMTs), in different parts of the plant (De Angeli, Thomine, & Frachisse, 2016; Ramesh et al., 2015). For example, GABA is thought to alter stomatal movement by blocking ALMTs (Mekonnen, Flügge, & Ludewig, 2016). Combined with the evidence that both GABA and ATP are key players in plant responses to various stresses including drought and salinity (Bown & Shelp, 2016; De Angeli, Thomine, & Frachisse, 2016; Kiani-Pouya et al., 2017; Li, Tester, et al., 2017; Li, Yu, et al., 2017; Ramesh et al., 2015), it further supports the importance of investigating their effects on xylem loading (Fig. 2).

4.3 Potassium

4.3.1 Transcriptional Changes

Given that the molecular identity of NORC channels is as yet unknown, most of the knowledge in the field comes from studies on Arabidopsis SKOR. Decreased SKOR transcript accumulation has been reported in response to K^+ deprivation, auxin and cytokinin treatments (Pilot et al., 2003) and importantly with regards to drought, ABA treatment (Gaymard et al., 1998; Pilot et al., 2003). This suggests that under a limited water supply, K^+ translocation to shoots is restricted to retain K^+ in the roots for osmotic adjustment (Pritchard, Jones, & Tomos, 1991; Sharp, Hsiao, & Silk, 1990) and maintenance of root turgor (Ahmad & Maathuis, 2014).

In contrast to drought, salinity increases SKOR transcript levels (Maathuis, 2006), but not systematically (Pilot et al., 2003). The resulting

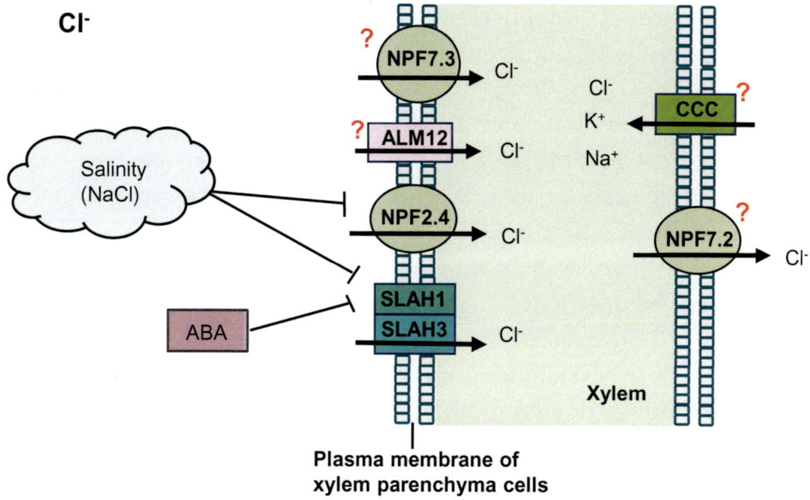

Fig. 2 Membrane transporters mediating xylem Cl^- loading under abiotic stress condition and signalling pathways involved in their control. *ALM*, Aluminium-Activated Malate Transporter; *CCC*, chloride-cation cotransporter; *NPF*, Nitrate/Peptide Family transporter; *SLAH*, Slow Anion efflux channel.

increase in SKOR activity could be crucial in maintaining and possibly restoring shoot K^+ levels under saline conditions, this function being vital in a plant's survival response to salt stress. Importantly, because SKOR is strongly K^+ selective, up-regulation would not increase Na^+ translocation to the shoot as it would if NORC expression/activity were increased—this channel shows equal selectivity for K^+ and Na^+. ABA which decreases SKOR transcript levels is probably not involved in the up-regulation of SKOR during salinity stress. ROS levels, which are known to increase under saline conditions (Jacoby, Taylor, & Millar, 2011; Schachtman & Shin, 2007), are more likely responsible for this upregulation. H_2O_2 is known to activate SKOR at the post-translational level (Garcia-Mata et al., 2010). Whether it also acts at the transcriptional level remains to be tested.

The expression of SKOR, together with that of NRT1.5/NPF7.3, is up-regulated by NO_3^- (Lin et al., 2008; Wang et al., 2004). This suggests that K^+ and NO_3^- transport into the xylem vessels may be co-ordinated via transcriptional regulation of several types of K^+ transporters (Li, Yu, et al., 2017). Indeed, SKOR seems to mediate root-to-shoot translocation of K^+ under high NO_3^- and low K^+ availability, while NRT1.5/NPF7.3 is more important for K^+ translocation under low NO_3^- availability, irrespective of the K^+ supply (Drechsler et al., 2015). Salinity may affect this balance. Under this stress condition, the chloride component of NaCl frequently reduces root NO_3^- uptake, leading to reduced availability in the plant. The strong induction of SKOR by salt stress may help mitigate this, especially as NRT1.5/NPF7.3 expression is down-regulated when potassium is replaced by sodium (Lin et al., 2008) as frequently occurs under saline conditions.

Very little is known about transcriptional regulation of CCCs. In rice, the expression of the *OsCCC1* gene is significantly enhanced by external KCl treatment (Kong et al., 2011), while in citrus and grapevine, *CCC* genes seem not to be transcriptionally regulated (Sorensson et al., 2012). The only report of transcriptional regulation in KUP7 is that it is not induced by K^+ and that any regulation is at the post-translational/post-transcriptional level (Han et al., 2016).

4.3.2 Post-translational Regulation and Signalling Pathways

Being an outward-rectifying channel, SKOR is activated upon membrane depolarisation. It is also sensitive to the external concentration of K^+, with its open probability decreasing when external K^+ concentration increases. SKOR is therefore in a closed state when the driving force for K^+ is inwardly directed and a far higher membrane potential is then required for its opening. This property minimises the risk of a K^+ influx into the parenchyma cell back from the xylem vessels (Johansson et al., 2006). The mechanism of external K^+ sensing involves interactions between the gating domain present in the sixth transmembrane segment (S6) of the channel polypeptide and the base of the pore helix (Johansson et al., 2006). In addition to this intrinsic regulation by external K^+, SKOR displays post-translational activation by H_2O_2. This process involves a cysteine residue present in the third transmembrane segment (S3) of the channel polypeptide. This H_2O_2 sensitive site is only accessible from the extracellular side of the membrane and when the channel is in an open conformation (Garcia-Mata et al., 2010). In agreement with this sensitivity to H_2O_2, the contribution of SKOR to root-to-shoot K^+ translocation is increased after H_2O_2 treatment (Garcia-Mata et al., 2010).

In addition to transcriptional regulation, ABA also acts post-translationally on SKOR (Gaymard et al., 1998; Roberts, 1998; Roberts & Snowman, 2000). In maize stellar cells, ABA triggers a rapid decrease in outward channel activity, which cannot result from a transcriptional regulation since it occurs within minutes (Roberts, 1998). This inhibition is rather post-translational and may be pH mediated; ABA would induce a decrease in cytosolic pH (Wilkinson & Davies, 2002), leading to inhibition of SKOR.

The SKOR channel indeed exhibits a strong pH-dependence. An internal pH decrease from 7.4 to 7.2 accounts for about 80% of the voltage independent decrease of the macroscopic SKOR current, and an external acidification from pH 7.4 to 6.4 has similar effects (Lacombe, Pilot, Gaymard, Sentenac, & Thibaud, 2000). Because the macroscopic gating parameters and the single channel conductance remain unchanged, it was proposed that internal and external pH adjustments regulate K^+ secretion to the xylem by altering the number of channels available for activation (Lacombe et al., 2000). Keeping in mind that stress-induced changes in the xylem pH are well reported (Bahrun, Jensen, Asch, & Mogensen, 2002; Wang, Liu, & Jensen, 2012), these findings suggest a causal link between xylem K^+ loading and plant adaptive responses to environmental stresses such as drought and salinity.

A number of regulatory partners have been identified that control the targeting to the plasma membrane, or the activity of plant Shaker channels (to which SKOR belongs). These include β-subunits (Zhang, Ma, & Berkowitz, 1999), 14-3-3 proteins (Sottocornola et al., 2006), different types of protein kinases (Xu et al., 2006; Zhao et al., 2013), phosphatases (Chérel et al., 2002; Lee et al., 2007), and SNAREs (Honsbein et al., 2009; Sutter, Campanoni, Tyrrell, & Blatt, 2006). However, compared to other, more conveniently-detectable Shaker channels, a few of such regulatory partners have been investigated for SKOR. Yet, it has been reported that the availability of phosphatidylinositol phosphates (PIPs) in the cell membrane is essential to the function of SKOR (Liu, Li, & Luan, 2005), indicating that PIPs are essential regulators for the voltage-dependent and independent activation of plant Shaker-type channels. This study also showed that PIPs are required for SKOR activation by voltage, suggesting that PIPs maintain the channel conformation in a 'ready state'.

The regulation of K^+ transport proteins by calcineurin B-like (CBL)-interacting protein kinases (CIPK) and their regulatory CBL partners has been characterised for a number of Shaker channels in Arabidopsis. Arabidopsis has 26 serine/threonine CIPKs and 10 CBLs, which form a signalling network that is involved in plant responses to environmental cues, including salt stress and K^+ deficiency (Luan, 2009; Wang & Wu, 2013).

Binding of Ca^{2+} to CBLs is required for CIPK-CBL interaction (Halfter, Ishitani, & Zhu, 2000; Shi, Quintero, Pardo, & Zhu, 2002). The CIPK23 protein, which is activated by CBL1 or CBL9 and subsequently phosphorylates the AKT1 channel, is well described (Lee et al., 2007; Xu et al., 2006). Similarly, the SOS3-SOS2-SOS1 pathway is well known (Gong, Guo, Schumaker, & Zhu, 2004; Halfter et al., 2000; Shi et al., 2002). Thus far, a similar regulatory pathway has not been found or even investigated for SKOR. Nonetheless, the ankyrin domain which serves as a CIPK interaction site in Shaker channels (Lee et al., 2007) is present in SKOR and therefore this type of regulation probably exists.

NORC channels show strong voltage dependence, are open at zero potential and are controlled by external and cytosolic Ca^{2+}, not by external salt concentration (Pottosin & Dobrovinskaya, 2014; Shabala, Cuin, & Pottosin, 2007). Due to their sensitivity to intracellular Ca^{2+}, it has been proposed that NORC can mediate massive salt loading to the xylem under any stress condition that invokes an increase in cytosolic Ca^{2+}. One such stress is salinity. Unfortunately, NORC is equally selective for K^+ and Na^+ (Wegner & Raschke, 1994), and the far higher Na^+ soil concentration under salinised conditions could result in large amounts of toxic Na^+ reaching the shoots. However, the potentially low K^+ content in the xylem could activate SKOR-mediated K^+ efflux and SKOR is known to be up-regulated by salinity (Maathuis, 2006). In addition, SKOR is activated by H_2O_2 as discussed above, and ROS levels increase under salinity (Jacoby et al., 2011; Schachtman & Shin, 2007). Thus, this putative dual regulation of SKOR and NORC activity has implications for salinity tolerance mechanisms, increasing the K^+/Na^+ selectivity of xylem loading, thereby improving the K^+ status, and as a consequence, the salinity tolerance of the plant.

An indirect effect of HKT1 on SKOR/NORC activity could also be present. Sodium absorption into xylem parenchyma cells via HKT transporters would depolarise xylem parenchyma cells. This could activate the depolarisation-activated SKOR and NORC, thus mediating K^+ efflux (Schroeder, Raschke, & Neher, 1987; Wegner & De Boer, 1997; Wegner & Raschke, 1994). Such a hypothesis corresponds with the frequently observed reduction in K^+ accumulation in the xylem and leaves of *athkt1.1* mutant plants alongside increases in the root K^+ content (Ren et al., 2005; Sunarpi et al., 2005), even though HKT1 is not a K^+ transporter (Fig. 3).

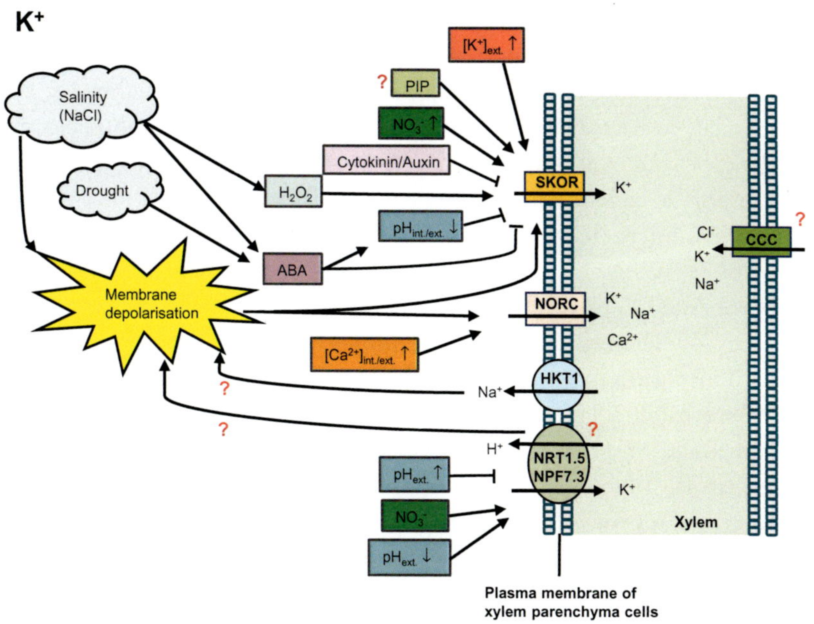

Fig. 3 Membrane transporters mediating xylem K^+ loading under abiotic stress condition and signalling pathways involved in their control. *CCC,* chloride-cation cotransporter; *HKT,* High Affinity Potassium Transporter; *NORC,* non-selective outward-rectifying channel; *SKOR,* Shaker-like outward rectifying K^+ channel.

5. IMPLICATIONS FOR PLANT BREEDING

As evident from the analysis above, control of xylem ion loading is essential for plant adaptive responses to abiotic stresses, including drought, salinity and flooding. Surprisingly, our search on the Web of Science has failed to reveal any papers reporting QTLs associated to any of above xylem parenchyma cell-based transporters mediating xylem Na^+, Cl^- and K^+ loading. Several reasons may explain this absence. First, plant xylem is not easily accessible, and the physiological markers that can be used to characterise the activity of these transporters are rather complicated and unsuited for the high throughput assays required for screening doubled haploid (DH) lines. Second, the molecular identity of NSCCs remains an enigma, which complicates any gene editing, and makes the direct proof of concept highly challenging. Finally, the kinetics of xylem ion loading is a highly dynamic process, hence the candidate genes need to be targeted at specific time points.

More optimistically, because the xylem loading trait has never been targeted in any breeding programmes, it is an unexplored opportunity for genetic improvement of plant germplasm. In most staple crops, the root stele can be isolated relatively simply by mechanical means. This enables electrophysiological studies and, specifically, allows measurement of the kinetics of ion transport in xylem parenchyma cells by non-invasive vibrating MIFE ion-selective microelectrodes (Shabala et al., 2010; Sun et al., 2017; Zhu et al., 2017). When applied to DH lines, this approach is likely to reveal the appropriate QTLs conferred by one or several transporters that mediate the process of xylem loading. More excitingly, key xylem-specific regulators of these transporters may be uncovered and/or validated. Our previous studies have revealed a large extent of variability in the xylem ion content between contrasting barley (Shabala et al., 2010; Zhu et al., 2017) and wheat (Zhu, Shabala, Shabala, Fan, & Zhou, 2016) varieties. These data can be used to create appropriate DH lines between contrasting parental varieties.

ACKNOWLEDGEMENTS

This work was supported by the Australian Research Council; Department of Industry, Innovation and Science; and China National Science Foundation Grants to S.S. N.B. is a recipient of the Marie Curie Fellowship (grant 700001).

REFERENCES

Ache, P., Becker, D., Ivashikina, N., Dietrich, P., Roelfsema, M. R. G., & Hedrich, R. (2000). GORK, a delayed outward rectifier expressed in guard cells of *Arabidopsis thaliana*, is a K^+-selective, K^+-sensing ion channel. *FEBS Letters*, *486*, 93–98.

Aguilar, E. A., Turner, D. W., Gibbs, D. J., Armstrong, W., & Sivasithamparam, K. (2003). Oxygen distribution and movement, respiration and nutrient loading in banana roots (*Musa spp*. L.) subjected to aerated and oxygen-depleted environments. *Plant and Soil*, *253*, 91–102.

Ahmad, I., & Maathuis, F. J. M. (2014). Cellular and tissue distribution of potassium: Physiological relevance, mechanisms and regulation. *Journal of Plant Physiology*, *171*, 708–714.

Almeida-Rodriguez, A. M., & Hacke, U. G. (2012). Cellular localization of aquaporin mRNA in hybrid poplar stems. *American Journal of Botany*, *99*, 1249–1254.

Assaha, D. V. M., Ueda, A., Saneoka, H., Al-Yahyai, R., & Yaish, M. W. (2017). The role of Na^+ and K^+ transporters in salt stress adaptation in glycophytes. *Frontiers in Physiology*, *8*, 509.

Bahrun, A., Jensen, C. R., Asch, F., & Mogensen, V. O. (2002). Drought-induced changes in xylem pH, ionic composition, and ABA concentration act as early signals in field-grown maize (*Zea mays* L.). *Journal of Experimental Botany*, *53*, 251–263.

Balagué, C., Lin, B. Q., Alcon, C., Flottes, G., Malmström, S., Köhler, C., et al. (2003). HLM1, an essential signaling component in the hypersensitive response, is a member of the cyclic nucleotide-gated channel ion channel family. *Plant Cell*, *15*, 365–379.

Bañuelos, M. A., Garciadeblas, B., Cubero, B., & Rodriguez-Navarro, A. (2002). Inventory and functional characterization of the HAK potassium transporters of rice. *Plant Physiology*, *130*, 784–795.

Barbier-Brygoo, H., De Angeli, A., Filleur, S., Frachisse, J.-M., Gambale, F., Thomine, S., et al. (2011). Anion channels/transporters in plants: From molecular bases to regulatory networks. *Annual Review of Plant Biology*, *62*, 25–51.

Berthomieu, P., Conejero, G., Nublat, A., Brackenbury, W. J., Lambert, C., Savio, C., et al. (2003). Functional analysis of AtHKT1 in Arabidopsis shows that Na^+ recirculation by the phloem is crucial for salt tolerance. *The EMBO Journal*, *22*, 2004–2014.

Bose, J., Shabala, L., Pottosin, I., Zeng, F. R., Velarde-Buendía, A. M., Massart, A., et al. (2014). Kinetics of xylem loading, membrane potential maintenance, and sensitivity of K^+-permeable channels to reactive oxygen species: Physiological traits that differentiate salinity tolerance between pea and barley. *Plant, Cell & Environment*, *37*, 589–600.

Bown, A. W., & Shelp, B. J. (2016). Plant GABA: Not just a metabolite. *Trends in Plant Science*, *21*, 811–813.

Büchsenschütz, K., Marten, I., Becker, D., Philippar, K., Ache, P., & Hedrich, R. (2005). Differential expression of K^+ channels between guard cells and subsidiary cells within the maize stomatal complex. *Planta*, *222*, 968–976.

Byrt, C. S., Platten, J. D., Spielmeyer, W., James, R. A., Lagudah, E. S., Dennis, E. S., et al. (2007). HKT1;5-like cation transporters linked to Na^+ exclusion loci in wheat, Nax2 and Knal. *Plant Physiology*, *143*, 1918–1928.

Byrt, C. S., Zhao, M. C., Kourghi, M., Bose, J., Henderson, S. W., Qiu, J. E., et al. (2017). Non-selective cation channel activity of aquaporin AtPIP2;1 regulated by Ca2 + and pH. *Plant, Cell & Environment*, *40*, 802–815.

Campbell, E. M., Birdsell, D. N., & Yool, A. J. (2012). The activity of human aquaporin 1 as a cGMP-gated cation channel is regulated by tyrosine phosphorylation in the carboxyl-terminal domain. *Molecular Pharmacology*, *81*, 97–105.

Chen, Z. C., Yamaji, N., Fujii-Kashino, M., & Ma, J. F. (2016). A cation-chloride cotransporter gene is required for cell elongation and osmoregulation in rice. *Plant Physiology*, *171*, 494–507.

Chérel, I., Michard, E., Platet, N., Mouline, K., Alcon, C., Sentenac, H., et al. (2002). Physical and functional interaction of the Arabidopsis K^+ channel AKT2 and phosphatase AtPP2CA. *Plant Cell*, *14*, 1133–1146.

Chin, K., Moeder, W., & Yoshioka, K. (2009). Biological roles of cyclic-nucleotide-gated ion channels in plants: What we know and don't know about this 20 member ion channel family. *Botany-Botanique*, *87*, 668–677.

Chinnusamy, V., Schumaker, K., & Zhu, J. K. (2004). Molecular genetic perspectives on cross-talk and specificity in abiotic stress signalling in plants. *Journal of Experimental Botany*, *55*, 225–236.

Chung, J. S., Zhu, J. K., Bressan, R. A., Hasegawa, P. M., & Shi, H. H. (2008). Reactive oxygen species mediate Na^+-induced SOS1 mRNA stability in Arabidopsis. *Plant Journal*, *53*, 554–565.

Clarkson, D. T., & Hanson, J. B. (1986). Proton fluxes and the activity of a stelar proton pump in onion roots. *Journal of Experimental Botany*, *37*, 1136–1150.

Colcombet, J., Lelievre, F., Thomine, S., Barbier-Brygoo, H., & Frachisse, J. M. (2005). Distinct pH regulation of slow and rapid anion channels at the plasma membrane of *Arabidopsis thaliana* hypocotyl cells. *Journal of Experimental Botany*, *56*, 1897–1903.

Colmenero-Flores, J. M., Martínez, G., Gamba, G., Vázquez, N., Iglesias, D. J., Brumós, J., et al. (2007). Identification and functional characterization of cation-chloride cotransporters in plants. *Plant Journal*, *50*, 278–292.

Colmer, T. D., & Greenway, H. (2011). Ion transport in seminal and adventitious roots of cereals during O_2 deficiency. *Journal of Experimental Botany*, *62*, 39–57.

Corratgé-Faillie, C., Jabnoune, M., Zimmermann, S., Véry, A. A., Fizames, C., & Sentenac, H. (2010). Potassium and sodium transport in non-animal cells: The Trk/Ktr/HKT transporter family. *Cellular and Molecular Life Sciences*, *67*, 2511–2532.

Cubero-Font, P., Maierhofer, T., Jaslan, J., Rosales, M. A., Espartero, J., Díaz-Rueda, P., et al. (2016). Silent S-type anion channel subunit SLAH1 gates SLAH3 open for chloride root-to-shoot translocation. *Current Biology*, *26*, 2213–2220.

Davenport, R. J., Munoz-Mayor, A., Jha, D., Essah, P. A., Rus, A., & Tester, M. (2007). The Na^+ transporter AtHKT1;1 controls retrieval of Na^+ from the xylem in *Arabidopsis*. *Plant, Cell & Environment*, *30*, 497–507.

Davies, W. J., & Ribaut, J. M. (2017). Stress resilience in crop plants: Strategic thinking to address local food production problems. *Food and Energy Security*, *6*, 12–18.

De Angeli, A., Thomine, S., & Frachisse, J.-M. (2016). Anion channel blockage by ATP as a means for membranes to perceive the energy status of the cell. *Molecular Plant*, *9*, 320–322.

De Boer, A. H., & Volkov, V. (2003). Logistics of water and salt transport through the plant: Structure and functioning of the xylem. *Plant, Cell & Environment*, *26*, 87–101.

Demidchik, V. (2014). Mechanisms and physiological roles of K^+ efflux from root cells. *Journal of Plant Physiology*, *171*, 696–707.

Demidchik, V., Davenport, R. J., & Tester, M. (2002). Nonselective cation channels in plants. *Annual Review of Plant Biology*, *53*, 67–107.

Demidchik, V., & Maathuis, F. J. M. (2007). Physiological roles of nonselective cation channels in plants: From salt stress to signalling and development. *The New Phytologist*, *175*, 387–404.

Demidchik, V., Shabala, S. N., & Davies, J. M. (2007). Spatial variation in H_2O_2 response of *Arabidopsis thaliana* root epidermal Ca^{2+} flux and plasma membrane Ca^{2+} channels. *Plant Journal*, *49*, 377–386.

Donaldson, L., Ludidi, N., Knight, M. R., Gehring, C., & Denby, K. (2004). Salt and osmotic stress cause rapid increases in *Arabidopsis thaliana* cGMP levels. *FEBS Letters*, *569*, 317–320.

D'Onofrio, C., & Lindberg, S. (2009). Sodium induces simultaneous changes in cytosolic calcium and pH in salt-tolerant quince protoplasts. *Journal of Plant Physiology*, *166*, 1755–1763.

Drechsler, N., Zheng, Y., Bohner, A., Nobmann, B., von Wiren, N., Kunze, R., et al. (2015). Nitrate-dependent control of shoot K homeostasis by the nitrate transporter1/peptide transporter family member NPF7.3/NRT1.5 and the Stelar K^+ outward rectifier SKOR in Arabidopsis. *Plant Physiology*, *169*, 2832–2847.

FAO. (2005). *Status of research and application of crop biotechnologies in developing countries*. http://www.fao.org/docrep/008/y5800e/Y5800E06.htm.

Felle, H. H. (2005). pH regulation in anoxic plants. *Annals of Botany*, *96*, 519–532.

Frachisse, J. M., Colcombet, J., Guern, J., & Barbier-Brygoo, H. (2008). Characterization of a nitrate-permeable channel able to mediate sustained anion efflux in hypocotyl cells from *Arabidopsis thaliana*. *Plant Journal*, *21*, 361–371.

Fraile-Escanciano, A., Kamisugi, Y., Cuming, A. C., Rodríguez-Navarro, A., & Benito, B. (2010). The SOS1 transporter of *Physcomitrella* patens mediates sodium efflux *in planta*. *The New Phytologist*, *188*, 750–761.

Garciadeblás, B., Benito, B., & Rodriguez-Navarro, A. (2002). Molecular cloning and functional expression in bacteria of the potassium transporters CnHAK1 and CnHAK2 of the seagrass *Cymodocea nodosa*. *Plant Molecular Biology*, *50*, 623–633.

Garciadeblás, B., Senn, M. E., Banuelos, M. A., & Rodriguez-Navarro, A. (2003). Sodium transport and HKT transporters: The rice model. *Plant Journal*, *34*, 788–801.

Garcia-Mata, C., Wang, J. W., Gajdanowicz, P., Gonzalez, W., Hills, A., Donald, N., et al. (2010). A minimal cysteine motif required to activate the SKOR K^+ channel of *Arabidopsis* by the reactive oxygen species H_2O_2. *Journal of Biological Chemistry*, *285*, 29286–29294.

Gaymard, F., Pilot, G., Lacombe, B., Bouchez, D., Bruneau, D., Boucherez, J., et al. (1998). Identification and disruption of a plant shaker-like outward channel involved in K^+ release into the xylem sap. *Cell, 94*, 647–655.

Geilfus, C.-M. (2018). Chloride: From nutrient to toxicant. *Plant and Cell Physiology, 59*, 877–886.

Gibbs, J., & Greenway, H. (2003). Mechanisms of anoxia tolerance in plants. I. Growth, survival and anaerobic catabolism. *Functional Plant Biology, 30*, 1–47.

Gilliham, M., & Tester, M. (2005). The regulation of anion loading to the maize root xylem. *Plant Physiology, 137*, 819–828.

Gobert, A., Park, G., Amtmann, A., Sanders, D., & Maathuis, F. J. M. (2006). *Arabidopsis thaliana* cyclic nucleotide gated channel 3 forms a non-selective ion transporter involved in germination and cation transport. *Journal of Experimental Botany, 57*, 791–800.

Gong, D. M., Guo, Y., Schumaker, K. S., & Zhu, J. K. (2004). The SOS3 family of calcium sensors and SOS2 family of protein kinases in Arabidopsis. *Plant Physiology, 134*, 919–926.

Guo, K. M., Babourina, O., Christopher, D. A., Borsic, T., & Rengel, Z. (2010). The cyclic nucleotide-gated channel AtCNGC10 transports Ca^{2+} and Mg^{2+} in *Arabidopsis*. *Physiologia Plantarum, 139*, 303–312.

Guo, K. M., Babourina, O., Christopher, D. A., Borsics, T., & Rengel, Z. (2008). The cyclic nucleotide-gated channel, AtCNGC10, influences salt tolerance in *Arabidopsis*. *Physiologia Plantarum, 134*, 499–507.

Halfter, U., Ishitani, M., & Zhu, J. K. (2000). The *Arabidopsis* SOS2 protein kinase physically interacts with and is activated by the calcium-binding protein SOS3. *Proceedings of the National Academy of Sciences of the United States of America, 97*, 3735–3740.

Hamamoto, S., Horie, T., Hauser, F., Deinlein, U., Schroeder, J. I., & Uozumi, N. (2015). HKT transporters mediate salt stress resistance in plants: From structure and function to the field. *Current Opinion in Biotechnology, 32*, 113–120.

Han, M., Wu, W., Wu, W. H., & Wang, Y. (2016). Potassium transporter KUP7 is involved in K^+ acquisition and translocation in *Arabidopsis* root under K^+-limited conditions. *Molecular Plant, 9*, 437–446.

Hedrich, R., & Geiger, D. (2017). Biology of SLAC1-type anion channels—From nutrient uptake to stomatal closure. *The New Phytologist, 216*, 46–61.

Henderson, W. S., Wege, S., & Gilliham, M. (2018). Plant cation-chloride cotransporters (CCC): Evolutionary origins and functional insights. *International Journal of Molecular Sciences, 19*, 492.

Honsbein, A., Sokolovski, S., Grefen, C., Campanoni, P., Pratelli, R., Paneque, M., et al. (2009). A tripartite SNARE-K^+ channel complex mediates in channel-dependent K^+ nutrition in *Arabidopsis*. *Plant Cell, 21*, 2859–2877.

Horie, T., Costa, A., Kim, T. H., Han, M. J., Horie, R., Leung, H. Y., et al. (2007). Rice OsHKT2;1 transporter mediates large Na^+ influx component into K^+-starved roots for growth. *The EMBO Journal, 26*, 3003–3014.

Horie, T., Hauser, F., & Schroeder, J. I. (2009). HKT transporter-mediated salinity resistance mechanisms in *Arabidopsis* and monocot crop plants. *Trends in Plant Science, 14*, 660–668.

Horie, T., Yoshida, K., Nakayama, H., Yamada, K., Oiki, S., & Shinmyo, A. (2001). Two types of HKT transporters with different properties of Na^+ and K^+ transport in *Oryza sativa*. *Plant Journal, 27*, 129–138.

Hu, J., Ma, Q., Kumar, T., Duan, H. R., Zhang, J. L., Yuan, H. J., et al. (2016). ZxSKOR is important for salinity and drought tolerance of *Zygophyllum xanthoxylum* by maintaining K^+ homeostasis. *Plant Growth Regulation, 80*, 195–205.

Huang, L., Raats, D., Sela, H., Klymiuk, V., Lidzbarsky, G., Feng, L. H., et al. (2016). Evolution and adaptation of wild emmer wheat populations to biotic and abiotic stresses. *Annual Review of Phytopathology, 54*, 279–301.

Isner, J. C., Nühse, T., & Maathuis, F. J. M. (2012). The cyclic nucleotide cGMP is involved in plant hormone signalling and alters phosphorylation of *Arabidopsis thaliana* root proteins. *Journal of Experimental Botany, 63*, 3199–3205.

Ivashikina, N., Becker, D., Ache, P., Meyerhoff, O., Felle, H. H., & Hedrich, R. (2001). K^+ channel profile and electrical properties of *Arabidopsis* root hairs. *FEBS Letters, 508*, 463–469.

Jabnoune, M., Espeout, S., Mieulet, D., Fizames, C., Verdeil, J. L., Conéjéro, G., et al. (2009). Diversity in expression patterns and functional properties in the rice HKT transporter family. *Plant Physiology, 150*, 1955–1971.

Jacoby, R. P., Taylor, N. L., & Millar, A. H. (2011). The role of mitochondrial respiration in salinity tolerance. *Trends in Plant Science, 16*, 614–623.

Jacquot, A., Li, Z., Gojon, A., Schulze, W., & Lejay, L. (2017). Post-translational regulation of nitrogen transporters in plants and microorganisms. *Journal of Experimental Botany, 68*, 2567–2580.

Jiang, C. F., Belfield, E. J., Cao, Y., Smith, J. A. C., & Harberd, N. P. (2013). An *Arabidopsis* soil-salinity-tolerance mutation confers ethylene-mediated enhancement of sodium/potassium homeostasis. *Plant Cell, 25*, 3535–3552.

Johansson, I., Wulfetange, K., Poree, F., Michard, E., Gajdanowicz, P., Lacombe, B., et al. (2006). External K^+ modulates the activity of the Arabidopsis potassium channel SKOR via an unusual mechanism. *Plant Journal, 46*, 269–281.

Julkowska, M. M., & Testerink, C. (2015). Tuning plant signaling and growth to survive salt. *Trends in Plant Science, 20*, 586–594.

Kaplan, B., Sherman, T., & Fromm, H. (2007). Cyclic nucleotide-gated channels in plants. *FEBS Letters, 581*, 2237–2246.

Katschnig, D., Bliek, T., Rozema, J., & Schat, H. (2015). Constitutive high-level SOS1 expression and absence of HKT1;1 expression in the salt-accumulating halophyte *Salicornia dolichostachya*. *Plant Science, 234*, 144–154.

Kiani-Pouya, A., Roessner, U., Jayasinghe, N. S., Lutz, A., Rupasinghe, T., Bazihizina, N., et al. (2017). Epidermal bladder cells confer salinity stress tolerance in the halophyte quinoa and Atriplex species. *Plant, Cell & Environment, 40*, 1900–1915.

Kim, H. Y., Choi, E. H., Min, M. K., Hwang, H., Moon, S. J., Yoon, I., et al. (2015). Differential gene expression of two outward-rectifying shaker-like potassium channels OsSKOR and OsGORK in rice. *Journal of Plant Biology, 58*, 230–235.

Kinnersley, A. M., & Turano, F. J. (2000). Gamma aminobutyric acid (GABA) and plant responses to stress. *Critical Reviews in Plant Sciences, 19*, 479–509.

Knight, H., Brandt, S., & Knight, M. R. (1998). A history of stress alters drought calcium signalling pathways in Arabidopsis. *Plant Journal, 16*, 681–687.

Köhler, B., & Raschke, K. (2000). The delivery of salts to the xylem. Three types of anion conductance in the plasmalemma of the xylem parenchyma of roots of barley. *Plant Physiology, 122*, 243.

Kong, X. Q., Gao, X. H., Sun, W., An, J., Zhao, Y. X., & Zhang, H. (2011). Cloning and functional characterization of a cation-chloride cotransporter gene OsCCC1. *Plant Molecular Biology, 75*, 567–578.

Kotula, L., Clode, P. L., Striker, G. G., Pedersen, O., Lauchli, A., Shabala, S., et al. (2015). Oxygen deficiency and salinity affect cell-specific ion concentrations in adventitious roots of barley (*Hordeum vulgare*). *The New Phytologist, 208*, 1114–1125.

Kourghi, M., Pei, J. V., De Ieso, M. L., Nourmohammadi, S., Chow, P. H., & Yool, A. J. (2018). Fundamental structural and functional properties of Aquaporin ion channels found across the kingdoms of life. *Clinical and Experimental Pharmacology and Physiology, 45*, 401–409.

Lacombe, B., Pilot, G., Gaymard, F., Sentenac, H., & Thibaud, J. B. (2000). pH control of the plant outwardly-rectifying potassium channel SKOR. *FEBS Letters, 466*, 351–354.

Lagarde, D., Basset, M., Lepetit, M., Conejero, G., Gaymard, F., & Astruc, S. (1996). Tissue-specific expression of *Arabidopsis AKT1* gene is consistent with a role in K^+ nutrition. *Plant Journal, 9*, 195–203.

Lam, H. M., Chiu, J., Hsieh, M. H., Meisel, L., Oliveira, I. C., Shin, M., et al. (1998). Glutamate-receptor genes in plants. *Nature, 396*, 125–126.

Laurie, S., Feeney, K. A., Maathuis, F. J. M., Heard, P. J., Brown, S. J., & Leigh, R. A. (2002). A role for HKT1 in sodium uptake by wheat roots. *Plant Journal, 32*, 139–149.

Lee, S. C., Lan, W. Z., Kim, B. G., Li, L. G., Cheong, Y. H., Pandey, G. K., et al. (2007). A protein phosphorylation/dephosphorylation network regulates a plant potassium channel. *Proceedings of the National Academy of Sciences of the United States of America, 104*, 15959–15964.

Leng, Q., Mercier, R. W., Hua, B. G., Fromm, H., & Berkowitz, G. A. (2002). Electrophysiological analysis of cloned cyclic nucleotide-gated ion channels. *Plant Physiology, 128*, 400–410.

Li, X. L., Borsics, T., Harrington, H. M., & Christopher, D. A. (2005). *Arabidopsis* AtCNGC10 rescues potassium channel mutants of *E. coli*, yeast and *Arabidopsis* and is regulated by calcium/calmodulin and cyclic GMP in *E. coli*. *Functional Plant Biology, 32*, 643–653.

Li, B., Byrt, C., Qiu, J., Baumann, U., Hrmova, M., Evrard, A., et al. (2016). Identification of a stelar-localized transport protein that facilitates root-to-shoot transfer of chloride in Arabidopsis. *Plant Physiology, 170*, 1014.

Li, J., Long, Y., Qi, G. N., Xu, Z. J., Wu, W. H., & Wang, Y. (2014). The Os-AKT1 channel is critical for K^+ uptake in rice roots and is modulated by the rice CBL1-CIPK23 complex. *Plant Cell, 26*, 3387–3402.

Li, B., Tester, M., & Gilliham, M. (2017). Chloride on the move. *Trends in Plant Science, 22*(3), 236–248.

Li, H., Yu, M., Du, X. Q., Wang, Z. F., Wu, W. H., Quintero, F. J., et al. (2017). NRT1.5/NPF7.3 functions as a proton-coupled H^+/K^+ antiporter for K^+ loading into the xylem in Arabidopsis. *Plant Cell, 29*, 2016–2026.

Lin, S. H., Kuo, H. F., Canivenc, G., Lin, C. S., Lepetit, M., Hsu, P. K., et al. (2008). Mutation of the *Arabidopsis NRT1.5* nitrate transporter causes defective root-to-shoot nitrate transport. *Plant Cell, 20*, 2514–2528.

Lind, C., Dreyer, I., López-Sanjurjo, E. J., von Meyer, K., Ishizaki, K., Kohchi, T., et al. (2015). Stomatal guard cells co-opted an ancient ABA-dependent desiccation survival system to regulate stomatal closure. *Current Biology, 25*, 928–935.

Lips, S. H. (1997). The role of inorganic nitrogen ions in plant adaptation processes. *Russian Journal of Plant Physiology, 44*, 421–431.

Liu, K., Li, L. G., & Luan, S. (2005). An essential function of phosphatidylinositol phosphates in activation of plant shaker-type K^+ channels. *Plant Journal, 42*, 433–443.

Liu, K. H., & Tsay, Y. F. (2003). Switching between the two action modes of the dual-affinity nitrate transporter CHL1 by phosphorylation. *The EMBO Journal, 22*, 1005.

Luan, S. (2009). The CBL-CIPK network in plant calcium signaling. *Trends in Plant Science, 14*, 37–42.

Maathuis, F. J. M. (2006). The role of monovalent cation transporters in plant responses to salinity. *Journal of Experimental Botany, 57*, 1137–1147.

Maathuis, F. J. M. (2014). Sodium in plants: Perception, signalling, and regulation of sodium fluxes. *Journal of Experimental Botany, 65*, 849–858.

Mäser, P., Eckelman, B., Vaidyanathan, R., Horie, T., Fairbairn, D. J., Kubo, M., et al. (2002). Altered shoot/root Na^+ distribution and bifurcating salt sensitivity in *Arabidopsis* by genetic disruption of the Na^+ transporter *AtHKT1*. *FEBS Letters*, *531*, 157–161.

Mäser, P., Gierth, M., & Schroeder, J. I. (2002). Molecular mechanisms of potassium and sodium uptake in plants. *Plant and Soil*, *247*, 43–54.

Maurel, C., Kado, R. T., Guern, J., & Chrispeels, M. J. (1995). Phosphorylation regulates the water channel activity of the seed-specific aquaporin α-tip. *The EMBO Journal*, *14*, 3028–3035.

Mekonnen, D. W., Flügge, U.-I., & Ludewig, F. (2016). Gamma-aminobutyric acid depletion affects stomata closure and drought tolerance of *Arabidopsis thaliana*. *Plant Science*, *245*, 25–34.

Miller, G., Shulaev, V., & Mittler, R. (2008). Reactive oxygen signaling and abiotic stress. *Physiologia Plantarum*, *133*, 481–489.

Mittler, R. (2002). Oxidative stress, antioxidants and stress tolerance. *Trends in Plant Science*, 7, 405–410.

Munns, R., & Tester, M. (2008). Mechanisms of salinity tolerance. *Annual Review of Plant Biology*, *59*, 651–681.

Nguyen, T. H., Huang, S. G., Meynard, D., Chaine, C., Michel, R., Roelfsema, M. R. G., et al. (2017). A dual role for the OsK5.2 ion channel in stomatal movements and K^+ loading into xylem sap. *Plant Physiology*, *174*, 2409–2418.

Nieves-Cordones, M., Martinez, V., Benito, B., & Rubio, F. (2016). Comparison between *Arabidopsis* and rice for main pathways of K^+ and Na^+ uptake by roots. *Frontiers in Plant Science*, 7, 14.

Núñez-Ramírez, R., Sánchez-Barrena, M. J., Villalta, I., Vega, J. F., Pardo, J. M., Quintero, F. J., et al. (2012). Structural insights on the plant salt-overly-sensitive 1 (SOS1) Na^+/H^+ antiporter. *Journal of Molecular Biology*, *424*, 283–294.

Oddo, E., Inzerillo, S., La Bella, F., Grisafi, F., Salleo, S., & Nardini, A. (2011). Short-term effects of potassium fertilization on the hydraulic conductance of *Laurus nobilis* L. *Tree Physiology*, *31*, 131–138.

Osakabe, Y., Arinaga, N., Umezawa, T., Katsura, S., Nagamachi, K., Tanaka, H., et al. (2013). Osmotic stress responses and plant growth controlled by potassium transporters in *Arabidopsis*. *Plant Cell*, *25*, 609–624.

Parent, B., Hachez, C., Redondo, E., Simonneau, T., Chaumont, F., & Tardieu, F. (2009). Drought and abscisic acid effects on aquaporin content translate into changes in hydraulic conductivity and leaf growth rate: A trans-scale approach. *Plant Physiology*, *149*, 2000–2012.

Pilot, G., Gaymard, F., Mouline, K., Cherel, I., & Sentenac, H. (2003). Regulated expression of *Arabidopsis* Shaker K^+ channel genes involved in K^+ uptake and distribution in the plant. *Plant Molecular Biology*, *51*, 773–787.

Pingali, P. L. (2012). Green revolution: Impacts, limits, and the path ahead. *Proceedings of the National Academy of Sciences of the United States of America*, *109*, 12302–12308.

Platten, J. D., Cotsaftis, O., Berthomieu, P., Bohnert, H., Davenport, R. J., Fairbairn, D. J., et al. (2006). Nomenclature for *HKT* transporters, key determinants of plant salinity tolerance. *Trends in Plant Science*, *11*, 372–374.

Pottosin, I., & Dobrovinskaya, O. (2014). Non-selective cation channels in plasma and vacuolar membranes and their contribution to K^+ transport. *Journal of Plant Physiology*, *171*, 732–742.

Prado, K., Boursiac, Y., Tournaire-Roux, C., Monneuse, J. M., Postaire, O., Da Ines, O., et al. (2013). Regulation of *Arabidopsis* leaf hydraulics involves light-dependent phosphorylation of aquaporins in veins. *Plant Cell*, *25*, 1029–1039.

Pritchard, J., Jones, R. G. W., & Tomos, A. D. (1991). Turgor, growth and rheological gradients of wheat roots following osmotic-stress. *Journal of Experimental Botany*, *42*, 1043–1049.

Qiu, J., Henderson, S. W., Tester, M., Roy, S. J., & Gilliham, M. (2016). SLAH1, a homologue of the slow type anion channel SLAC1, modulates shoot Cl^- accumulation and salt tolerance in *Arabidopsis thaliana*. *Journal of Experimental Botany*, *67*, 4495–4505.

Quan, R. D., Lin, H. X., Mendoza, I., Zhang, Y. G., Cao, W. H., Yang, Y. Q., et al. (2007). SCABP8/CBL10, a putative calcium sensor, interacts with the protein kinase SOS2 to protect *Arabidopsis* shoots from salt stress. *Plant Cell*, *19*, 1415–1431.

Quintero, F. J., Martinez-Atienza, J., Villalta, I., Jiang, X. Y., Kim, W. Y., Ali, Z., et al. (2011). Activation of the plasma membrane Na/H antiporter Salt-Overly-Sensitive 1 (SOS1) by phosphorylation of an auto-inhibitory C-terminal domain. *Proceedings of the National Academy of Sciences of the United States of America*, *108*, 2611–2616.

Ramesh, S. A., Tyerman, S. D., Xu, B., Bose, J., Kaur, S., Conn, V., et al. (2015). GABA signalling modulates plant growth by directly regulating the activity of plant-specific anion transporters. *Nature Communications*, *6*, 7879.

Ren, Z. H., Gao, J. P., Li, L. G., Cai, X. L., Huang, W., Chao, D. Y., et al. (2005). A rice quantitative trait locus for salt tolerance encodes a sodium transporter. *Nature Genetics*, *37*, 1141–1146.

Roberts, S. K. (1998). Regulation of K^+ channels in maize roots by water stress and abscisic acid. *Plant Physiology*, *116*, 145–153.

Roberts, S. K. (2006). Plasma membrane anion channels in higher plants and their putative functions in roots. *The New Phytologist*, *169*, 647–666.

Roberts, S. K., & Snowman, B. N. (2000). The effects of ABA on channel-mediated K^+ transport across higher plant roots. *Journal of Experimental Botany*, *51*, 1585–1594.

Rodrigues, O., Reshetnyak, G., Grondin, A., Saijo, Y., Leonhardt, N., Maurel, C., et al. (2017). Aquaporins facilitate hydrogen peroxide entry into guard cells to mediate ABA- and pathogen-triggered stomatal closure. *Proceedings of the National Academy of Sciences of the United States of America*, *114*, 9200.

Rubio, F., Aleman, F., Nieves-Cordones, M., & Martinez, V. (2010). Studies on Arabidopsis *athak5*, *atakt1* double mutants disclose the range of concentrations at which AtHAK5, AtAKT1 and unknown systems mediate K^+ uptake. *Physiologia Plantarum*, *139*, 220–228.

Rubio, F., Gassmann, W., & Schroeder, J. I. (1995). Sodium-driven potassium uptake by the plant potassium transporter HKT1 and mutations conferring salt tolerance. *Science*, *270*, 1660–1663.

Rubio, F., Schwarz, M., Gassmann, W., & Schroeder, J. I. (1999). Genetic selection of mutations in the high affinity K^+ transporter HKT1 that define functions of a loop site for reduced Na^+ permeability and increased Na^+ tolerance. *Journal of Biological Chemistry*, *274*, 6839–6847.

Rus, A., Lee, B. H., Muňoz-Mayor, A., Sharkhuu, A., Miura, K., Zhu, J. K., et al. (2004). AtHKT1 facilitates Na^+ homeostasis and K^+ nutrition in planta. *Plant Physiology*, *136*, 2500–2511.

Sakr, S., Alves, G., Morillon, R. L., Maurel, K., Decourteix, M., Guilliot, A., et al. (2003). Plasma membrane aquaporins are involved in winter embolism recovery in walnut tree. *Plant Physiology*, *133*, 630–641.

Sánchez-Barrena, M. J., Martínez-Ripoll, M., & Albert, A. (2013). Structural biology of a major signaling network that regulates plant abiotic stress: The CBL-CIPK mediated pathway. *International Journal of Molecular Sciences*, *14*, 5734–5749.

Schachtman, D. P., & Schroeder, J. I. (1994). Structure and transport mechanism of a high-affinity potassium uptake transporter from higher-plants. *Nature*, *370*, 655–658.

Schachtman, D. P., & Shin, R. (2007). Nutrient sensing and signaling: NPKS. *Annual Review of Plant Biology*, *58*, 47–69.

Schroeder, J. I., Raschke, K., & Neher, E. (1987). Voltage dependence of K^+ channels in guard-cell protoplasts. *Proceedings of the National Academy of Sciences of the United States of America*, *84*, 4108–4112.

Secchi, F., Pagliarani, C., & Zwieniecki, M. A. (2017). The functional role of xylem parenchyma cells and aquaporins during recovery from severe water stress. *Plant, Cell & Environment*, *40*, 858–871.
Shabala, S. (2013). Learning from halophytes: Physiological basis and strategies to improve abiotic stress tolerance in crops. *Annals of Botany*, *112*, 1209–1221.
Shabala, S., Bose, J., Fuglsang, A. T., & Pottosin, I. (2016). On a quest for stress tolerance genes: Membrane transporters in sensing and adapting to hostile soils. *Journal of Experimental Botany*, *67*, 1015–1031.
Shabala, S., Cuin, T. A., & Pottosin, I. (2007). Polyamines prevent NaCl-induced K^+ efflux from pea mesophyll by blocking non-selective cation channels. *FEBS Letters*, *581*, 1993–1999.
Shabala, S., Shabala, S., Cuin, T. A., Pang, J., Percey, W., Chen, Z., et al. (2010). Xylem ionic relations and salinity tolerance in barley. *Plant Journal*, *61*, 839–853.
Shabala, S., White, R. G., Djordjevic, M. A., Ruan, Y. L., & Mathesius, U. (2016). Root-to-shoot signalling: Integration of diverse molecules, pathways and functions. *Functional Plant Biology*, *43*, 87–104.
Shabala, S., Wu, H. H., & Bose, J. (2015). Salt stress sensing and early signalling events in plant roots: Current knowledge and hypothesis. *Plant Science*, *241*, 109–119.
Sharp, R. E., Hsiao, T. C., & Silk, W. K. (1990). Growth of the maize primary root at low water potentials. II. Role of growth and deposition of hexose and potassium in osmotic adjustment. *Plant Physiology*, *93*, 1337–1346.
Shi, H. Z., Quintero, F. J., Pardo, J. M., & Zhu, J. K. (2002). The putative plasma membrane Na^+/H^+ antiporter SOS1 controls long-distance Na^+ transport in plants. *Plant Cell*, *14*, 465–477.
Shkolnik-Inbar, D., Adler, G., & Bar-Zvi, D. (2013). *ABI4* downregulates expression of the sodium transporter *HKT1;1* in Arabidopsis roots and affects salt tolerance. *Plant Journal*, *73*, 993–1005.
Siebrecht, S., Herdel, K., Schurr, U., & Tischner, R. (2003). Nutrient translocation in the xylem of poplar—Diurnal variations and spatial distribution along the shoot axis. *Planta*, *217*, 783–793.
Sorensson, C., Lenman, M., Veide-Vilg, J., Schopper, S., Ljungdahl, T., & Grotli, M. (2012). Determination of primary sequence specificity of *Arabidopsis* MAPKs MPK3 and MPK6 leads to identification of new substrates. *Biochemical Journal*, *446*, 271–278.
Sottocornola, B., Visconti, S., Orsi, S., Gazzarrini, S., Giacometti, S., Olivari, C., et al. (2006). The potassium channel KAT1 is activated by plant and animal 14-3-3 proteins. *Journal of Biological Chemistry*, *281*, 35735–35741.
Sun, Y., Lindberg, S., Shabala, L., Morgan, S., Shabala, S., & Jacobsen, S. E. (2017). A comparative analysis of cytosolic Na^+ changes under salinity between halophyte quinoa (*Chenopodium quinoa*) and glycophyte pea (*Piswn sativwn*). *Environmental and Experimental Botany*, *141*, 154–160.
Sunarpi, Horie, T., Motoda, J., Kubo, M., Yang, H., Yoda, K., et al. (2005). Enhanced salt tolerance mediated by AtHKT1 transporter-induced Na^+ unloading from xylem vessels to xylem parenchyma cells. *Plant Journal*, *44*, 928–938.
Sung, J., Sonn, Y., Lee, Y., Kang, S., Ha, S., Krishnan, H. B., et al. (2015). Compositional changes of selected amino acids, organic acids, and soluble sugars in the xylem sap of N, P, or K-deficient tomato plants. *Journal of Plant Nutrition and Soil Science*, *178*, 792–797.
Sutter, J. U., Campanoni, P., Tyrrell, M., & Blatt, M. R. (2006). Selective mobility and sensitivity to SNAREs is exhibited by the *Arabidopsis* KAT1 K^+ channel at the plasma membrane. *Plant Cell*, *18*, 935–954.
Swarbreck, S. M., Colaço, R., & Davies, J. M. (2013). Plant calcium-permeable channels. *Plant Physiology*, *163*, 514–522.

Teakle, N. L., & Tyerman, S. D. (2010). Mechanisms of Cl^- transport contributing to salt tolerance. *Plant, Cell & Environment, 33*, 566–589.

Tester, M., & Davenport, R. J. (2003). Na^+ tolerance and Na^+ transport in higher plants. *Annals of Botany, 91*, 503–527.

Tester, M., & Langridge, P. (2010). Breeding technologies to increase crop production in a changing world. *Science, 327*, 818–822.

Tounsi, S., Feki, K., Hmidi, D., Masmoudi, K., & Brini, F. (2017). Salt stress reveals differential physiological, biochemical and molecular responses in T-monococcum and T-durum wheat genotypes. *Physiology and Molecular Biology of Plants, 23*, 517–528.

Véry, A. A., Nieves-Cordones, M., Daly, M., Khan, I., Fizames, C., & Sentenac, H. (2014). Molecular biology of K^+ transport across the plant cell membrane: What do we learn from comparison between plant species? *Journal of Plant Physiology, 171*, 748–769.

Vincill, E. D., Clarin, A. E., Molenda, J. N., & Spalding, E. P. (2013). Interacting glutamate receptor-like proteins in phloem regulate lateral root initiation in *Arabidopsis*. *Plant Cell, 25*, 1304–1313.

Wang, T. B., Gassmann, W., Rubio, F., Schroeder, J. I., & Glass, A. D. M. (1998). Rapid up-regulation of *HKT1*, a high-affinity potassium transporter gene, in roots of barley and wheat following withdrawal of potassium. *Plant Physiology, 118*, 651–659.

Wang, C., Hu, H., Qin, X., Zeise, B., Xu, D., Rappel, W.-J., et al. (2016). Reconstitution of CO_2 regulation of SLAC1 anion channel and function of CO_2-permeable PIP2;1 aquaporin as CARBONIC ANHYDRASE4 interactor. *Plant Cell, 28*, 568.

Wang, Y. S., Liu, F. L., & Jensen, C. R. (2012). Comparative effects of deficit irrigation and alternate partial root-zone irrigation on xylem pH, ABA and ionic concentrations in tomatoes. *Journal of Experimental Botany, 63*, 1907–1917.

Wang, R. C., Tischner, R., Gutierrez, R. A., Hoffman, M., Xing, X. J., Chen, M. S., et al. (2004). Genomic analysis of the nitrate response using a nitrate reductase-null mutant of Arabidopsis. *Plant Physiology, 136*, 2512–2522.

Wang, Y., & Wu, W. H. (2013). Potassium transport and signaling in higher plants. *Annual Review of Plant Biology, 64*, 451–476.

Watson, R., Pritchard, J., & Malone, M. (2001). Direct measurement of sodium and potassium in the transpiration stream of salt-excluding and non-excluding varieties of wheat. *Journal of Experimental Botany, 52*, 1873–1881.

Wegner, L. H. (2017). Cotransport of water and solutes in plant membranes: The molecular basis, and physiological functions. *AIMS Biophysics, 4*, 192–209.

Wegner, L. H., & De Boer, A. H. (1997). Properties of two outward-rectifying channels in root xylem parenchyma cells suggest a role in K^+ homeostasis and long-distance signaling. *Plant Physiology, 115*, 1707–1719.

Wegner, L. H., & Raschke, K. (1994). Ion channels in the xylem parenchyma of barley roots—A procedure to isolate protoplasts from this tissue and a patch-clamp exploration of salt passageways into xylem vessels. *Plant Physiology, 105*, 799–813.

Wegner, L. H., Stefano, G., Shabala, L., Rossi, M., Mancuso, S., & Shabala, S. (2011). Sequential depolarization of root cortical and stelar cells induced by an acute salt shock—Implications for Na^+ and K^+ transport into xylem vessels. *Plant, Cell & Environment, 34*, 859–869.

White, P. J., & Broadley, M. R. (2001). Chloride in soils and its uptake and movement within the plant: A review. *Annals of Botany, 88*, 967–988.

Wilkinson, S., & Davies, W. J. (2002). ABA-based chemical signalling: The co-ordination of responses to stress in plants. *Plant, Cell & Environment, 25*, 195–210.

Xu, J., Li, H. D., Chen, L. Q., Wang, Y., Liu, L. L., He, L., et al. (2006). A protein kinase, interacting with two calcineurin B-like proteins, regulates K^+ transporter AKT1 in Arabidopsis. *Cell, 125*, 1347–1360.

Yang, T. Y., Zhang, S., Hu, Y. B., Wu, F. C., Hu, Q. D., Chen, G., et al. (2014). The role of a potassium transporter OsHAK5 in potassium acquisition and transport from roots to shoots in rice at low potassium supply levels. *Plant Physiology*, *166*, 945–959.

Zepeda-Jazo, I., Velarde-Buendía, A. M., Enríquez-Figueroa, R., Bose, J., Shabala, S., Muñiz-Murguía, J., et al. (2011). Polyamines interact with hydroxyl radicals in activating Ca^{2+} and K^{+} transport across the root epidermal plasma membranes. *Plant Physiology*, *157*, 2167–2180.

Zhang, X. A., Ma, J. O., & Berkowitz, G. A. (1999). Evaluation of functional interaction between K^{+} channel alpha- and beta-subunits and putative inactivation gating by co-expression in *Xenopus laevis* oocytes. *Plant Physiology*, *121*, 995–1002.

Zhang, W. D., Wang, P., Bao, Z. L. T., Ma, Q., Duan, L. J., Bao, A. K., et al. (2017). SOS1, HKT1;5, and NHX1 synergistically modulate Na^{+} homeostasis in the halophytic grass *Puccinellia tenuiflora*. *Frontiers in Plant Science*, *8*, 576.

Zhao, L. N., Shen, L. K., Zhang, W. Z., Zhang, W., Wang, Y., & Wu, W. H. (2013). Ca^{2+}-dependent protein kinase11 and 24 modulate the activity of the inward rectifying K^{+} channels in *Arabidopsis* pollen tubes. *Plant Cell*, *25*, 649–661.

Zhao, F. G., Song, C. P., He, J. Q., & Zhu, H. (2007). Polyamines improve K^{+}/Na^{+} homeostasis in barley seedlings by regulating root ion channel activities. *Plant Physiology*, *145*, 1061–1072.

Zhu, J. K. (2003). Regulation of ion homeostasis under salt stress. *Current Opinion in Plant Biology*, *6*, 441–445.

Zhu, M., Shabala, S., Shabala, L., Fan, Y., & Zhou, M. X. (2016). Evaluating predictive values of various physiological indices for salinity stress tolerance in wheat. *Journal of Agronomy and Crop Science*, *202*, 115–124.

Zhu, M., Zhou, M., Shabala, L., & Shabala, S. (2017). Physiological and molecular mechanisms mediating xylem Na^{+} loading in barley in the context of salinity stress tolerance. *Plant, Cell & Environment*, *40*, 1009–1020.

FURTHER READING

Barrett-Lennard, E. G. (2003). The interaction between waterlogging and salinity in higher plants: Causes, consequences and implications. *Plant and Soil*, *253*, 35–54.

Bose, J., Rodrigo-Moreno, A., & Shabala, S. (2014). ROS homeostasis in halophytes in the context of salinity stress tolerance. *Journal of Experimental Botany*, *65*, 1241–1257.

Choudhury, F. K., Rivero, R. M., Blumwald, E., & Mittler, R. (2016). Reactive oxygen species, abiotic stress and stress combination. *Plant Journal*, *90*, 856–867.

Henderson, S. W., Wege, S., Qiu, J., Blackmore, D. H., Walker, A. R., Tyerman, S. D., et al. (2015). Grapevine and Arabidopsis cation-chloride cotransporters localize to the Golgi and trans-Golgi network and indirectly influence long-distance ion transport and plant salt tolerance. *Plant Physiology*, *169*, 2215.

Hsu, F.-C., Chou, M.-Y., Peng, H.-P., Chou, S.-J., & Shih, M.-C. (2011). Insights into hypoxic systemic responses based on analyses of transcriptional regulation in *Arabidopsis*. *PLoS One*, *6*, e28888.

Hunte, C., Screpanti, E., Venturi, M., Rimon, A., Padan, E., & Michel, H. (2005). Structure of a Na^{+}/H^{+} antiporter and insights into mechanism of action and regulation by pH. *Nature*, *435*, 1197–1202.

Kawaguchi, R., & Bailey-Serres, J. (2005). mRNA sequence features that contribute to translational regulation in Arabidopsis. *Nucleic Acids Research*, *33*, 955–965.

Mullan, D. J., Colmer, T. D., & Francki, M. G. (2007). *Arabidopsis*-rice-wheat gene orthologues for Na^{+} transport and transcript analysis in wheat-*L. elongatum* aneuploids under salt stress. *Molecular Genetics and Genomics*, *277*, 199–212.

Niu, L., & Liao, W. (2016). Hydrogen peroxide signaling in plant development and abiotic responses: Crosstalk with nitric oxide and calcium. *Frontiers in Plant Science*, *7*, 230.

Shabala, S., Shabala, L., Barcelo, J., & Poschenrieder, C. (2014). Membrane transporters mediating root signalling and adaptive responses to oxygen deprivation and soil flooding. *Plant, Cell & Environment*, *37*, 2216–2233.

Shi, H. Z., Ishitani, M., Kim, C. S., & Zhu, J. K. (2000). The *Arabidopsis thaliana* salt tolerance gene *SOS1* encodes a putative Na^+/H^+ antiporter. *Proceedings of the National Academy of Sciences of the United States of America*, *97*, 6896–6901.

Teakle, N. L., Flowers, T. J., Real, D., & Colmer, T. D. (2007). *Lotus tenuis* tolerates the interactive effects of salinity and waterlogging by 'excluding' Na^+ and Cl^- from the xylem. *Journal of Experimental Botany*, *58*, 2169–2180.

Vinothkumar, K. R., Smits, S. H. J., & Kuhlbrandt, W. (2005). pH-induced structural change in a sodium/proton antiporter from Methanococcus jannaschii. *The EMBO Journal*, *24*, 2720–2729.

CHAPTER TEN

The Role of Plant Transporters in Mycorrhizal Symbioses

Carmen Guerrero-Galán*[,2], Gabriella Houdinet*, Monica Calvo-Polanco*, Katia E. Bonaldi†, Kevin Garcia‡, Sabine Dagmar Zimmermann*[,1]

*Biochimie et Physiologie Moléculaire des Plantes (BPMP), Université de Montpellier, CNRS, INRA, SupAgro, Montpellier, France
†Division of Biological Sciences, University of California, San Diego, CA, United States
‡Department of Crop and Soil Sciences, North Carolina State University, Raleigh, NC, United States
[1]Corresponding author: e-mail address: sabine.zimmermann@cnrs.fr

Contents

Abstract

Membrane transport systems are crucial elements for plant nutrition and development as they play a key role in the absorption of mineral nutrients and water at the root level but also in the translocation within the plant. Moreover, membrane transport is involved in signalling and communication e.g. to adapt and interact with the environment. Most

[2] *Present address*: Centro de Biotecnología y Genómica de Plantas, Instituto Nacional de Investigación y Tecnología Agraria y Alimentación (INIA), Universidad Politécnica de Madrid (UPM), 28223 Pozuelo de Alarcón, Spain.

Advances in Botanical Research, Volume 87
ISSN 0065-2296
https://doi.org/10.1016/bs.abr.2018.09.012

plants live in tight contact with beneficial soil microbes, such as bacteria and mycorrhizal fungi, which contribute to plant nutrition in part through modulation of the expression and functioning of plant transporter systems, as ion channels and transporters. In addition, mycorrhizal fungi largely increase the absorption surface of roots thereby promoting plant's access to soil resources as minerals and water. In turn, plants "reward" mycorrhizal fungi with sugars and/or lipids. This "fair trade" requires specific communication and a series of exchanges between the two symbiotic partners enabled by the adaptability and plasticity of their transporters. Here, we summarize recent advances allowing molecular insight in the impact of mycorrhizal symbiosis on the plant "transportome". We highlight results obtained in ecto- and endomycorrhizal associations for plant transporters involved in the absorption of mineral nutrients and water released by the fungus at the symbiotic interface, and molecular players responsible for carbon and lipid nutrition of the fungal partner. We focus also on plant membrane transport systems implicated in early communication between plant and fungal partners.

1. INTRODUCTION

Plants form mutualistic associations with specialized fungi that modulate their ability to acquire water and nutrients from the soil. Mycorrhiza is a new root structure that is formed by both symbiotic partners, and which provides a continuum between soil, fungi and plant roots. Among the different mycorrhizal associations, arbuscular mycorrhizal (AM) fungi interact with most land plants, while ectomycorrhizal (ECM) fungi colonize more specifically woody plant species, mostly from temperate and boreal forests (Smith & Read, 2008). There are also some types of mycorrhizal symbioses more specific to plant families, such as Ericaceae and Orchidaceae. These beneficial fungi can be seen as a kind of prolongation of the plant root system, enabling access to water and nutrients located outside of depletion zones around the roots (Fig. 1). Other benefits provided by mycorrhizal symbioses include protection against biotic and abiotic stresses (Miransari, 2014). In exchange, mycorrhizal fungi are fed by the plants with carbon substrates originating from photosynthesis, thus implicating regulation of plant sugar transport (Fig. 1). Nowadays, lipid transfer has also been shown in AM symbiosis (Keymer et al., 2017).

Regarding AM symbiosis, research on the mechanisms governing nutritional exchanges between the symbiotic partners has mainly focused on the plant host due to the complexity of the fungal genomes and the difficulty for handling AM fungi (Garcia, Doidy, Zimmermann, Wipf, & Courty, 2016). In contrast, studies on ECM symbiosis are more advanced on the fungal partners than on the plants due to limited tools for woody plants, with

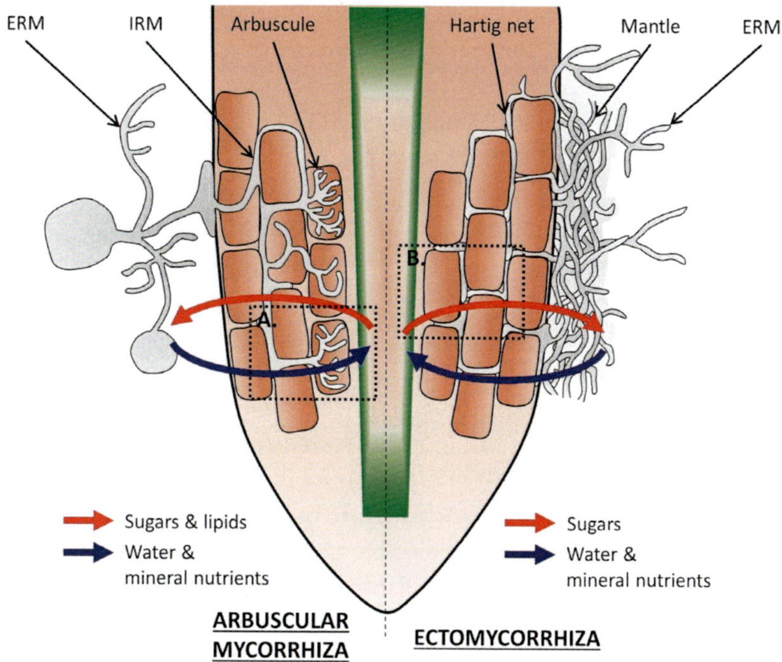

Fig. 1 Nutritional exchanges modulate the functioning of arbuscular mycorrhizal and ectomycorrhizal associations. In arbuscular mycorrhizae, the fungus absorbs water and mineral nutrients from the soil solution with its extraradical mycelium (ERM). These nutriments are transferred within the root through the intraradical mycelium (IRM) and finally transmitted to the host plant through structures within root cells, called arbuscules, that are formed by fungal and plant membranes. The plant controls the system of "reciprocal rewards" established with the fungus via the allocation of carbon compounds obtained by photosynthesis. In ectomycorrhizae, the wide ERM extends the root absorbing surface and allows a better mineral and water nutrition of the plant, which is also rewarded by the transfer of sugars to the fungus. These exchanges take place at the level of the Hartig net. Boxes (A) and (B) are further detailed in Fig. 2.

the exception of poplar, and easier handling of ECM fungi. Genomic projects such as The JGI Mycorrhizal Genomics Initiative (https://jgi.doe.gov/the-mycorrhizal-genomics-initiative-or-exploring-tree-fungal-symbioses/) recently extended our knowledge on the fungal side, which has been particularly useful in the case of ECM associations (Kohler et al., 2015). Nevertheless, ongoing researches are releasing key essential information on both the fungal and plant sides regarding the functioning of nutritional exchanges in different types of mycorrhizal symbiosis (Garcia et al., 2016).

Traditionally, there has been a great research interest on plant membrane transport systems, ion channels and transporters (cf. this book), due to their impact on whole plant physiology, signalling, nutrient uptake and translocation, and agricultural outputs (Schroeder et al., 2013). At first, they have been extensively studied in the model plant *Arabidopsis thaliana*. Translational research is expanding this field to crop species, such as *Medicago* sp., tomato, potato, rice or maize, all of them being known to interact with arbuscular mycorrhizal fungi. In the case of mycorrhiza involving tree species, there has been much less research efforts on characterizing membrane transport systems on the plant side, besides poplar (Courty, Doidy, Garcia, Wipf, & Zimmermann, 2017), partly because of the large size of tree genomes (Loth-Pereda et al., 2011; Neale et al., 2014). This is one of the reasons why the fungal side is more explored in this symbiotic interaction.

First molecular evidence for a tight regulation of plant transporters during mycorrhizal interactions has been found for a phosphate (Pi) transporter from tomato (Rausch et al., 2001). In the meantime, recent transcriptomic studies and further detailed analyses have underlined such a tight control of plant membrane transport systems by symbiotic fungal partners, revealing the plasticity of the plant "transportome". The importance and implication of membrane transport for communication and nutritional exchange (Fig. 1) between fungi and plant roots become more and more recognized. Within this chapter, recent knowledge on this topic is summarized and discussed.

2. ECTOMYCORRHIZAL SYMBIOSIS REQUIRES TIGHTLY REGULATED PLANT MEMBRANE TRANSPORT

ECM symbioses are mutualistic interactions between roots of ligneous plants and soil fungi mainly belonging to the Basidiomycota and Ascomycota phyla (Tedersoo & Smith, 2013). They characteristically develop around short thick lateral roots forming three structures: (i) an extraradical mycelium (ERM) in contact with the substrate, (ii) a dense fungal mantle covering the root, and (iii) the so-called Hartig net formed by the hyphae that enter the root and develop in layers around the cortex and epidermal cells (Smith & Read, 2008). The ERM explores the soil and is specialized in nutrient and water absorption, functionally substituting the root hairs, whose development is inhibited. Remarkably, ECM fungi in symbiosis with trees form an extensive hyphal network, contributing to carbon storage in the soil (Clemmensen et al., 2013, 2015). This extraradical mycelium can mobilize soil nutrients and water even under extremely unfavourable soil conditions (for example, Berruti, Lumini, Balestrini, &

Bianciotto, 2015; Landeweert, Hoffland, Finlay, Kuyper, & Van Breemen, 2001), thus improving plant tolerance to challenging environments. The fungal mantle can store nutrients and serves as a buffering compartment against environmental changes or stresses (for example, Bücking & Heyser, 2000; Bücking, Kuhn, Schröder, & Heyser, 2002; Muhsin & Zwiazek, 2002). Finally, the Hartig net hyphae have been reported to be responsible for nutritional exchanges between the fungus and the plant (Kottke & Oberwinkler, 1987). Due to the small and rather simple genomes of ECM fungi compared to those of the interacting plants, genetic and molecular research in this field is far more developed on the fungal side, whereas the plant transportome involved in symbiotic exchange largely remains to be explored.

2.1 Plant Root's Uptake of Mineral Nutrients and Water Transferred From Symbiotic Fungi

2.1.1 Plant Phosphate Nutrition

ECM fungi improve plant inorganic orthophosphate (Pi) nutrition through several mechanisms including the expression of membrane transport proteins at the uptake side from the soil and at the release side between the Hartig net and plant cortical cells (reviewed in Becquer, Trap, Irshad, Ali, & Plassard, 2014; Nehls & Plassard, 2018). This beneficial effect is of great importance with respect to the limiting availability of phosphorus (P) representing an increasing problem in agroecosystems. Thanks to a higher exploratory capacity provided by the fungal structure and the smaller size of fungal hyphae compared to root systems, mycorrhizal plants have the ability to access soil P in a more efficient manner than plants *per se*. In addition, fungi release Pi from the substrate through the acidification of the environment and secretion of phosphatases (Plassard & Dell, 2010). Pi absorbed by the extraradical hyphae can be transferred to the plant in the Hartig net, where as yet unknown transporters mediate its release into the apoplasm (Torres-Aquino et al., 2017). Recently, a fungal Pi transporter from *Hebeloma cylindrosporum* with a dual function likely involved in Pi absorption from the soil and in Pi secretion towards the symbiotic interface has been described (Becquer et al., 2018). However, molecular knowledge on Pi uptake by the root cells at the symbiotic interface is still missing. Loth-Pereda et al. (2011) studied several Pi transporters from ECM poplars, finding that two genes (*PtPT9* and *PtPT12*) were strongly upregulated when interacting with *Laccaria bicolor* and *Paxillus involutus*. They proposed that the corresponding transporters could perform Pi uptake in mycorrhizal conditions, but suggested that the expression increase could be biased by the lower Pi concentrations in the symbiotic space. Similar results have been obtained by Zhang, Wen, and Ding (2017), who found four phosphate transporter genes (*PmPT1–4*) with a higher expression in *Pinus massoniana–Boletus edulis/Pisolithus tinctorius* mycorrhizae than in non-inoculated roots, especially under low Pi conditions. These authors also considered the possibility that low Pi availability was responsible for the overexpression of *PmPT1–4* genes in ectomycorrhizae. It is likely that the combination of both factors, low Pi and symbiosis, enhances the beneficial role of ECM fungi, which could be otherwise masked under high Pi conditions.

2.1.2 Ectomycorrhizal Contribution to Nitrogen Nutrition

ECM fungi contribute largely to nitrogen (N) uptake and assimilation via nitrate, ammonium, peptide and amino acid transporters (Garcia et al., 2016; Nehls & Plassard, 2018). The molecular mechanisms responsible for plant N uptake in ECM associations have been partly studied in model angiosperms and gymnosperms. In mycorrhizal poplar (*Populus trichocarpa*), the high-affinity ammonium transporter *PttAMT1.2* was upregulated when the plant was inoculated with *Amanita muscaria* in *in vitro* cultures or in field conditions when naturally associated with wild species (Selle et al., 2005) or when inoculated with *P. involutus* (Couturier et al., 2007). This transporter is believed to take up ammonium released by the fungus at the symbiotic interface. In mycorrhizal association, the related tree *Populus tremula* × *tremuloides* also undergoes an upregulation of ammonium transporters, including *PoptrAMT1.2b* which is the closest *PttAMT1.2* ortholog, as well as *PoptrAMT1.3* and *PoptrAMT1.4*. In ECM *Pinus pinaster*, ammonium uptake is believed to be mediated by *PpAMT1.3*, a high-affinity and high-capacity transporter whose transcription is induced in mycorrhizal roots (Castro-Rodríguez et al., 2016). Also, in the mycorrhizal model *Quercus robur-Piloderma croceum*, colonization induced the overexpression of one AMT (Tarkka et al., 2013).

Strikingly, a rise in the transcript abundance of the putative nitrate transporter *PttNRT2.5B* has been observed during ECM association of *P. tremula* × *tremuloides* and *A. muscaria* (Willmann, Thomfohrde, Haensch, & Nehls, 2014). Because nitrate is commonly not considered as a major form of transported N from the fungus to the host plant (Chalot, Blaudez, & Brun, 2006), it has been proposed that this transporter uses other substrates than nitrate (Willmann et al., 2014).

So far, little transcriptomic data are available on trees, whether or not they are engaged in a symbiotic interaction with an ECM fungus. The most up-to-date information led to an incomplete N model where ammonium transporters play the main role (Garcia et al., 2016). Further research is needed to confirm this function and to complete our view about the other nitrogenous compounds (nitrate, amino acids or peptides) in the N nutrition of the ECM plant.

2.1.3 Potassium

Indisputably, N and P are the main mineral nutrients acquired by mycorrhizal associations and regulating mycorrhizal establishment and functioning. However, other elements are presumably involved as well, particularly when they are scarce in the environment. For instance, some ECM plants have a better potassium (K) nutrition than non-mycorrhizal controls under shortage conditions (Garcia et al., 2014; Garcia & Zimmermann, 2014; Jentschke, Brandes, Kuhn, Schröder, & Godbold, 2001; Jourand et al., 2014), indicating a beneficial effect of this symbiotic interaction. Although the mechanisms involved in K nutrition have been well described and characterized in herbaceous plants (for a review, see Nieves-Cordones, Alemán, Martínez, & Rubio, 2014), little is known about K transport systems of ECM plants (Zhang, Yin, & Xia, 2010). These transport systems are likely to be distributed among the Shaker-like channel family (Dreyer & Uozumi, 2011) as well as the Trk/Ktr/HKT and KT/KUP/HAK transporter families (Corratgé-Faillie et al., 2010; Grabov, 2007). To date, there is only one mention of the regulation of these genes during ECM symbiosis in literature. Peter et al. (2016) found an induction of the putative Shaker-like channel *SKOR* (Stellar K^+ Outward-Rectifier) of *Pinus sylvestris* upon mycorrhization with *Cenococcum geophilum*, which could be involved in an increased xylem loading during symbiosis. Once again, most data available so far are on the fungal partner where transport proteins were identified and characterized (Corratgé et al., 2007; Garcia et al., 2014; Guerrero-Galán, Delteil, et al., 2018; Guerrero-Galán, Garcia, Houdinet, & Zimmermann, 2018). Efforts are still needed on the plant side.

2.1.4 Ectomycorrhizal Fungi Modify Root Water Transport in Plants

The association of ECM fungi with plant roots modifies deeply water uptake in the host trees. The absorption of water from the soil by plant roots is mainly controlled by radial water transport following two main paths, the apoplastic and the cell-to-cell pathways (Steudle & Peterson,

1998). The apoplastic pathway is formed by the cell wall continuum and controlled by the development of Casparian strips and suberin at the exodermis and endodermis (Barberon & Geldner, 2014). The cell-to-cell pathway occurs mainly through plasmodesmata or plasma membranes by specialized water channel proteins called aquaporins, which mediate passive water transport using the driving force due to water potential gradients. Aquaporins are membrane intrinsic proteins (MIPs) found in all plant species (Maurel et al., 2015) and in fungi (Dietz, von Bülow, Beitz, & Nehls, 2011; Nehls & Dietz, 2014; Xu et al., 2015). They are distributed among five subfamilies (PIPs, TIPs, NIPs, XIPs and SIPs) with different substrate selectivities and localizations (Chaumont & Tyerman, 2014). Among them, the most important PIP and TIP subfamilies have a prominent role in whole plant water transport (Maurel et al., 2015).

In general, the presence of ECM fungi enhances the plant's root hydraulic conductivity (Lee, Calvo-Polanco, & Zwiazek, 2010; Xu et al., 2015), although other results showed no effect (Calvo-Polanco, Jones, & Zwiazek, 2009; Xu, Cooke, Kemppainen, Pardo, & Zwiazek, 2016). The increase of root water transport in the presence of ECM fungi has been mainly attributed to their effect on the symplastic water transport through upregulation of PIP aquaporins, as shown in the following associations: *A. muscaria-poplar* (Marjanović et al., 2005), *Tuber melanosporum*-poplar; (Hacquard et al., 2013), *Tuber clavery-Helianthemum almeriense* (Navarro-Rodenas et al., 2013), *L. bicolor-spruce* (Xu et al., 2015).

The effect of ECM fungi on plant water transport is fungal/host dependent. In addition, other parameters, that are not yet unravelled and well studied, might contribute to upregulation or deregulation of plant aquaporins. In particular, proper phosphorylation of aquaporins, as in AM symbiosis (Vandeleur et al., 2014), as well as their precise localization within the roots can be critical for water uptake both under control conditions and when plants are exposed to different soil stresses.

Due to the role of extraradical fungal hyphae as prolongation of the plant roots, fungal aquaporins are of importance for the uptake and transfer of water towards the plant. In line with this hypothesis, results obtained after artificial downregulation or overexpression of aquaporins from *L. bicolor* support their role in ectomycorrhizae in *Populus tremuloides* seedlings (Navarro-Ródenas, Xu, Kemppainen, Pardo, & Zwiazek, 2015) and in *Picea glauca* (Xu et al., 2015). Furthermore, fungal aquaporins have been shown to be transcriptionally regulated in ECM fungi under mycorrhizal conditions, such as in *C. geophilum* in symbiosis with *P. sylvestris* (Peter et al., 2016).

These observations support the hypothesis that fungal aquaporins are required for transport of water during symbiosis and that their expression is regulated by water needs from the host plant.

2.2 Delivering Carbon Food for the Fungal Partner

Carbon supply by the plant to the fungal partner is essential to balance the relationship between partners engaged in the mycorrhizal symbiosis. ECM fungi often derive from ancestral saprotrophic fungi (Veneault-Fourrey & Martin, 2013), which, however, have lost most genes coding for plant cell wall-degrading enzymes (Kohler et al., 2015). Therefore, their nutrition mostly relies on sugars and simple carbon compounds exuded by the plant into the root apoplasm.

Although the transport of carbohydrates in ectomycorrhizae has not been deeply explored, it has been pointed out that three families of plant transporters could be involved in this process: sucrose transporters (SUTs), monosaccharide transporters (MSTs), and sugar SWEET transporters ("sugars will eventually be exported transporter" family) (Garcia et al., 2016). Upregulation of plant SWEET members has been reported in transcriptomic studies of several ECM couples such as *Q. robur-P. croceum* (Tarkka et al., 2013), *P. trichocarpa-L. bicolor* (Plett et al., 2015) and *P. sylvestris-C. geophilum/Suillus granulatus* ectomycorrhizae (Peter et al., 2016). However, the precise process by which SWEET transporters undergo this regulation is still unknown. Initially, it has been hypothesized that sucrose would be the main substrate released by the plant to the fungus even if this latter cannot absorb sucrose. Assimilation of carbon compounds might then rely on plant invertases that hydrolyse sucrose into glucose and fructose; these two carbohydrate forms being taken up by the fungus transport systems (Nehls, Grunze, Willmann, Reich, & Küster, 2007).

To sustain a balanced sharing of resources, mechanisms controlling plant-fungal C allocation have been proposed. Grunze, Willmann, and Nehls (2004) found that *PttMST3.1*, a *P. tremula* × *tremuloides* MST gene likely involved in hexose import from the root apoplasm, is induced in *A. muscaria* ectomycorrhizae. By expressing this transporter, the plant would compete for sugars with the fungus in the symbiotic interface, thereby regulating the benefit balance of the mutualism.

To sum up, Nehls et al. (2007) suggested that the plant reaches an equilibrium in the interaction thanks to carbohydrate allocation in three ways: (i) the exudation of sucrose through a plant transporter, (ii) its hydrolysis (in glucose and fructose) by invertases, and (iii) the competition for hexoses in the Hartig net between root and fungal cells, also mediated by plant transporters. However, this relative simple model will have to be implemented to include other substrates and mechanisms.

2.3 Communication Between Symbiotic Partners

Mutual recognition between host plants and their fungal partners constitutes the first step in the establishment of ECM symbiosis (Martin et al., 2001). Plant roots secrete a large amount of primary and secondary metabolites in the rhizosphere to attract symbiotic microbes, stimulating growth and spore germination of ECM fungi (Ali & Jackson, 1988; Fries, Serck-Hanssen, Dimberg, & Theander, 1987; Melin & Rama Das, 1954). For example, seven flavonoids exuded by *Pinus densiflora* roots stimulate germination of *Suillus bovinus* spores (Kikuchi, Matsushita, Suzuki, & Hogetsu, 2007), and rutin secreted by *Eucalyptus globulus* roots enhances the hyphal growth of two *Pisolithus* strains (Lagrange, Jay-Allemand, & Lapeyrie, 2001). Rutin and quercitin from poplar roots trigger the expression of an effector protein in the ECM fungus *L. bicolor* (Plett & Martin, 2012). Carbon-derived compounds are also exuded by the plant and can be taken up by ECM fungi growing at the root vicinity (Nehls, 2008). Although it was thought for a long time that root exudates are passively exported from roots into the rhizosphere, it becomes clear that an active transport involving plant transporters also occurs (Baetz & Martinoia, 2014; Sasse, Martinoia, & Northen, 2018). For example, ABC-type transporters allow exudation of secondary metabolites in the non-mycorrhizal plant *A. thaliana* (Baudri et al., 2008; Ziegler, Schmidt, Strehmel, Scheel, & Abel, 2017), or flavonoids in soybean (Sugiyama, Shitan, & Yazaki, 2007). Although plant transporters obviously play a key role in root exudation processes, only a few of these transporters have been reported and characterized so far. In particular, no data are currently available regarding their role in exudation of compounds that stimulate the germination and growth of ECM fungi.

It is also hypothesized that nutrients themselves play an important role for the early recognition between host plants and ECM fungi (Garcia, Delaux, Cope, & Ané, 2015). The dual trophic and signalling role of nutrients was already observed in AM symbiosis (cf. Section 3.1; Carbonnel & Gutjahr, 2014). In ECM plants, it has been demonstrated that nutrient availability affects the establishment of the symbiotic interaction (Arnebrant, 1994; Plassard, Bonafos, & Touraine, 2000; Plassard & Dell, 2010), but the molecular players involved still remain unknown. Becquer et al. (2018) recently characterized a Pi transporter from the ECM fungus *H. cylindrosporum* that is primarily involved in Pi acquisition from the soil and its transfer towards the plant. Excitingly, it was observed that downregulation of *HcPT2* inhibits ECM formation, suggesting that a less cooperative fungus can be sanctioned by the host and that nutrients themselves are key regulators of ECM symbiosis. On the plant side, although Pi transporters are differentially regulated upon ECM colonization in poplar (Loth-Pereda et al., 2011), reverse genetic experiments are still missing to validate their role in both Pi transport and signalization in symbiosis.

3. ARBUSCULAR MYCORRHIZAL FUNGI CONTROL PLANT ION CHANNELS AND TRANSPORTERS

Arbuscular mycorrhizal symbiosis is largely widespread and occurs between a wide range of land plants and AM fungi belonging to the Glomeromycota group (Smith & Read, 2008). The principle of their relationship relies on mutual exchanges like the ones described for the ECM symbiosis, the fungi providing water and minerals, mainly phosphorus and nitrogen, while getting carbon resources from the plant (MacLean, Bravo, & Harrison, 2017). However, in contrast to the ECM association (cf. Section 2), the interface for exchange between AM fungi and the plant is not formed between the root cells. The AM fungi rather enter the plant cells and produce an expanded membrane surface within the root cell. During the symbiotic interaction, AM fungi grow in two different structures: the extraradical mycelium (ERM) and the intraradical mycelium (IRM).

The role of the ERM is to explore the soil by establishing an extensive hyphal network to mobilize and acquire soil water and nutrients, mainly phosphorus and nitrogen (Bompadre et al., 2013; Bonfante & Genre, 2010). Thus, establishment of AM symbiosis enables the host plants to resist short- and long-term environmental constraints, increasing the plant capacity to adapt and survive unfavourable conditions (Al-Karaki & Clark, 1998; Ruíz-Lozano, Porcel, Azcón, & Aroca, 2012; Smith & Read, 2008). The success of the interaction and the beneficial effects of AM fungi on plants vary according to the AM fungal strain and plant cultivars used, as well as to the biochemistry and microbial composition of the soil where the interaction takes place (Burns, Anacker, Strauss, & Burke, 2015; Calvo-Polanco, Sánchez-Castro, et al., 2016; Öpik, Metsis, Daniell, Zobel, & Moora, 2009; Varela-Cevero et al., 2015). Furthermore, recent studies have demonstrated that AM symbiosis regulates plant transport systems thus allowing a tight control of nutrient exchange between both symbiotic partners as well as uptake from the soil.

The IRM creates arbuscules, and eventually vesicules, within the host's roots where molecule exchanges between symbiotic partners take place. These arbuscules are formed by hyphae crossing the cell wall of root cells but not their plasma membranes. Thus, nutrients have to move across both the fungal and the plant membranes to reach their final destination. As a result of root cell stimulation by the fungal colonization, the host plant develops a new membrane surrounding the arbuscules, called periarbuscular membrane (Aloui et al., 2018; Bravo, Brands, Wewer, Dörmann, & Harrison, 2017; Gaude, Bortfeld, Duensing, Lohse, & Krajinski, 2012). Some transport proteins are exclusively expressed here, like phosphate transporters *MtPT4* in *Medicago truncatula* and *OsPT11* in rice (Harrison, Dewbre, & Liu, 2002; Javot, Penmetsa, Terzaghi, Cook, & Harrison, 2007; Kobae & Hata, 2010).

3.1 Plant Phosphate Transporters Are Key Elements for Symbiotic Functioning

As mentioned above, acquiring nutrients allocated by the AM fungus requires specific expression and regulation of dozens of transporters in cortical cells containing an arbuscule (reviewed in Garcia et al., 2016). To date, most researches have focused on inorganic phosphate (Pi) and nitrogen (N; cf. Section 3.2.1) since these two elements play a key role in AM symbiosis due to their dual trophic and signalling roles. Several studies reported that root Pi transporters are upregulated upon AM colonization. For example, a specific upregulation of *MtPT4, OsPT11, StPt3, LePt1* and *ZmPt9*, all these genes coding for Pi transporters, was observed in *M. truncatula*, rice, potato, tomato and maize roots, respectively, after colonization by AM fungi (Harrison et al., 2002; Kobae & Hata, 2010; Liu et al., 2018; Rausch et al., 2001; Rosewarne, Barker, Smith, Smith, & Schachtman, 1999). Plant Pi transporters belong to the Pht1, Pht2 and Pht3 families and drive the transport of negatively charged orthophosphates (Pi) using an energy-driven mechanism (Rausch & Bucher, 2002; Smith, Mudge, Rae, & Glassop, 2003). Krajinski et al. (2014) revealed that the energy required for uptake of fungal Pi in *M. truncatula* is provided by an

H^+-ATPase. Although Na^+/Pi transporters are also suspected to be involved in Pi transport during ECM symbiosis (Casieri et al., 2013; Garcia et al., 2016), only low- and high-affinity H^+/Pi symporters clustered in two subgroups of the plant Pht1 family have been described so far during AM associations (Harrison et al., 2002; Liu et al., 2018; Yang et al., 2012). While AM-specific Pi transporters are upregulated, expression is reduced for many other transporters, particularly those that are involved in AM-independent Pi acquisition. However, the causal link between those two observations is not yet elucidated. Indeed, downregulations could simply be due to a general repression of genes upon colonization, regardless of Pi availability in soil and regulation of AM-specific Pi transporters. Although the involvement of plant transporters in Pi acquisition during AM symbiosis has been clearly established over the last decades, post-transcriptional and post-translational regulations of these transport proteins are still largely unknown and need further investigation.

A key role of Pi during mycorrhizal associations has emerged more recently, due to its dual trophic and signalling role (Carbonnel & Gutjahr, 2014; Garcia et al., 2015). Indeed, premature degeneration of arbuscules was observed in loss-of-function *mtpt4* mutants (Javot et al., 2007), indicating that the absence of Pi fluxes originating from the fungus affects the proper formation and functioning of AM symbiosis. In addition, the expression of *LjPT4* observed in root tips of *Lotus japonicus* without any fungal colonization revealed that transporters suspected to be AM-specific can also be a key part of the root Pi-sensing machinery (Volpe, Giovannetti, Sun, Fiorilli, & Bonfante, 2015).

3.2 Root Mineral Nutrient Transport Adapts to Mycorrhizal Interaction

3.2.1 Nitrogen Acquisition by the Plant in Arbuscular Mycorrhizae

Nitrogen is transported from the soil to colonized roots in organic and inorganic forms (Courty et al., 2017; Courty, Smith, Koegel, Redecker, & Wipf, 2015). Govindarajulu et al. (2005) developed a model for N assimilation that seems to be confirmed by recent research (Bücking & Kafle, 2015; Gomez et al., 2009). AM fungi absorb N mainly as ammonium, although they can also use nitrate and organic compounds. According to this model, N is incorporated into the arginine metabolic pathway and broken down in the urea cycle of the fungus to release ammonium into the periarbuscular space, where it is absorbed by the plant.

Plant AM-specific transporters involved in N uptake from fungal arbuscules have been described in several host plants. Ammonium transporters of the AMT family were detected on the branch domain of periarbuscular membranes, but not on the trunk region where nutrient release is probably less active (Kobae, Tamura, Takai, Banba, & Hata, 2010; Koegel et al., 2013). In addition, transcriptomic studies have demonstrated an induction of *AMT* genes in arbuscular mycorrhizal roots of several plant species (Courty et al., 2017). However, studies characterizing these genes are rather scarce. The *L. japonicus AMT2;2* is strongly upregulated in mycorrhizal roots, particularly in arbusculated cells (Guether, Balestrini, et al., 2009). The transporter encoded by this gene is able to take up ammonium in the periarbuscular space, deprotonates it and transfers ammonia inside the cell (Guether, Neuhäuser, et al., 2009). Thus, the activity of *Lj*AMT2;2 helps to reduce the energetic cost of the transport and to maintain the proton gradient necessary for other transport processes. Similar gene expression patterns and transport activity have been identified for its soybean orthologue *Gm*AMT4;1 (Kobae et al., 2010). The induction of a rice gene (Koegel et al., 2017), that is phylogenetically close to *Lj*AMT2;2, underlines a conserved mechanism between monocots and dicots. However, the loss-of-function of this gene had no effect on mycorrhizal rice indicating a minor role in symbiosis.

Specific studies in grasses have highlighted the importance of mycorrhiza-induced *AMT* genes in sorghum (*Sorghum bicolor*; *SbAMT3;1* and *SbAMT4*), and other grasses (Koegel et al., 2013; Pérez-Tienda, Corrêa, Azcón-Aguilar, & Ferrol, 2014). The two *sorghum* transporters were specifically expressed in cells containing arbuscules or being adjacent to arbuscule-containing cells. In addition, the *Sb*AMT3;1 protein was observed in arbuscule-developing cells. In addition, a broader analysis investigated these two mycorrhiza-induced genes, AMT3;1 and AMT4, in five species of the *Poaceae* family; *sorghum*, rice, brome grass (*Brachypodium distachyon*), maize (*Zea mays*) and millet (*Setaria italica*) (Koegel et al., 2017). These transporters, whose activity was measured by yeast complementation, were found to be upregulated in all species analysed, suggesting that they have evolved from a common ancestor of current grasses. Furthermore, the overexpression of *AMT3;1* orthologues in mycorrhizal roots of dicots (*GmAMT3;1* in soybean; Kobae et al., 2010) indicated the conservation of this symbiosis-induced gene in both monocot and dicot clades. In rice, *OsAMT3;1* knock-down mutants had an impaired N and P nutrition and plant growth was not stimulated by mycorrhizal fungi (Koegel et al., 2017). In contrast, this gene does not seem to be essential for symbiosis establishment or arbuscule development.

It has been hypothesized that some AMT transporters could have a signalling role, rather than a transport activity *per se*. For instance, *Mt*AMT2;3 does not complement yeasts deficient for ammonium transport but is required to prevent premature arbuscule degeneration (Breuillin-Sessoms et al., 2015). In addition to the AMT transporters, other transporter families could be involved in ammonium/ammonia transport in arbuscular mycorrhizal symbiosis. This is the case for the NIP1 (Nod 26-like intrinsic protein) aquaporin of *M. truncatula*. This channel protein is specifically expressed in arbusculated cells where it allows the entry of ammonia molecules into the cell (Hogekamp et al., 2011; Uehlein et al., 2007).

Nitrate and organic compounds are other forms of N that might be transferred but that are believed to be less significant for AM symbiosis. It is unlikely that nitrate plays a major role in plant N nutrition in this context, because of the energetic cost required for its subsequent reduction to ammonium. Although some plant NRT transporters exhibit a symbiotic regulation (Drechsler, Courty, Brulé, & Kunze, 2018; Guether, Balestrini, et al., 2009; Hildebrandt, Schmelzer, & Bothe, 2002) and a specific expression in mycorrhizal structures, it seems that they are also modulated by N and P concentrations (Chen et al., 2018; Hohnjec, Vieweg, Pühler, Becker, & Küster, 2005). This indicates a more complex regulation mechanism and suggests that plant and fungal nutritional status may affect the symbiotic assimilation of nitrate. However, plant nitrate assimilation in arbuscular mycorrhizae should not be dismissed until more thorough studies are conducted.

Among the three families of putative nitrate transporters, NPF (NRT1/PTR) members are also able to transport various N-containing compounds, including oligopeptides (Corratgé-Faillie & Lacombe, 2017; Léran et al., 2014). AM roots and arbusculated cells show specific upregulation of genes coding for di- and tripeptide transporters (PTR; Garcia et al., 2016). Transcriptomic analyses of mycorrhizal *L. japonicus* roots have identified nine putative NPF (PTR) transporters upregulated by interaction with fungi, of which the most upregulated one is exclusively expressed in arbusculated cells at the transcript level (Guether, Balestrini, et al., 2009). Likewise, several NPF (PTR) candidates of *M. truncatula* respond positively to fungal colonization, and at least one is specific of arbuscule-containing cells according to Gomez et al. (2009) and Benedito et al. (2010). Moreover, Hogekamp et al. (2011) found 11 members of this family that could also be related to plant N nutrition in arbuscular mycorrhizae, two of which were present in arbusculated cells.

In conclusion, most knowledge on N compounds in AM symbiosis is about the transcriptional regulation of genes in colonized roots and symbiotic structures. The lack of information on the functional role of these membrane transporters hampers the development of a complete model for N nutrition in this particular symbiosis. The model proposed by Govindarajulu et al. (2005) and mentioned above was recently reviewed and refined by Bücking and Kafle (2015) and Chen et al. (2018). This model provides hints regarding the path that should be followed to elucidate the complete role of AM fungi in plant N nutrition.

3.2.2 Potassium Transport in Arbuscular Mycorrhizae

K nutrition under mycorrhizal symbioses has received little attention so far, but there are increasing evidence illustrating its importance for plant performance (Garcia & Zimmermann, 2014). Recent research highlighted the need for specific nutritional conditions, such as nutrient deprivation or environmental stress, to amplify the benefit gain by the host plant from the AM fungi (Garcia, Chasman, Roy, & Ane, 2017; Zhang, Wei, Hu, Xiao, & Tang, 2017). For instance, *M. truncatula* plants colonized with *Rhizophagus irregularis* accumulated more biomass than control plants when exposed to K deprivation (Garcia et al., 2017). A transcriptomic analysis uncovered one upregulated putative K^+/H^+ exchanger (CHX) that could be responsible for this increased K absorption, as well as a protein phosphatase that is likely regulating the function of K channels. In *Lycium barbarum*, Zhang, Wei, et al. (2017) identified two Shaker-like channels, *LbKT1* and *LbSKOR*, whose expression is induced in *R. irregularis* mycorrhizal roots. *Lb*KT1 is an orthologue of the AKT1 transporter involved in K uptake in roots (Nieves-Cordones et al., 2014), whereas *Lb*SKOR is closely related to SKOR (Ahmad & Maathuis, 2014; Gaymard et al., 1998), known to be responsible for the translocation of K to the xylem sap in *A. thaliana*. An unknown K transporter was also found to be strongly upregulated in *L. japonicus-Gigaspora margarita* mycorrhizae (Guether, Balestrini, et al., 2009). In conclusion, proteins responsible for K uptake from the periarbuscular space have not been characterized yet, but promising candidates are to explore.

3.2.3 Transport of Metal Nutrients

The transport of other nutrients, such as metals, has been given less attention due to the importance of P and N transport in arbuscular mycorrhizae, and remains largely unexplored. Some researchers have identified putative symbiosis-related genes possibly involved in metal transport, but so far none of these transport proteins has been studied in the context of symbiosis. A thorough summary on the knowledge accumulated on putative metal transporters has been published recently by Ferrol, Tamayo, and Vargas (2016). See also chapter "Metal transport in the developing plant seed" by Eroglu, in this volume.

3.3 Water Transport Is Regulated by Symbiotic Partners

As previously reported for ECM fungi, the symbiotic association between plants and AM fungi regulates the uptake of water both under control conditions and in the presence of different soil stresses. Overall, AM symbioses modify the root hydraulic conductivity of their host plants (Bárzana, Aroca, Bienert, Chaumont, & Ruíz-Lozano, 2014; Calvo-Polanco, Molina, Zamarreno, Garcia-Mina, & Aroca, 2014; Calvo-Polanco, Sánchez-Castro, et al., 2016; Calvo-Polanco, Sánchez-Romera, et al., 2016; Quiroga et al., 2018; Sánchez-Romera et al., 2018; Sánchez-Romera, Ruíz-Lozano, Zamarreño, García-Mina, & Aroca, 2016) and result in altered rates of water transfer in-and-out of the host plants (Augé, 2001).

In particular, AM fungi have been shown to increase plant resistance to several soil stresses such as drought (Azcón, Gomez, & Tobar, 1996; Bárzana et al., 2012; Calvo-Polanco, Sánchez-Romera, et al., 2016; Porcel, Azcón, & Ruíz-Lozano, 2004; Quiroga et al., 2018; Sheng et al., 2008), salt (Ruíz-Lozano et al., 2012; Saxena, Shukla, & Giri, 2017) or flooding (Calvo-Polanco et al., 2014). Thus, increase in tolerance to these conditions causes improvement of plant productivity (Abbaspour, Saeidi-Sar, Afshari, & Abdel-Wahhab, 2012; Navarro-Fernández, Aroca, & Barea, 2011) and plant nutrient status (Farzaneh, Vierheilig, Lössl, & Kaul, 2011; Lee, Muneer, Avice, Jin, & Kim, 2012). This effect has been largely attributed to the action of AM fungi on water transport and particularly on aquaporins. Indeed, the presence of AM fungi triggers specific changes in aquaporin gene expression and protein abundance in many host plants (Aroca, Porcel, & Ruíz-Lozano, 2007; Bárzana et al., 2014; Calvo-Polanco et al., 2014; Calvo-Polanco, Sánchez-Castro, et al., 2016; Calvo-Polanco, Sánchez-Romera, et al., 2016; Chitarra et al., 2016; Jahromi, Aroca, Porcel, & Ruíz-Lozano, 2008; Sánchez-Romera et al., 2016, 2018). Moreover, the effects of AM fungi on plant aquaporins are specific to the environmental conditions plants are exposed to (El-Mesbahi, Azcon, Ruíz-Lozano, & Aroca, 2012). Hence, aquaporins belonging to the same subgroup may exhibit distinct expression patterns when exposed to similar circumstances (Aroca et al., 2007) or may be similarly regulated under various stress conditions (Bárzana et al., 2014; Guo et al., 2006). The high propensity of aquaporins to adapt their expression and function according to the environmental context is likely accounting for differences in described results. In fact, discrepancies are commonly observed in studies where aquaporins are differentially expressed during AM even when the same combinations of plant and AM fungal species are being used (Aroca et al., 2007; Sánchez-Romera et al., 2016).

Furthermore, posttranslational regulations of aquaporins, such as phosphorylation of specific serine residues, are necessary to adjust their activity in response to fast environmental changes (Vandeleur et al., 2014). These regulations trigger conformational changes that act on the gating of aquaporins (Johansson et al., 1998) or determine their subcellular localization and membrane targeting (Prak et al., 2008). The presence of AM fungi within roots seems to be critical for these phosphorylation events under different stresses (Calvo-Polanco, Sánchez-Romera, et al., 2016; Quiroga et al., 2018; Sánchez-Romera et al., 2018), and are believed to be a key factor during plant adaptation to stress. Conversely, de-phosphorylation of aquaporins in the presence of AM is believed to prevent water loss, giving a main role to the apoplastic pathway for water transport (Bárzana et al., 2012).

AM fungal aquaporins have been related to water transport in the extraradical mycelium, and in the periarbuscular membrane (Li et al., 2013). Thus, in AM plants, the enhanced root hydraulic conductivity could be due to the activity of the associated fungal aquaporins (Bárzana et al., 2014; Bárzana, Aroca, & Ruiz-Lozano, 2015).

It has also been suggested that water could be absorbed by the external AM mycelium and delivered to the cortical apoplast, at the symbiotic interfaces, where it would join water taken up via the root apoplastic pathway (Bárzana et al., 2012; Smith, Facelli, Pope, & Smith, 2010). Hyphal water uptake and transfer to the host plants have been demonstrated in several studies (Khalvati, Hu, Mozafar, & Schmidhalter, 2005; Marulanda, Azcón, & Ruíz-Lozano, 2003; Ruth, Khalvati, & Schmidhalter, 2011). Increased water uptake by mycorrhizal plants under drought has been related to the increased absorbing surface of growing hyphae, and mycorrhizal ability to take up water from soil pores inaccessible to roots, as AM hyphae represent a low-resistance pathway for water movement until root cells (Allen, 2009; Ruíz-Lozano, 2003).

3.4 Sugars and Lipids Are Delivered to the Fungal Partner

As discussed above, plant mineral and water nutrition by AM symbiosis is part of a kind of "fair trade" between fungi and plants including a "reciprocal reward" delivered by the host plant towards its mutualistic partner in form of carbon substrates (Fellbaum et al., 2012, 2014; Hammer, Pallon, Wallander, & Olsson, 2011; Kiers et al., 2011). In contrast to ECM fungi, AM fungi are dependent on their plant hosts and cannot be cultivated alone. It is well established that plants transfer a part of their photosynthetically produced sugars to their associated fungi. Very recently, lipids have been added to the range of substrates delivered by the plant to the fungi. In fact, the two symbiotic partners have to produce so many membranes at their root cell interface that it is not astonishing that the costs are shared between the two partners. Here, we highlight transport processes involved in fungal feeding by AM plants. Recently, Aloui et al. (2018) demonstrated that 82 plasma membrane-associated proteins are responsive to mycorrhization, and some of these are sugar transporters or proteins putatively involved in lipid biosynthesis.

3.4.1 Plant Carbohydrates Are Feeding AM Fungi

During arbuscular mycorrhization, the development of the mycelium by fungi and the development of the periarbuscular membrane by plants are costly processes requiring energy in the form of carbohydrates (Bago et al., 2003, 2002; Gaude, Schulze, Franken, & Krajinski, 2012). AM fungi are obligate biotrophs that completely rely on carbon supply by their host to grow and develop (Luginbuehl et al., 2017; Rich, Nouri, Courty, & Reinhardt, 2017; Smith & Read, 2008). Sugar transfer from plant roots to fungi has been demonstrated using ^{14}C tracer studies (Řezáčová, Konvalinková, & Jansa, 2017). In fact, during AM symbiosis, up to 20% of plant photosynthates are transferred to the fungus (Bago, Pfeffer, & Shachar-Hill, 2000), preferentially in the form of hexoses, the main substrate being glucose (Pfeffer, Douds, Bécard, & Shachar-Hill, 1999; Shachar-Hill et al., 1995). This glucose is probably released by the plant through the periarbuscular membrane where it becomes available to the fungal partner (Rich et al., 2017). Fungal transporters involved in sugar transfer from the plant towards the AM fungi have been reviewed recently (Courty et al., 2017; Garcia et al., 2016). The sugar fluxes are mediated by several transport systems described in plants, comprising sucrose transporters (SUTs), monosaccharide transporters (MSTs), and SWEETs

(Doidy, Grace, et al., 2012). The monosaccharide transporter MST2 of *Glomus* sp. is a fungal hexose transporter known to be an important actor of sugar absorption, arbuscule development and root colonization (Bago et al., 2000; Helber et al., 2011; Pfeffer et al., 1999; Rich et al., 2017). Thus, glucose is certainly released by the plant through the periarbuscular membrane to the fungus including the cleavage of plant sucrose (Rich et al., 2017). Correspondingly, several studies have focused on the analysis of plant MST expression to possibly pinpoint differentially regulated candidates in tomato and *M. truncatula* (Ge, Sun, Chen, Kapulnik, & Xu, 2008; Harrison, 1996). Differential regulation of such MSTs (MtSt1 and MtHext1) was also found in non-colonized cells and neighbouring arbuscular cells, suggesting that non-colonized adjacent cells might feed the arbuscular cells with carbon through plasmodesmata (Gaude, Bortfeld, et al., 2012).

Due to the intense transit of sugars from the plant to the fungus, mycorrhizal roots represent particularly strong carbon sinks (Gaude, Schulze, et al., 2012; Rich et al., 2017). Implication of transporters of the SUT and SWEET families in symbiotic carbon transfer has been shown for *M. truncatula* (Kafle et al., 2018). Expression of all SUT types is regulated in leaves and in colonized roots of tomato and *M. truncatula* when inoculated with AM fungi (Boldt et al., 2011; Doidy, van Tuinen, et al., 2012).

SWEET transporters, characterized as exporters of both sucrose and monosaccharides using nanosensors (Chen, 2014), were found to be specifically expressed in arbusculated cortical cells of potato roots (Manck-Götzenberger & Requena, 2016), and differently regulated in *M. truncatula* roots in tripartite interaction with the AM fungus *R. irregularis* and the N-fixing bacteria *E. meliloti* (Kafle et al., 2018). The fact that mycorrhizae represent such an important carbon sink implies that a number of plant sugar transport systems are likely to be involved in sugar translocation and transport throughout the plant body, to support the delivery function of root sugar transporters.

3.4.2 Plant Lipids Are Needed to Establish and Maintain AM Symbiosis

Other important forms for carbon transport and storage in AM fungi are lipids, in the form of triacylglycerols (Bago et al., 2003, 2002; Beilby & Kidby, 1980; Jabaji-hare, 1988; Trépanier et al., 2005). AM fungi can use hexoses to synthetize lipids by their own. This *de novo* biosynthesis of fungal fatty acids (FAs) takes place only at the IRM and not in ERM or spores (Bago et al., 1999; Pfeffer et al., 1999; Trépanier et al., 2005). Nevertheless, AM fungi are not able to complete all the lipid biosynthesis by themselves.

Indeed, recent studies showed that the type I fatty acid synthase (FAS-I), an enzyme with a key role in the initiation of *de novo* synthesis of most FAs, is absent in some AM fungi such as *R. irregularis* or *Gigaspora rosea* indicating that enzymes originating from the plant partner are required to overcome this lack (Tang et al., 2016; Wewer, Brands, & Dörmann, 2014). However, subsequent steps of the FAs synthesis *per se* (e.g. FAs elongation, desaturation) can be done by fungi alone (Trépanier et al., 2005). Interestingly, many genes involved in lipid biosynthesis in the host plant are activated and/or upregulated, especially in arbuscocytes, arbuscule-containing plant cells (Gaude, Bortfeld, et al., 2012; Gaude, Schulze, et al., 2012; Gomez et al., 2009; Hogekamp & Küster, 2013; Tang et al., 2016; Tisserant et al., 2012).

Recent studies have revealed several genes being putatively related to lipid metabolism and transport in AM fungi-colonized plants (Bravo et al., 2017; Keymer et al., 2017; Luginbuehl et al., 2017; Rich et al., 2017). These genes, involved in both lipid metabolism and transport, are FatM (fatty acid synthase), RAM1 (Required for Arbuscular Mycorrhization 1), RAM2, STR1 (Stunted Arbuscule) and STR2.

Regarding lipid metabolism, the FAs biosynthesis takes place in the plastids of plant cells (Bravo et al., 2017; Keymer et al., 2017). Involvement of a Fat gene in AM symbiosis could be established for *FatM* in *M. truncatula*, by showing that *fatm* mutants are affected in the development of arbuscules within the root cells (Bravo, York, Pumplin, Mueller, & Harrison, 2016). The *RAM2* gene encodes a GPAT enzyme (glycerol 3-phosphate acyl transferase) and is close to *A. thaliana GPAT6* (Keymer et al., 2017). As GPAT6, RAM2 is likely to produce *sn*-2-monoacylglycerol (MAG) especially of one type, βMAG (Bravo et al., 2017; Keymer et al., 2017; Yang et al., 2010). Furthermore, the *ram2* mutation in *M. truncatula* impairs arbuscule development as the *fatm* mutation (Bravo et al., 2017; Gobbato et al., 2013; Wang et al., 2012).

Finally, initial steps have been taken in elucidating the travel of the plant lipids towards the fungal hyphae. STR1 and STR2 are ABC-type transporters specific of the periarbuscular membrane, that work together and are required for arbuscule formation (Gutjahr et al., 2012; Zhang, Blaylock, & Harrison, 2010). Their role(s) and substrate(s) remain still to be further characterized, but two research teams suggest independently that they may transport lipids, probably βMAG, from plants to fungi (Bravo et al., 2017; Keymer et al., 2017). For further information on ABC transporters, see also chapter "The ABC of ABC transporters" by Pierman, Boutry, and Lefèvre in this volume.

Moreover, RAM1 is a GRAS-domain transcription factor that can regulate many genes involved in lipid biosynthesis and transport such as FatM, RAM2 and ABC transporters (Luginbuehl et al., 2017). RAM1 was shown to be crucial for arbuscule development (Gobbato et al., 2012; Park, Floss, Levesque-Tremblay, Bravo, & Harrison, 2015).

Taken together, all of these studies indicate that plant RAM1, FatM, RAM2, STR1 and STR2 genes can form a functional unit/operational module that is regulated in AM associations and serves for lipid biosynthesis and transport particularly in arbuscocytes (Bravo et al., 2017; Keymer et al., 2017; Luginbuehl et al., 2017; Rich et al., 2017). Once more, the role of plant transporters in the functioning of the symbiotic interaction is underlined by the involvement of ABC-type transporters.

3.5 Membrane Transport Is Needed for Early Signalling to Establish Symbiosis

Besides their trophic role as described above, plant transporters can regulate the AM symbiosis by sensing nutrient availability and therefore affecting the lifespan of the symbiotic structures (Breuillin-Sessoms et al., 2015; Javot et al., 2007). Although not well described, these mechanisms are thought to play a key role in the functioning of the mutualistic association (Lanfranco, Fiorilli, & Gutjahr, 2018). Indeed, specific plant transporters appear to be crucial during the steps preceding the formation of fungal arbuscules, and even before the root colonization by hyphae. Thus, a molecular dialogue engaging the fungus and host roots must occur before any physical contact. Plant roots excrete the plant hormone strigolactone that is perceived by fungal spores, triggering their germination followed by hyphal branching (Akiyama, Matsuzaki, & Hayashi, 2005; Besserer et al., 2006). On the other side, AM fungi release a mixture of signaling molecules, including lipochitooligosaccharides and short chitooligosaccharides (Genre et al., 2013; Maillet et al., 2011), that are perceived by host plant LysM-receptor-like kinases. The signal is then transduced through various proteins, including membrane transport systems, resulting in the expression of AM-specific transcription factors (recently reviewed in Charpentier, 2018). All currently known transport proteins needed for this signalling cascade are localized on the nuclear envelope, at either or both the inner and outer nuclear membranes. Finely coordinated movements of potassium and calcium ions through the nuclear envelope are required to decode signals coming from the perception of fungal molecules. In response to unknown secondary messenger(s), calcium is released into the nucleus through calcium channels recently described in *M. truncatula* and belonging to the CNGC family (CNGC15; Charpentier et al., 2016). In order to balance transmembrane charges due to calcium transport, the DMI1 channel (DOES NOT MAKE INFECTION 1) participates in potassium movements between the outer and inner nuclear membranes (Ané et al., 2004). A SERCA-type calcium ATPase (MCA8 in *M. truncatula*) also participates in these movements of ions (Capoen et al., 2011). It is worth noting that nucleoporins anchored in the nuclear envelope, through both membranes, were also described in *L. japonicus* to participate in calcium oscillations (Kanamori et al., 2006; Saito et al., 2007). Because all these transport proteins are localized at both inner and outer nuclear membranes, it is still unclear if calcium oscillations occur inside or outside the nucleus, or on both sides (Charpentier, 2018). The resulting calcium oscillations are finally decoded by a calcium and calmodulin-dependent kinase (CCaMK), transducing the signal to specific transcription factors and resulting in fungal colonization.

4. FIRST STEPS IN THE STUDY OF PLANT NUTRITION IN ORCHID MYCORRHIZAE

To date, little research has been done on the molecular players responsible for nutritional exchanges in orchid mycorrhizal symbiosis (Dearnaley, Perrotto, & Selosse, 2017), although C, N and P transfers have long been known in this context (e.g. Cameron, Johnson, Leake, & Read, 2007; Cameron, Leake, & Read, 2006). Recent advances in "omics" techniques have helped unravel key genes putatively involved in N transport in orchid mycorrhizae. Comparisons in gene expression in *Serapias vomeracea* protocorms colonized or not by the cosmopolitan fungus *Tulasnella calospora* have revealed interesting facts about N dynamics in the first stages of orchid development (Fochi et al., 2016).

Because it lacks nitrate transporters, it is very likely that the fungus *T. calospora* absorbs N as ammonium and/or amino acids, as suggested by the transcriptomic analysis. In the symbiotic interface, the fungus would release amino acids as a source of N for the orchid plant (*S. vomeracea*), at least in the first stages of plant development, as suggested by comparative transcriptomic analysis in which the arginine-urea pathways are not modulated. In line with such a model, two plant amino acid permeases (*Sv*AAP1 and *Sv*AAP2) and a lysine histidine transporter 1 (*Sv*LHT1) that are strongly upregulated by symbiosis are probably responsible for plant N nutrition originated from the symbiotic fungus (Fochi, Falla, Girlanda, Perotto, & Balestrini, 2017; Fochi et al., 2016).

In general, it is believed that ammonium does not play an important role in the nutrition of *S. vomeracea*, because its ammonium transporters (*Sv*AMT1 and *Sv*AMT2) have low expression levels and only one of them, *Sv*AMT1, is slightly upregulated in symbiosis. In addition, the fungal AMT transporters (*Tc*AMT1 and *Tc*AMT2) are also expressed in the interacting organs, competing for this N resource with the plant. In a model proposed by Dearnaley and Cameron (2017), the first stages of orchid development are characterized by a loop in N transport, the fungus providing the plant with amino acids and the plant releasing NH_4^+ in return in the symbiotic interface. Apart from the already mentioned transport systems, members of the OligoPeptide Transporters (*Sv*OPT1 and *Sv*OPT2) and Peptide Transport (*Sv*PTR1 and *Sv*PTR2) families could be involved in the recovery of larger N-containing molecules by the plant (Fochi et al., 2017). In addition, it has also been proposed that intracellular fungal hyphae forming coils that increase the interface between fungus and orchid root cells, called pelotons, degrade, thus releasing nutrients in the root cell.

Finally, orchids have typically a mycoheterotrophic phase after germination in which they require a mycorrhizal fungus to obtain C, thanks to a symbiotic partner with a simultaneous saprotrophic or ECM lifestyle (Jacquemyn et al., 2017; Selosse, Bocayuva, Kasuya, & Courty, 2017). In this situation, the supply of amino acids from the fungus to the plant would also be a source of C towards the growing orchid.

In conclusion, although we have many clues on element fluxes in orchid mycorrhizae thanks to physiological studies, many aspects regarding nutritional exchanges and their molecular mechanisms are still ignored. The discovery of these systems will be necessary to understand this fascinating symbiosis essential for the life cycle of the largest plant family on Earth.

5. CONCLUDING REMARKS AND PERSPECTIVES

To sum up, the contribution of mycorrhizal fungi to plant nutrition and water transport becomes possible in part by the adaptation and plasticity of both fungal (reviewed recently, Garcia et al., 2016) and plant transportomes (Fig. 2). Beneficial effects of mycorrhizal fungi have been described at plant physiological level since long time. Mineral and water uptake by the fungus, delivery towards and absorption by colonized root cells, and release of plant carbohydrates and lipids to the fungus require a

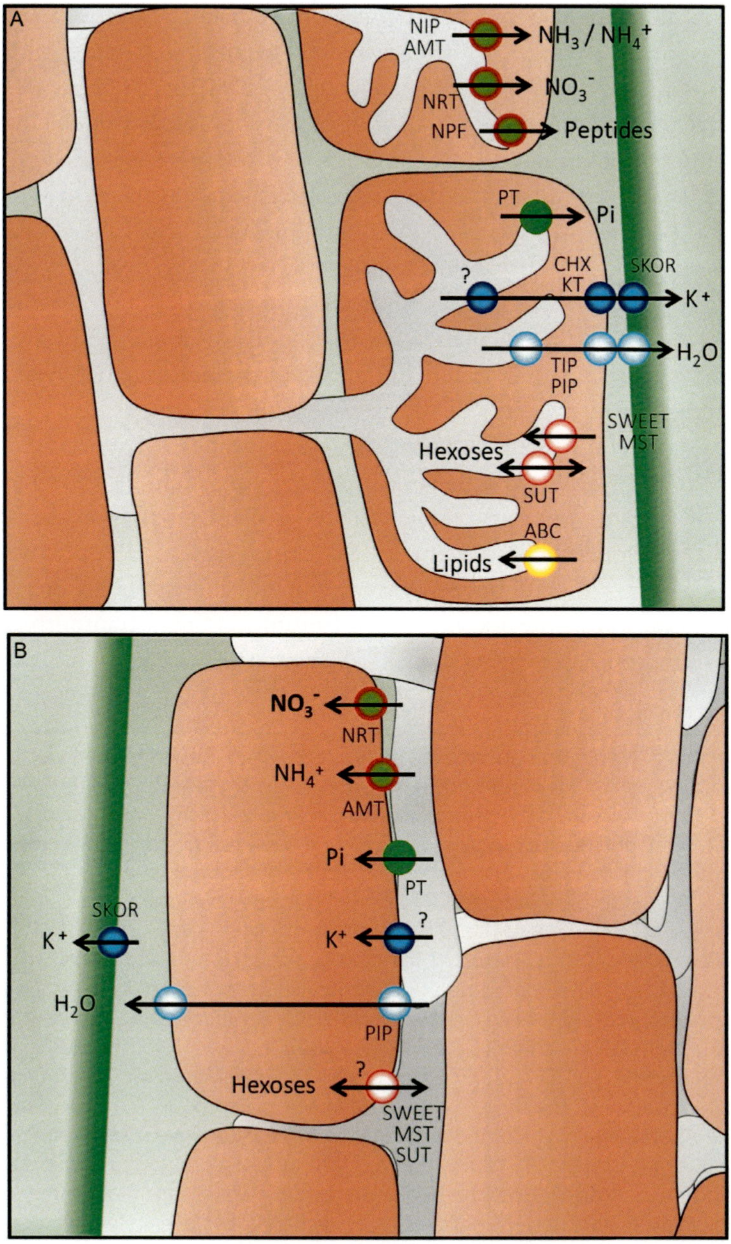

Fig. 2 Plant transporters are involved in nutritional exchanges between symbiotic fungi and root cells in arbuscular mycorrhizae (A) and ectomycorrhizae (B). A close-up from Fig. 1 shows plant transport systems that might be directly regulated and involved in nutrient exchange with mycorrhizal fungi. In addition, membrane transport is also *(Continued)*

tight coordination of membrane transporters from early communication to actual resource exchanges. Molecular knowledge of all steps contributing to such an elaborate mutualistic association is increasing thanks to the number of available genomes and accumulation of transcriptomic studies. Insights into transcriptional regulation of the different transport systems involved need to be supported by functional validations to unravel the *in situ* roles of such identified candidates. Moreover, molecular mechanisms behind the polarized expression patterns and/or specific localization of symbiotic transport systems, which overall control directional transport, have still to be dissected. In particular, membrane microdomains might be responsible for a tight regulation of secretion and uptake at the interface between symbiotic fungi and host plants. Very local conditions with respect to e.g. pH or calcium could also make the difference with respect to transport directions and activities. In addition, activity of proton pumps might be closely linked to that of other transporters (Krajinski et al., 2014; Lanfranco et al., 2018). Many of these questions regarding molecular mechanisms are still awaiting answers. Progress is being made by a computational cell biology approach, which allows simulating the network of proton-coupled transporters and proton pumps in the context of AM symbiosis (Schott et al., 2016).

Fig. 2—Cont'd probably implicated in molecular communication between both symbiotic partners and in signalling. Known transport systems on the fungal side have been reported before and are not indicated here. (A) In arbuscular mycorrhizae, some plant transporters are upregulated and/or specifically expressed and could be involved in the uptake of mineral nutrients and water released by the fungus at the periarbuscular space. Some studies have indicated a role related to the absorption of N compounds in the forms of ammonium (AMT and NIP families), nitrate or peptides (NRT, NPF), phosphate (PT family) and K^+ (CHX and KT families) or its translocation to the xylem (SKOR members). It has also been proposed that arbuscular mycorrhizal fungi enhance water transport through the cell-to-cell pathway, increasing the expression and activation of plant aquaporins (TIPs and PIPs). On the release side, it has been hypothesized that hexoses (through MST, SWEET and SUT transporters) and lipids (through STR proteins) are transported to nourish the fungal partner. (B) In ectomycorrhizae, the host plant may also take up N-containing compounds, mainly ammonium (AMTs) but maybe also nitrate (NRT); phosphate (PT family) and K^+, through transport systems that are still to be elucidated. Possibly due to a higher demand for K^+ translocation to the xylem, as provoked by an enhanced K^+ nutrition, a SKOR channel has been related to the functioning of ectomycorrhizae. It has also been demonstrated that ectomycorrhizal fungi have a stimulating effect on cell-to-cell water transport (induction of PIP aquaporins). The plant is thought to secrete and take up hexoses (MST, SWEET and SUT families) from the apoplasm to feed the fungus and also to control the amount of C allocated to the fungus.

Moreover, the reality of multiple interactions under natural and field conditions is much more complex than the study of a given interaction under more or less determined artificial laboratory conditions. Such studies, taking in account e.g. tripartite interactions or different abiotic stress are coming up, and start to reveal an adapted expression of plant transporters in response to different symbionts (Kafle et al., 2018). Excitingly, it was very recently shown that the AM fungus *R. irregularis* can even feed Pi-solubilizing bacteria (*Rahnella aquatilis*) with the carbon substrate fructose to stimulate phytate mineralization (Zhang, Feng, & Declerck, 2018).

Altogether, the study of these beneficial mycorrhizal associations deserves more attention because of their ecological importance for plant health and biomass production in agriculture and agroforestry. Last, but not least, knowledge on molecular mechanisms governing beneficial associations is urgently needed with respect to plant growth under increasing environmental stress and challenging climate and soil conditions.

ACKNOWLEDGEMENTS

Thanks to all students ever participating in our own research project, thanks to CNRS for acceptance and support, thanks to all worldwide colleagues for sharing stimulating discussions during all the last years. We apologize for interesting work perhaps not cited here. Thanks for patience to my family for "burning the midnight oil". C.G.G. and G.H. acknowledge the *Ecole doctorale* "*GAIA*" (University Montpellier) for PhD fellowships financed by the French Ministry of Higher Education and Research. K.G. acknowledges support of the North Carolina Agriculture Research Service (NCARS).

REFERENCES

Abbaspour, H., Saeidi-Sar, S., Afshari, H., & Abdel-Wahhab, M. A. (2012). Tolerance of mycorrhiza infected pistachio (*Pistacia vera* L.) seedling to drought stress under glasshouse conditions. *Journal of Plant Physiology*, *169*, 704–709.

Ahmad, I., & Maathuis, F. J. M. (2014). Cellular and tissue distribution of potassium: Physiological relevance, mechanisms and regulation. *Journal of Plant Physiology*, *171*, 708–714.

Akiyama, K., Matsuzaki, K., & Hayashi, H. (2005). Plant sesquiterpenes induce hyphal branching in arbuscular mycorrhizal fungi. *Nature*, *435*, 824–827.

Ali, N. A., & Jackson, R. M. (1988). Effects of plant roots and their exudates on germination of spores of ectomycorrhizal fungi. *Transactions of the British Mycological Society*, *91*, 253–260.

Al-Karaki, G. N., & Clark, R. B. (1998). Growth, mineral acquisition and water use by mycorrhizal wheat grown under water stress. *Journal of Plant Nutrition*, *21*, 263–276.

Allen, M. F. (2009). Bidirectional water flows through the soil-fungal-plant mycorrhizal continuum. *The New Phytologist*, *182*, 290–293.

Aloui, A., Recorbet, G., Lemaître-Guillier, C., Mounier, A., Balliau, T., Zivy, M., et al. (2018). The plasma membrane proteome of *Medicago truncatula* roots as modified by arbuscular mycorrhizal symbiosis. *Mycorrhiza*, *28*, 1–16.

Ané, J.-M., Kiss, G. B., Riely, B. K., Penmetsa, R. V., Oldroyd, G. E. D., Ayax, C., et al. (2004). *Medicago truncatula* DMI1 required for bacterial and fungal symbioses in legumes. *Science, 303*, 1364–1367.

Arnebrant, K. (1994). Nitrogen amendments reduce the growth of extramatrical ectomycorrhizal mycelium. *Mycorrhiza, 5*, 7–15.

Aroca, R., Porcel, R., & Ruíz-Lozano, J. M. (2007). How does arbuscular mycorrhizal symbiosis regulate root hydraulic properties and plasma membrane aquaporins in *Phaseolus vulgaris* under drought, cold or salinity stresses? *The New Phytologist, 173*, 808–816.

Augé, R. M. (2001). Water relations, drought and vesicular-arbuscular mycorrhizal symbiosis. *Mycorrhiza, 11*, 3–42.

Azcón, R., Gomez, M., & Tobar, R. (1996). Physiological and nutritional responses by *Lactuca sativa* L. to nitrogen sources and mycorrhizal fungi under drought conditions. *Biology and Fertility of Soils, 22*, 156–161.

Baetz, U., & Martinoia, E. (2014). Root exudates: The hidden part of plant defense. *Trends in Plant Science, 19*, 90–98.

Bago, B., Pfeffer, P. E., Abubaker, J., Jun, J., Allen, J. W., Brouillette, J., et al. (2003). Carbon export from arbuscular mycorrhizal roots involves the translocation of carbohydrate as well as lipid. *Plant Physiology, 131*, 1496–1507.

Bago, B., Pfeffer, P. E., Douds, D. D., Brouillette, J., Bécard, G., & Shachar-Hill, Y. (1999). Carbon metabolism in spores of the arbuscular mycorrhizal fungus *Glomus intraradices* as revealed by nuclear magnetic resonance spectroscopy. *Plant Physiology, 121*, 263–271.

Bago, B., Pfeffer, P. E., & Shachar-Hill, Y. (2000). Carbon metabolism and transport in arbuscular mycorrhizas. *Plant Physiology, 124*, 949–957.

Bago, B., Zipfel, W., Williams, R. M., Jun, J., Arreola, R., Lammers, P. J., et al. (2002). Translocation and utilization of fungal storage lipid in the arbuscular mycorrhizal symbiosis. *Plant Physiology, 128*, 108–124.

Barberon, M., & Geldner, N. (2014). Radial transport of nutrients: The plant root as a polarized epithelium. *Plant Physiology, 166*, 528–537.

Bárzana, G., Aroca, R., Bienert, G. P., Chaumont, F., & Ruíz-Lozano, J. M. (2014). New insights into the regulation of aquaporins by the arbuscular mycorrhizal symbiosis in maize plants under drought stress and possible implications for plant performance. *Molecular Plant-Microbe Interactions, 27*, 349–363.

Bárzana, G., Aroca, R., Paz, J. A., Chaumont, F., Martinez-Ballesta, M., Carvajal, M., et al. (2012). Arbuscular mycorrhizal symbiosis increases relative apoplastic water flow in roots of the host plant under both well-watered and drought stress conditions. *Annals of Botany, 109*, 1009–1017.

Bárzana, G., Aroca, R., & Ruiz-Lozano, J. M. (2015). Localized and non-localized effects of arbuscular mycorrhizal symbiosis on accumulation of osmolytes and aquaporins and on antioxidant systems in maize plants subjected to total or partial root drying. *Plant, Cell & Environment, 38*, 1613–1627.

Baudri, D. V., Loyola-Vargas, V. M., Broeckling, C. D., De-la-Peña, C., Jasinski, M., Santelia, D., et al. (2008). Altered profile of secondary metabolites in the root exudates of *Arabidopsis* ATP-binding cassette transporter mutants. *Plant Physiology, 146*, 762–771.

Becquer, A., Garcia, K., Amenc, L., Rivard, C., Doré, J., Trives-Segura, C., et al. (2018). The *Hebeloma cylindrosporum Hc*PT2 Pi transporter plays a key role in the ectomycorrhizal symbiosis. *The New Phytologist*. https://doi.org/10.1111/nph.15281 [in press].

Becquer, A., Trap, J., Irshad, U., Ali, M. A., & Plassard, C. (2014). From soil to plant, the journey of P through trophic relationships and ectomycorrhizal association. *Frontiers in Plant Science, 5*, 548.

Beilby, J. P., & Kidby, D. K. (1980). Biochemistry of ungerminated and germinated spores of the vesicular-arbuscular mycorrhizal fungus, *Glomus caledonius*: Changes in neutral and polar lipids. *Journal of Lipid Research, 21*, 739–750.

Benedito, V. A., Li, H., Dai, X., Wandrey, M., He, J., Kaundal, R., et al. (2010). Genomic inventory and transcriptional analysis of *Medicago truncatula* transporters. *Plant Physiology*, *152*, 1716–1730.

Berruti, A., Lumini, E., Balestrini, R., & Bianciotto, V. (2015). Arbuscular mycorrhizal fungi as natural biofertilizers: Let's benefit from past successes. *Frontiers in Microbiology*, *6*, 1559.

Besserer, A., Puech-Pagès, V., Kiefer, P., Gomez-Roldan, V., Jauneau, A., Roy, S., et al. (2006). Strigolactones stimulate arbuscular mycorrhizal fungi by activating mitochondria. *PLoS Biology*, *4*, e226.

Boldt, K., Pors, Y., Haupt, B., Bitterlich, M., Kuhn, C., Grimm, B., et al. (2011). Photochemical processes, carbon assimilation and RNA accumulation of sucrose transporter genes in tomato arbuscular mycorrhiza. *Journal of Plant Physiology*, *168*, 1256–1263.

Bompadre, M. J., Rios de Molina, M. C., Colombo, R. P., Bidondo, L. F., Silvani, V. A., Pardo, A. G., et al. (2013). Differential efficiency of two strains of the arbuscular mycorrhizal fungus *Rhizophagus irregularis* on olive (*Olea europaea*) plants under two water regimes. *Symbiosis*, *61*, 105–112.

Bonfante, P., & Genre, A. (2010). Mechanisms underlying beneficial plant-fungus interactions in mycorrhizal symbiosis. *Nature Communications*, *1*, 48.

Bravo, A., Brands, M., Wewer, V., Dörmann, P., & Harrison, M. J. (2017). Arbuscular mycorrhiza-specific enzymes FatM and RAM2 fine-tune lipid biosynthesis to promote development of arbuscular mycorrhiza. *The New Phytologist*, *214*, 1631–1645.

Bravo, A., York, T., Pumplin, N., Mueller, L. A., & Harrison, M. J. (2016). Genes conserved for arbuscular mycorrhizal symbiosis identified through phylogenomics. *Nature Plants*, *2*, 15208.

Breuillin-Sessoms, F., Floss, D. S., Gomez, S. K., Pumplin, N., Ding, Y., Levesque-Tremblay, V., et al. (2015). Suppression of arbuscule degeneration in *Medicago truncatula* phosphate transporter4 mutants is dependent on the ammonium transporter 2 family protein AMT2;3. *The Plant Cell*, *27*, 1352–1366.

Bücking, H., & Heyser, W. (2000). Subcellular compartmentation of elements in non-mycorrhizal and mycorrhizal roots of *Pinus sylvestris*: An X-ray microanalytical study. II. The distribution of calcium, potassium and sodium. *The New Phytologist*, *145*, 321–331.

Bücking, H., & Kafle, A. (2015). Role of arbuscular mycorrhizal fungi in the nitrogen uptake of plants: Current knowledge and research gaps. *Agronomy*, *5*, 587–612.

Bücking, H., Kuhn, A. J., Schröder, W. H., & Heyser, W. (2002). The fungal sheath of ectomycorrhizal pine roots: An apoplastic barrier for the entry of calcium, magnesium, and potassium into the root cortex? *Journal of Experimental Botany*, *53*, 1659–1669.

Burns, J. H., Anacker, B. L., Strauss, S. Y., & Burke, D. J. (2015). Soil microbial community variation correlates most strongly with plant species identity, followed by soil chemistry, spatial location and plant genus. *AoB Plants*, 7. plv030.

Calvo-Polanco, M., Jones, M. D., & Zwiazek, J. J. (2009). Effects of pH on NaCl tolerance of American elm (*Ulmus americana*) seedlings inoculated with *Hebeloma crustuliniforme* and *Laccaria bicolor*. *Acta Physiologiae Plantarum*, *31*, 515–522.

Calvo-Polanco, M., Molina, S., Zamarreno, A. M., Garcia-Mina, J. M., & Aroca, R. (2014). The symbiosis with the arbuscular mycorrhizal fungus *Rhizophagus irregularis* drives root water transport in flooded tomato plants. *Plant & Cell Physiology*, *55*, 1017–1029.

Calvo-Polanco, M., Sánchez-Castro, I., Cantos, M., García, J. L., Azcón, R., Ruíz-Lozano, J. M., et al. (2016). Effects of different arbuscular mycorrhizal fungal backgrounds and soils on olive plants growth and water relation properties under well-watered and drought conditions. *Plant, Cell & Environment*, *39*, 2498–2514.

Calvo-Polanco, M., Sánchez-Romera, B., Aroca, R., Asins, M. J., Declerck, S., Dodd, I. C., et al. (2016). Exploring the use of recombinant inbred lines in combination with beneficial microbial inoculants (AM fungus and PGPR) to improve drought stress tolerance in tomato. *Environmental and Experimental Botany*, *131*, 47–57.

Cameron, D. D., Johnson, I., Leake, J. R., & Read, D. J. (2007). Mycorrhizal acquisition of inorganic phosphorus by the green-leaved terrestrial orchid *Goodyera repens*. *Annals of Botany*, *99*, 831–834.

Cameron, D. D., Leake, J. R., & Read, D. J. (2006). Mutualistic mycorrhiza in orchids: Evidence from plant–fungus carbon and nitrogen transfers in the green-leaved terrestrial orchid *Goodyera repens*. *The New Phytologist*, *171*, 405–416.

Capoen, W., Sun, J., Wysham, D., Otegui, M. S., Venkateshwaran, M., Hirsch, S., et al. (2011). Nuclear membranes control symbiotic calcium signaling of legumes. *Proceedings of the National Academy of Sciences of the United States of America*, *108*, 14348–14353.

Carbonnel, S., & Gutjahr, C. (2014). Control of arbuscular mycorrhiza development by nutrient signals. *Frontiers in Plant Science*, *5*, 462.

Casieri, L., Lahmidi, N. A., Doidy, J., Veneault-Fourrey, C., Migeon, A., Bonneau, L., et al. (2013). Biotrophic transportome in mutualistic plant–fungal interactions. *Mycorrhiza*, *23*, 597–625.

Castro-Rodríguez, V., Assaf-Casals, I., Pérez-Tienda, J., Fan, X., Avila, C., Miller, A., et al. (2016). Deciphering the molecular basis of ammonium uptake and transport in maritime pine. *Plant, Cell & Environment*, *39*, 1669–1682.

Chalot, M., Blaudez, D., & Brun, A. (2006). Ammonia: A candidate for nitrogen transfer at the mycorrhizal interface. *Trends in Plant Science*, *11*, 263–266.

Charpentier, M. (2018). Calcium signals in the plant nucleus: Origin and function. *Journal of Experimental Botany*, *69*, 4165–4173.

Charpentier, M., Sun, J., Martins, T. V., Radhakrishnan, G. V., Findlay, K., Soumpourou, E., et al. (2016). Nuclear-localized cyclic nucleotide-gated channels mediate symbiotic calcium oscillations. *Science*, *352*, 1102–1105.

Chaumont, F., & Tyerman, S. D. (2014). Aquaporins: Highly regulated channels controlling plant water relations. *Plant Physiology*, *164*, 1600–1618.

Chen, L.-Q. (2014). SWEET sugar transporters for phloem transport and pathogen nutrition. *The New Phytologist*, *201*, 1150–1155.

Chen, A., Gu, M., Wang, S., Chen, J., Xu, G., Aiqun, C., et al. (2018). Transport properties and regulatory roles of nitrogen in arbuscular mycorrhizal symbiosis. *Seminars in Cell & Developmental Biology*, *74*, 80–88.

Chitarra, W., Pagliarani, C., Maserti, B., Lumini, E., Siciliano, I., Cascone, P., et al. (2016). Insights on the impact of arbuscular mycorrhizal symbiosis on tomato tolerance to water stress. *Plant Physiology*, *171*, 1009–1023.

Clemmensen, K. E., Bahr, A., Ovaskainen, O., Dahlberg, A., Ekblad, A., Wallander, H., et al. (2013). Roots and associated fungi drive long-term carbon sequestration in boreal forest. *Science*, *339*, 1615–1618.

Clemmensen, K. E., Finlay, R. D., Dahlberg, A., Stenlid, J., Wardle, D. A., & Lindahl, B. D. (2015). Carbon sequestration is related to mycorrhizal fungal community shifts during long-term succession in boreal forests. *The New Phytologist*, *205*, 1525–1536.

Corratgé, C., Zimmermann, S., Lambilliotte, R., Plassard, C., Marmeisse, R., Thibaud, J.-B., et al. (2007). Molecular and functional characterization of a Na^+-K^+ transporter from the Trk family in the ectomycorrhizal fungus *Hebeloma cylindrosporum*. *The Journal of Biological Chemistry*, *282*, 26057–26066.

Corratgé-Faillie, C., Jabnoune, M., Zimmermann, S., Véry, A. A., Fizames, C., & Sentenac, H. (2010). Potassium and sodium transport in non-animal cells: The Trk/Ktr/HKT transporter family. *Cellular and Molecular Life Sciences*, *67*, 2511–2532.

Corratgé-Faillie, C., & Lacombe, B. (2017). Substrate (un)specificity of Arabidopsis NRT1/PTR FAMILY (NPF) proteins. *Journal of Experimental Botany*, *68*, 3107–3113.

Courty, P.-E., Doidy, J., Garcia, K., Wipf, D., & Zimmermann, S. D. (2017). The transportome of mycorrhizal systems. In F. Martin (Ed.), *Molecular mycorrhizal symbiosis* (pp. 239–256). Hoboken, NJ: John Wiley & Sons, Inc., Wiley-Blackwell Press.

Courty, P. E., Smith, P., Koegel, S., Redecker, D., & Wipf, D. (2015). Inorganic nitrogen uptake and transport in beneficial plant root-microbe interactions. *Critical Reviews in Plant Sciences*, *34*, 4–16.

Couturier, J., Montanini, B., Martin, F., Brun, A., Blaudez, D., & Chalot, M. (2007). The expanded family of ammonium transporters in the perennial poplar plant. *The New Phytologist*, *174*, 137–150.

Dearnaley, J. D. W., & Cameron, D. D. (2017). Nitrogen transport in the orchid mycorrhizal symbiosis—Further evidence for a mutualistic association. *The New Phytologist*, *213*, 10–12.

Dearnaley, J. D. W., Perrotto, S., & Selosse, M.-A. (2017). Structure and development of orchid mycorrhizas. In F. Martin (Ed.), *Molecular mycorrhizal symbiosis* (pp. 63–86). Hoboken, NJ: John Wiley & Sons, Inc., Wiley-Blackwell Press.

Dietz, S., von Bülow, J., Beitz, E., & Nehls, U. (2011). The aquaporin gene family of the ectomycorrhizal fungus *Laccaria bicolor*: Lessons for symbiotic functions. *The New Phytologist*, *190*, 927–940.

Doidy, J., Grace, E., Kuhn, C., Simon-Plas, F., Casieri, L., & Wipf, D. (2012). Sugar transporters in plants and in their interactions with fungi. *Trends in Plant Science*, *17*, 413–422.

Doidy, J., van Tuinen, D., Lamotte, O., Corneillat, M., Alcaraz, G., & Wipf, D. (2012). The *Medicago truncatula* sucrose transporter family. Characterization and implication of key members in carbon partitioning towards arbuscular mycorrhizal fungi. *Molecular Plant*, *5*, 1346–1358.

Drechsler, N., Courty, P.-E., Brulé, D., & Kunze, R. (2018). Identification of arbuscular mycorrhiza-inducible nitrate transporter 1/peptide transporter family (NPF) genes in rice. *Mycorrhiza*, *28*, 93–100.

Dreyer, I., & Uozumi, N. (2011). Potassium channels in plant cells. *The FEBS Journal*, *278*, 4293–4303.

El-Mesbahi, M. N., Azcon, R., Ruíz-Lozano, J. M., & Aroca, R. (2012). Plant potassium content modifies the effects of arbuscular mycorrhizal symbiosis on root hydraulic properties in maize plants. *Mycorrhiza*, *22*, 555–564.

Farzaneh, M., Vierheilig, H., Lössl, A., & Kaul, H. P. (2011). Arbuscular mycorrhiza enhances nutrient uptake in chickpea. *Plant, Soil and Environment*, *57*, 465–470.

Fellbaum, C. R., Gachomo, E. W., Beesetty, Y., Choudhari, S., Strahan, G. D., Pfeffer, P. E., et al. (2012). Carbon availability triggers fungal nitrogen uptake and transport in arbuscular mycorrhizal symbiosis. *Proceedings of the National Academy of Sciences of the United States of America*, *109*, 2666–2671.

Fellbaum, C. R., Mensah, J. A., Cloos, A. J., Strahan, G. E., Pfeffer, P. E., Kiers, E. T., et al. (2014). Fungal nutrient allocation in common mycorrhizal networks is regulated by the carbon source strength of individual host plants. *The New Phytologist*, *203*, 646–656.

Ferrol, N., Tamayo, E., & Vargas, P. (2016). The heavy metal paradox in arbuscular mycorrhizas: From mechanisms to biotechnological applications. *Journal of Experimental Botany*, *67*, 6253–6265.

Fochi, V., Falla, N., Girlanda, M., Perotto, S., & Balestrini, R. (2017). Cell-specific expression of plant nutrient transporter genes in orchid mycorrhizae. *Plant Science*, *263*, 39–45.

Fochi, V., Walter, C., Kohler, A., Voyron, S., Singan, V. R., Lindquist, E. A., et al. (2016). Fungal and plant gene expression in the *Tulasnella calospora–Serapias vomeracea* symbiosis provides clues about nitrogen pathways in orchid mycorrhizas. *The New Phytologist*, *213*, 365–379.

Fries, N., Serck-Hanssen, K., Dimberg, L. H., & Theander, O. (1987). Abietic acid, and activator of basidiospore germination in ectomycorrhizal species of the genus *Suillus* (*Boletaceae*). *Experimental Mycology*, *11*, 360–363.

Garcia, K., Chasman, D., Roy, S., & Ane, J.-M. (2017). Physiological responses and gene co-expression network of mycorrhizal roots under K^+ deprivation. *Plant Physiology*, *173*, 1811–1823.

Garcia, K., Delaux, P.-M., Cope, K. R., & Ané, J.-M. (2015). Molecular signals required for the establishment and maintenance of ectomycorrhizal symbioses. *The New Phytologist*, *208*, 79–87.

Garcia, K., Delteil, A., Conéjéro, G., Becquer, A., Plassard, C., Sentenac, H., et al. (2014). Potassium nutrition of ectomycorrhizal *Pinus pinaster*: Overexpression of the *Hebeloma cylindrosporum HcTrk1* transporter affects the translocation of both K^+ and phosphorus in the host plant. *The New Phytologist*, *201*, 951–960.

Garcia, K., Doidy, J., Zimmermann, S. D., Wipf, D., & Courty, P.-E. (2016). Take a trip through the plant and fungal transportome of mycorrhiza. *Trends in Plant Science*, *21*, 937–950.

Garcia, K., & Zimmermann, S. D. (2014). The role of mycorrhizal associations in plant potassium nutrition. *Frontiers in Plant Science*, *5*, 1–9.

Gaude, N., Bortfeld, S., Duensing, N., Lohse, M., & Krajinski, F. (2012). Arbuscule-containing and non-colonized cortical cells of mycorrhizal roots undergo extensive and specific reprogramming during arbuscular mycorrhizal development. *The Plant Journal*, *69*, 510–528.

Gaude, N., Schulze, W. X., Franken, P., & Krajinski, F. (2012). Cell type-specific protein and transcription profiles implicate periarbuscular membrane synthesis as an important carbon sink in the mycorrhizal symbiosis. *Plant Signaling & Behavior*, 7, 461–464.

Gaymard, F., Pilot, G., Lacombe, B., Bouchez, D., Bruneau, D., Boucherez, J., et al. (1998). Identification and disruption of a plant Shaker-like outward channel involved in K^+ release into the xylem sap. *Cell*, *4*, 647–655.

Ge, L., Sun, S. B., Chen, A., Kapulnik, Y., & Xu, G. (2008). Tomato sugar transporter genes associated with mycorrhiza and phosphate. *Plant Growth Regulation*, *55*, 115–123.

Genre, A., Chabaud, M., Balzergue, C., Puech-Pagès, V., Novero, M., Rey, T., et al. (2013). Short-chain chitin oligomers from arbuscular mycorrhizal fungi trigger nuclear Ca^{2+} spiking in *Medicago truncatula* roots and their production is enhanced by strigolactone. *The New Phytologist*, *198*, 190–202.

Gobbato, E., Marsh, J. F., Vernié, T., Wang, E., Maillet, F., Kim, K., et al. (2012). A GRAS-type transcription factor with a specific function in mycorrhizal signaling. *Current Biology*, *22*, 2236–2241.

Gobbato, E., Wang, E., Higgins, G., Bano, S. A., Henry, C., Schultze, M., et al. (2013). *RAM1* and *RAM2* function and expression during arbuscular mycorrhizal symbiosis and *Aphanomyces euteiches* colonization. *Plant Signaling & Behavior*, *8*, e26049.

Gomez, S. K., Javot, H., Deewatthanawong, P., Torres-Jerez, I., Tang, Y., Blancaflor, E. B., et al. (2009). *Medicago truncatula* and *Glomus intraradices* gene expression in cortical cells harboring arbuscules in the arbuscular mycorrhizal symbiosis. *BMC Plant Biology*, *9*, 10.

Govindarajulu, M., Pfeffer, P. E., Jin, H., Abubaker, J., Douds, D. D., Allen, J. W., et al. (2005). Nitrogen transfer in the arbuscular mycorrhizal symbiosis. *Nature*, *435*, 819–823.

Grabov, A. (2007). Plant KT/KUP/HAK potassium transporters: Single family—Multiple functions. *Annals of Botany*, *99*, 1035–1041.

Grunze, N., Willmann, M., & Nehls, U. (2004). The impact of ectomycorrhiza formation on monosaccharide transporter gene expression in poplar roots. *The New Phytologist*, *164*, 147–155.

Guerrero-Galán, C., Delteil, A., Garcia, K., Houdinet, G., Conéjéro, G., Gaillard, I., et al. (2018). Plant potassium nutrition in ectomycorrhizal symbiosis: Properties and roles of the three fungal TOK potassium channels in *Hebeloma cylindrosporum*. *Environmental Microbiology*, *20*, 1873–1887.

Guerrero-Galán, C., Garcia, K., Houdinet, G., & Zimmermann, S. D. (2018). HcTOK1 participates in the maintenance of K^+ homeostasis in the ectomycorrhizal fungus *Hebeloma cylindrosporum*, which is essential for the symbiotic K^+ nutrition of *Pinus pinaster*. *Plant Signaling & Behavior, 13*(6). https://doi.org/10.1080/15592324.2018.1480845.

Guether, M., Balestrini, R., Hannah, M., He, J., Udvardi, M. K., & Bonfante, P. (2009). Genome-wide reprogramming of regulatory networks, transport, cell wall and membrane biogenesis during arbuscular mycorrhizal symbiosis in *Lotus japonicus*. *The New Phytologist, 182*, 200–212.

Guether, M., Neuhäuser, B., Balestrini, R., Dynowski, M., Ludewig, U., & Bonfante, P. (2009). A mycorrhizal-specific ammonium transporter from *Lotus japonicus* acquires nitrogen released by arbuscular mycorrhizal fungi. *Plant Physiology, 150*, 73–83.

Guo, L., Wang, Z. Y., Lin, H., Cui, W. E., Chen, J., Liu, M. H., et al. (2006). Expression and functional analysis of the rice plasma-membrane intrinsic protein gene family. *Cell Research, 16*, 277–286.

Gutjahr, C., Radovanovic, D., Geoffroy, J., Zhang, Q., Siegler, H., Chiapello, M., et al. (2012). The half-size ABC transporters STR1 and STR2 are indispensable for mycorrhizal arbuscule formation in rice. *The Plant Journal, 69*, 906–920.

Hacquard, S., Tisserant, E., Brun, A., Legué, V., Martin, F., & Kohler, A. (2013). Laser microdissection and microarray analysis of *Tuber melanosporum* ectomycorrhizas reveal functional heterogeneity between mantle and Hartig net compartments. *Environmental Microbiology, 5*, 1853–1869.

Hammer, E. C., Pallon, J., Wallander, H., & Olsson, P. A. (2011). Tit for tat? A mycorrhizal fungus accumulates phosphorus under low plant carbon availability. *FEMS Microbiology Ecology, 76*, 236–244.

Harrison, M. J. (1996). A sugar transporter from *Medicago truncatula*: Altered expression pattern in roots during vesicular-arbuscular (VA) mycorrhizal associations. *The Plant Journal, 9*, 491–503.

Harrison, M. J., Dewbre, G. R., & Liu, J. (2002). A phosphate transporter from *Medicago truncatula* involved in the acquisition of phosphate released by arbuscular mycorrhizal fungi. *The Plant Cell, 14*, 2413–2429.

Helber, N., Wippel, K., Sauer, N., Schaarschmidt, S., Hause, B., & Requena, N. (2011). A versatile monosaccharide transporter that operates in the arbuscular mycorrhizal fungus *Glomus* sp. is crucial for the symbiotic relationship with plants. *The Plant Cell, 23*, 3812–3823.

Hildebrandt, U., Schmelzer, E., & Bothe, H. (2002). Expression of nitrate transporter genes in tomato colonized by an arbuscular mycorrhizal fungus. *Physiologia Plantarum, 115*, 125–136.

Hogekamp, C., Arndt, D., Pereira, P. A., Becker, J. D., Hohnjec, N., & Küster, H. (2011). Laser microdissection unravels cell-type-specific transcription in arbuscular mycorrhizal roots, including CAAT-Box transcription factor gene expression correlating with fungal contact and spread. *Plant Physiology, 157*, 2023–2043.

Hogekamp, C., & Küster, H. (2013). A roadmap of cell-type specific gene expression during sequential stages of the arbuscular mycorrhiza symbiosis. *BMC Genomics, 14*, 306.

Hohnjec, N., Vieweg, M. F., Pühler, A., Becker, A., & Küster, H. (2005). Overlaps in the transcriptional profiles of *Medicago truncatula* roots inoculated with two different *Glomus* fungi provide insights into the genetic program activated during arbuscular mycorrhiza. *Plant Physiology, 137*, 1283–1301.

Jabaji-hare, S. (1988). Lipid and fatty acid profiles of some vesicular-arbuscular mycorrhizal fungi: Contribution to taxonomy. *Mycologia, 80*, 622.

Jacquemyn, H., Waud, M., Brys, R., Lallemand, F., Courty, P.-E., Robionek, A., et al. (2017). Mycorrhizal associations and trophic modes in coexisting orchids: An ecological continuum between auto- and mixotrophy. *Frontiers in Plant Science, 8*, 1497.

Jahromi, F., Aroca, R., Porcel, R., & Ruíz-Lozano, J. M. (2008). Influence of salinity on the in vitro development of *Glomus intraradices* and on the *in vivo* physiological and molecular responses of mycorrhizal lettuce plants. *Microbial Ecology*, *55*, 45–53.

Javot, H., Penmetsa, R. V., Terzaghi, N., Cook, D. R., & Harrison, M. J. (2007). A *Medicago truncatula* phosphate transporter indispensable for the arbuscular mycorrhizal symbiosis. *Proceedings of the National Academy of Sciences of the United States of America*, *104*, 1720–1725.

Jentschke, G., Brandes, B., Kuhn, A. J., Schröder, W. H., & Godbold, D. L. (2001). Interdependence of phosphorus, nitrogen, potassium and magnesium translocation by the ectomycorrhizal fungus *Paxillus involutus*. *The New Phytologist*, *149*, 327–337.

Johansson, I., Karlsson, M., Shukla, V. K., Chrispeels, M. J., Larsson, C., & Kjellbom, P. (1998). Water transport activity of the plasma membrane aquaporin PM28A is regulated by phosphorylation. *The Plant Cell*, *10*, 451–459.

Jourand, P., Hannibal, L., Majorel, C., Mengant, S., Ducousso, M., & Lebrun, M. (2014). Ectomycorrhizal *Pisolithus albus* inoculation of *Acacia spirorbis* and *Eucalyptus globulus* grown in ultramafic topsoil enhances plant growth and mineral nutrition while limits metal uptake. *Journal of Plant Physiology*, *171*, 164–172.

Kafle, A., Garcia, K., Wang, X., Pfeffer, P. E., Strahan, G. D., & Bücking, H. (2018). Nutrient demand and fungal access to resources control the carbon allocation to the symbiotic partners in tripartite interactions of *Medicago truncatula*. *Plant, Cell & Environment*. https://doi.org/10.1111/pce.13359 [in press].

Kanamori, N., Madsen, L. H., Radutoiu, S., Frantescu, M., Quistgaard, E. M. H., Miwa, H., et al. (2006). A nucleoporin is required for induction of Ca^{2+} spiking in legume nodule development and essential for rhizobial and fungal symbiosis. *Proceedings of the National Academy of Sciences of the United States of America*, *103*, 359–364.

Keymer, A., Pimprikar, P., Wewer, V., Huber, C., Brands, M., Bucerius, S. L., et al. (2017). Lipid transfer from plants to arbuscular mycorrhiza fungi. *eLife*, *6*, e29107.

Khalvati, M. A., Hu, Y., Mozafar, A., & Schmidhalter, U. (2005). Quantification of water uptake by arbuscular mycorrhizal hyphae and its significance for leaf growth, water relations, and gas exchange of barley subjected to drought stress. *Plant Biology*, *7*, 706–712.

Kiers, E. T., Duhamel, M., Beesetty, Y., Mensah, J., Franken, O., Verbruggen, E., et al. (2011). Reciprocal rewards stabilize cooperation in the mycorrhizal symbiosis. *Science*, *333*, 880–882.

Kikuchi, K., Matsushita, N., Suzuki, K., & Hogetsu, T. (2007). Flavonoids induce germination of basidiospores of the ectomycorrhizal fungus *Suillus bovinus*. *Mycorrhiza*, *17*, 563–570.

Kobae, Y., & Hata, S. (2010). Dynamics of periarbuscular membranes visualized with a fluorescent phosphate transporter in arbuscular mycorrhizal roots of rice. *Plant & Cell Physiology*, *51*, 341–353.

Kobae, Y., Tamura, Y., Takai, S., Banba, M., & Hata, S. (2010). Localized expression of arbuscular mycorrhiza-inducible ammonium transporters in soybean. *Plant & Cell Physiology*, *51*, 1411–1415.

Koegel, S., Ait Lahmidi, N., Arnould, C., Chatagnier, O., Walder, F., Ineichen, K., et al. (2013). The family of ammonium transporters (AMT) in *Sorghum bicolor*: Two AMT members are induced locally, but not systemically in roots colonized by arbuscular mycorrhizal fungi. *The New Phytologist*, *198*, 853–865.

Koegel, S., Mieulet, D., Baday, S., Chatagnier, O., Lehmann, M. F., Wiemken, A., et al. (2017). Phylogenetic, structural, and functional characterization of AMT3;1, an ammonium transporter induced by mycorrhization among model grasses. *Mycorrhiza*, *27*, 695–708.

Kohler, A., Kuo, A., Nagy, L. G., Morin, E., Barry, K. W., Buscot, F., et al. (2015). Convergent losses of decay mechanisms and rapid turnover of symbiosis genes in mycorrhizal mutualists. *Nature Genetics*, *47*, 410–415.

Kottke, I., & Oberwinkler, F. (1987). The cellular structure of the Hartig net: Coenocytic and transfer cell-like organization. *Nordic Journal of Botany*, 7, 85–95.

Krajinski, F., Courty, P.-E., Sieh, D., Franken, P., Zhang, H., Bucher, M., et al. (2014). The H^+-ATPase HA1 of *Medicago truncatula* is essential for phosphate transport and plant growth during arbuscular mycorrhizal symbiosis. *The Plant Cell*, *26*, 1808–1817.

Lagrange, H., Jay-Allemand, C., & Lapeyrie, F. (2001). Rutin, the phenolglycoside from eucalyptus root exudates, stimulates *Pisolithus* hyphal growth at picomolar concentrations. *The New Phytologist*, *149*, 349–355.

Landeweert, R., Hoffland, E., Finlay, R. D., Kuyper, T. W., & van Breemen, N. (2001). Linking plants to rocks: Ectomycorrhizal fungi mobilize nutrients from minerals. *Trends in Ecology & Evolution*, *279*, 2106–2110.

Lanfranco, L., Fiorilli, V., & Gutjahr, C. (2018). Partner communication and role of nutrients in the arbuscular mycorrhizal symbiosis. *The New Phytologist*. https://doi.org/10.1111/nph.15230.

Lee, S. H., Calvo-Polanco, M., & Zwiazek, J. J. (2010). Role of aquaporins in root water transport of ectomycorrhizal jack pine (*Pinus banksiana*) seedlings exposed to NaCl and fluoride. *Plant, Cell & Environment*, *33*, 769–780.

Lee, B.-R., Muneer, S., Avice, J.-C., Jin, W., & Kim, T.-H. (2012). Mycorrhizal colonization and P supplement effects on N uptake and N assimilation in perennial ryegrass under well-watered and drought-stressed conditions. *Mycorrhiza*, *22*, 525–534.

Léran, S., Varala, K., Boyer, J. C., Chiurazzi, M., Crawford, N., Daniel-Vedele, F., et al. (2014). A unified nomenclature of nitrate transporter 1/peptide transporter family members in plants. *Trends in Plant Science*, *19*, 5–9.

Li, T., Hu, Y. J., Hao, Z. P., Li, H., Wang, Y. S., & Chen, B. D. (2013). First cloning and characterization of two functional aquaporin genes from an arbuscular mycorrhizal fungus *Glomus intraradices*. *The New Phytologist*, *197*, 617–630.

Liu, F., Xu, Y., Han, G., Wang, W., Li, X., & Cheng, B. (2018). Identification and functional characterization of a maize phosphate transporter induced by mycorrhiza formation. *Plant & Cell Physiology*, *59*, 1683–1694.

Loth-Pereda, V., Orsini, E., Courty, P.-E., Lota, F., Kohler, A., Diss, L., et al. (2011). Structure and expression profile of the phosphate Pht1 transporter gene family in mycorrhizal *Populus trichocarpa*. *Plant Physiology*, *156*, 2141–2154.

Luginbuehl, L. H., Menard, G. N., Kurup, S., Van Herp, H., Radhakrishnan, G. V., Breakspear, A., et al. (2017). Fatty acids in arbuscular mycorrhizal fungi are synthesized by the host plant. *Science*, *356*, 1175–1178.

MacLean, A. M., Bravo, A., & Harrison, M. J. (2017). Plant signaling and metabolic pathways enabling arbuscular mycorrhizal symbiosis. *The Plant Cell*, *29*, 2319–2335.

Maillet, F., Poinsot, V., Andre, O., Puech-Pages, V., Haouy, A., Gueunier, M., et al. (2011). Fungal lipochitooligosaccharide symbiotic signals in arbuscular mycorrhiza. *Nature*, *469*, 58–63.

Manck-Götzenberger, J., & Requena, N. (2016). Arbuscular mycorrhiza symbiosis induces a major transcriptional reprogramming of the potato *SWEET* sugar transporter family. *Frontiers in Plant Science*, 7, 487.

Marjanović, Ž., Uehlein, N., Kaldenhoff, R., Zwiazek, J. J., Weiss, M., Hampp, R., et al. (2005). Aquaporins in poplar: What a difference a symbiont makes! *Planta*, *222*, 258–268.

Martin, F., Duplessis, S., Ditengou, F., Lagrange, H., Voiblet, S., & Lapeyrie, F. (2001). Developmental cross talking in the ectomycorrhizal symbiosis: Signals and communication genes. *The New Phytologist*, *151*, 145–154.

Marulanda, A., Azcón, R., & Ruíz-Lozano, J. M. (2003). Contribution of six arbuscular mycorrhizal fungal isolates to water uptake by *Lactuca sativa* plants under drought stress. *Physiologia Plantarum*, *119*, 526–533.

Maurel, C., Boursiac, Y., Luu, D. T., Santoni, V., Shahzad, Z., & Verdoucq, L. (2015). Aquaporins in plants. *Physiological Reviews*, *95*, 1321–1358.

Melin, E., & Rama Das, V. (1954). Influence of root-metabolites on the growth of tree mycorrhizal fungi. *Physiologia Plantarum*, 7, 851–858.

Miransari, M. (Ed.), (2014). *Use of microbes for the alleviation of soil stresses: Vol. 1.* New York, NY: Springer New York.

Muhsin, T. M., & Zwiazek, J. J. (2002). Colonization with *Hebeloma crustuliniforme* increases water conductance and limits shoot sodium uptake in white spruce (*Picea glauca*) seedlings. *Plant and Soil*, *238*, 217–225.

Navarro-Fernández, C. M., Aroca, R., & Barea, J. M. (2011). Influence of arbuscular mycorrhizal fungi and water regime on the development of endemic *Thymus* species in dolomitic soils. *Applied Soil Ecology*, *48*, 31–37.

Navarro-Rodenas, A., Barzana, G., Nicolas, E., Carra, A., Schubert, A., & Morte, A. (2013). Expression analysis of aquaporins from desert truffle mycorrhizal symbiosis reveals a fine-tuned regulation under drought. *Molecular Plant-Microbe Interactions*, *26*, 1068–1078.

Navarro-Ródenas, A., Xu, H., Kemppainen, M., Pardo, A. G., & Zwiazek, J. J. (2015). *Laccaria bicolor* aquaporin LbAQP1 is required for Hartig net development in trembling aspen (*Populus tremuloides*). *Plant, Cell & Environment*, *38*, 2475–2486.

Neale, D. B., Wegrzyn, J. L., Stevens, K. A., Zimin, A. V., Puiu, D., Crepeau, M. W., et al. (2014). Decoding the massive genome of loblolly pine using haploid DNA and novel assembly strategies. *Genome Biology*, *15*, R59.

Nehls, U. (2008). Mastering ectomycorrhizal symbiosis: The impact of carbohydrates. *Journal of Experimental Botany*, *59*, 1097–1108.

Nehls, U., & Dietz, S. (2014). Fungal aquaporins: Cellular functions and ecophysiological perspectives. *Applied Microbiology and Biotechnology*, *98*, 8835–8851.

Nehls, U., Grunze, N., Willmann, M., Reich, M., & Küster, H. (2007). Sugar for my honey: Carbohydrate partitioning in ectomycorrhizal symbiosis. *Phytochemistry*, *68*, 82–91.

Nehls, U., & Plassard, C. (2018). Nitrogen and phosphate metabolism in ectomycorrhizas. *The New Phytologist*, [in press]. https://doi.org/10.1111/nph.15257.

Nieves-Cordones, M., Alemán, F., Martínez, V., & Rubio, F. (2014). K^+ uptake in plant roots. The systems involved, their regulation and parallels in other organisms. *Journal of Plant Physiology*, *171*, 688–695.

Öpik, M., Metsis, M., Daniell, T. J., Zobel, M., & Moora, M. (2009). Large-scale parallel 454 sequencing reveals host ecological group specificity of arbuscular mycorrhizal fungi in a boreonemoral forest. *The New Phytologist*, *184*, 424–437.

Park, H.-J., Floss, D. S., Levesque-Tremblay, V., Bravo, A., & Harrison, M. J. (2015). Hyphal branching during arbuscule development requires Reduced Arbuscular Mycorrhiza 1. *Plant Physiology*, *169*, 2774–2788.

Pérez-Tienda, J., Corrêa, A., Azcón-Aguilar, C., & Ferrol, N. (2014). Transcriptional regulation of host NH_4^+ transporters and GS/GOGAT pathway in arbuscular mycorrhizal rice roots. *Plant Physiology and Biochemistry*, *75*, 1–8.

Peter, M., Kohler, A., Ohm, R. A., Kuo, A., Krützmann, J., Morin, E., et al. (2016). Ectomycorrhizal ecology is imprinted in the genome of the dominant symbiotic fungus *Cenococcum geophilum*. *Nature Communications*, 7, 12662.

Pfeffer, P. E., Douds, D. D., Bécard, G., & Shachar-Hill, Y. (1999). Carbon uptake and the metabolism and transport of lipids in an arbuscular mycorrhiza. *Plant Physiology*, *120*, 587–598.

Plassard, C., Bonafos, B., & Touraine, B. (2000). Differential effects of mineral and organic N sources, and of ectomycorrhizal infection by *Hebeloma cylindrosporum*, on growth and N utilization in *Pinus pinaster*. *Plant, Cell & Environment*, *23*, 1195–1205.

Plassard, C., & Dell, B. (2010). Phosphorus nutrition of mycorrhizal trees. *Tree Physiology*, *30*, 1129–1239.

Plett, J. M., & Martin, F. (2012). Poplar root exudates contain compounds that induce the expression of MiSSP7 in *Laccaria bicolor*. *Plant Signaling & Behavior*, *7*, 12–15.

Plett, J. M., Tisserant, E., Brun, A., Morin, E., Grigoriev, I. V., Kuo, A., et al. (2015). The mutualist *Laccaria bicolor* expresses a core gene regulon during the colonization of diverse host plants and a variable regulon to counteract host-specific defenses. *Molecular Plant-Microbe Interactions*, *28*, 261–273.

Porcel, R., Azcón, R., & Ruíz-Lozano, J. M. (2004). Evaluation of the role of genes encoding for Δ^1-pyrroline-5-carboxylate synthetase (P5CS) during drought stress in arbuscular mycorrhizal *Glycine max* and *Lactuca sativa* plants. *Physiological and Molecular Plant Pathology*, *65*, 211–221.

Prak, S., Hem, S., Boudet, J., Viennois, G., Sommerer, N., Rossignol, M., et al. (2008). Multiple phosphorylations in the C-terminal tail of plant plasma membrane aquaporins: Role in subcellular trafficking of AtPIP2;1 in response to salt stress. *Molecular & Cellular Proteomics*, *7*, 1019–1030.

Quiroga, G., Erice, G., Aroca, R., Zamarreño, A. M., García-Mina, J. M., & Ruíz-Lozano, J. M. (2018). Arbuscular mycorrhizal symbiosis and salicylic acid regulate aquaporins and root hydraulic properties in maize plants subjected to drought. *Agricultural Water Management*, *202*, 271–284.

Rausch, C., & Bucher, M. (2002). Molecular mechanisms of phosphate transport in plants. *Planta*, *216*, 23–37.

Rausch, C., Daram, P., Brunner, S., Jansa, J., Laloi, M., Leggewie, G., et al. (2001). A phosphate transporter expressed in arbuscule-containing cells in potato. *Nature*, *414*, 462–470.

Řezáčová, V., Konvalinková, T., & Jansa, J. (2017). Carbon fluxes in mycorrhizal plants. In A. Varma, R. Prasad, & N. Tuteja (Eds.), *Mycorrhiza—Eco-physiology, secondary metabolites, nanomaterials* (pp. 1–21). Springer International Publishing. [chapter 1].

Rich, M. K., Nouri, E., Courty, P.-E., & Reinhardt, D. (2017). Diet of arbuscular mycorrhizal fungi: Bread and butter? *Trends in Plant Science*, *22*, 652–660.

Rosewarne, G. M., Barker, S. J., Smith, S. E., Smith, F. A., & Schachtman, D. P. (1999). A *Lycopersicon esculentum* phosphate transporter (*Le*PT1) involved in phosphorus uptake from a vesicular–arbuscular mycorrhizal fungus. *The New Phytologist*, *144*, 507–516.

Ruíz-Lozano, J. M. (2003). Arbuscular mycorrhizal symbiosis and alleviation of osmotic stress. New perspectives for molecular studies. *Mycorrhiza*, *13*, 309–317.

Ruíz-Lozano, J. M., Porcel, R., Azcón, R., & Aroca, R. (2012). Regulation by arbuscular mycorrhizae of the integrated physiological response to salinity in plants: New challenges in physiological and molecular studies. *Journal of Experimental Botany*, *63*, 4033–4044.

Ruth, B., Khalvati, M., & Schmidhalter, U. (2011). Quantification of mycorrhizal water uptake via high-resolution on-line water content sensors. *Plant and Soil*, *342*, 459–468.

Saito, K., Yoshikawa, M., Yano, K., Miwa, H., Uchida, H., Asamizu, E., et al. (2007). NUCLEOPORIN 85 is required for calcium spiking, fungal and bacterial symbioses, and seed production in *Lotus japonicus*. *The Plant Cell*, *19*, 610–624.

Sánchez-Romera, B., Calvo-Polanco, M., Ruíz-Lozano, J. M., Zamarreño, A. M., Arbona, V., García-Mina, J. M., et al. (2018). Involvement of the *def-1* mutation in the response of tomato plants to arbuscular mycorrhizal symbiosis under well-watered and drought conditions. *Plant & Cell Physiology*, *59*, 248–261.

Sánchez-Romera, B., Ruíz-Lozano, J. M., Zamarreño, Á. M., García-Mina, J. M., & Aroca, R. (2016). Arbuscular mycorrhizal symbiosis and methyl jasmonate avoid the inhibition of root hydraulic conductivity caused by drought. *Mycorrhiza*, *26*, 111–122.

Sasse, J., Martinoia, E., & Northen, T. (2018). Feed your friends: Do plant exudates shape the root microbiome? *Trends in Plant Science*, *23*, 25–41.

Saxena, B., Shukla, K., & Giri, B. (2017). Arbuscular mycorrhizal fungi and tolerance of salt stress in plants. In Q. S. Wu (Ed.), *Arbuscular mycorrhizas and stress tolerance of plants.* Singapore: Springer.

Schott, S., Valdebenito, B., Bustos, D., Gomez-Porras, J. L., Sharma, T., & Dreyer, I. (2016). Cooperation through competition—Dynamics and microeconomics of a minimal nutrient trade system in arbuscular mycorrhizal symbiosis. *Frontiers in Plant Sciences*, 7, 912.

Schroeder, J. I., Delhaize, E., Frommer, W. B., Guerinot, M. L., Harrison, M. J., Herrera-Estrella, L., et al. (2013). Using membrane transporters to improve crops for sustainable food production. *Nature*, *497*, 60–66.

Selle, A., Willmann, M., Grunze, N., Gessler, A., Weiss, M., & Nehls, U. (2005). The high-affinity poplar ammonium importer *Ptt*AMT1.2 and its role in ectomycorrhizal symbiosis. *The New Phytologist*, *168*, 697–706.

Selosse, M.-A., Bocayuva, M. F., Kasuya, M. C. M., & Courty, P.-E. (2017). Mixotrophy in mycorrhizal plants: Extracting carbon from mycorrhizal networks. In F. Martin (Ed.), *Molecular mycorrhizal symbiosis* (pp. 451–471). Hoboken, NJ: John Wiley & Sons, Inc., Wiley-Blackwell Press.

Shachar-Hill, Y., Pfeffer, P. E., Douds, D., Osman, S. F., Doner, L. W., & Ratcliffe, G. (1995). Partitioning of intermediary carbon metabolism in vesicular-arbuscular mycorrhizal leek. *Plant Physiology*, *108*, 7–15.

Sheng, M., Tang, M., Chen, H., Yang, B. W., Zhang, F. F., & Huang, Y. H. (2008). Influence of arbuscular mycorrhizae on photosynthesis and water status of maize plants under salt stress. *Mycorrhiza*, *18*, 287–296.

Smith, S. E., Facelli, E., Pope, S., & Smith, F. A. (2010). Plant performance in stressful environments: Interpreting new and established knowledge of the roles of arbuscular mycorrhizas. *Plant and Soil*, *326*, 3–20.

Smith, F., Mudge, S., Rae, A., & Glassop, D. (2003). Phosphate transport in plants. *Plant and Soil*, *248*, 71–83.

Smith, S. E., & Read, D. (2008). *Mycorrhizal symbiosis* (3rd ed.). New York: Academic Press.

Steudle, E., & Peterson, C. A. (1998). How does water get through roots? *Journal of Experimental Botany*, *49*, 775–788.

Sugiyama, A., Shitan, N., & Yazaki, K. (2007). Involvement of a soybean ATP-binding cassette-type transporter in the secretion of genistein, a signal flavonoid in legume-*Rhizobium* symbiosis. *Plant Physiology*, *144*, 2000–2008.

Tang, N., San Clemente, H., Roy, S., Bécard, G., Zhao, B., & Roux, C. (2016). A survey of the gene repertoire of *Gigaspora rosea* unravels conserved features among Glomeromycota for obligate biotrophy. *Frontiers in Microbiology*, 7, 1–16.

Tarkka, M. T., Herrmann, S., Wubet, T., Feldhahn, L., Recht, S., Kurth, F., et al. (2013). OakContigDF159.1, a reference library for studying differential gene expression in *Quercus robur* during controlled biotic interactions: Use for quantitative transcriptomic profiling of oak roots in ectomycorrhizal symbiosis. *The New Phytologist*, *199*, 529–540.

Tedersoo, L., & Smith, M. E. (2013). Lineages of ectomycorrhizal fungi revisited: Foraging strategies and novel lineages revealed by sequences from belowground. *Fungal Biology Reviews*, *27*, 83–99.

Tisserant, E., Kohler, A., Dozolme-Seddas, P., Balestrini, R., Benandellah, K., Croll, D., et al. (2012). The transcriptome of the arbuscular mycorrhizal fungus *Glomus intraradices* (DAOM 197198) reveals functional tradeoffs in an obligate symbiont. *The New Phytologist*, *193*, 755–769.

Torres-Aquino, M., Becquer, A., Le Guernevé, C., Louche, J., Amenc, L. K., Staunton, S., et al. (2017). The host plant *Pinus pinaster* exerts specific effects on phosphate efflux and polyphosphate metabolism of the ectomycorrhizal fungus *Hebeloma cylindrosporum*: A radiotracer, cytological staining and ^{31}P NMR spectroscopy study. *Plant, Cell & Environment*, *40*, 190–202.

Trépanier, M., Bécard, G., Moutoglis, P., Willemot, C., Gagné, S., Avis, T. J., et al. (2005). Dependence of arbuscular-mycorrhizal fungi on their plant host for palmitic acid synthesis. *Applied and Environmental Microbiology, 71*, 5341–5347.

Uehlein, N., Fileschi, K., Eckert, M., Bienert, G. P., Bertl, A., & Kaldenhoff, R. (2007). Arbuscular mycorrhizal symbiosis and plant aquaporin expression. *Phytochemistry, 68*, 122–129.

Vandeleur, R. K., Sullivan, W., Athman, A., Jordans, C., Gilliham, M., Kaiser, B. N., et al. (2014). Rapid shoot-to-root signalling regulates root hydraulic conductance via aquaporins. *Plant, Cell & Environment, 37*, 520–538.

Varela-Cevero, S., Vasar, M., Davison, J., Barea, J. M., Öpik, M., & Azcón-Aguilar, C. (2015). The composition of arbuscular mycorrhizal fungal communities differs among the roots, spores and extraradical mycelia associated with five Mediterranean plant species. *Environmental Microbiology, 17*, 2882–2895.

Veneault-Fourrey, C., & Martin, F. (2013). 10 New insights into ectomycorrhizal symbiosis evolution and function. In F. Kempken & Agricultural Applications (Eds.), *The mycota—A comprehensive treatise on fungi as experimental systems for basic and applied research*, (pp. 273–293). Berlin, Heidelberg: Springer.

Volpe, V., Giovannetti, M., Sun, X. G., Fiorilli, V., & Bonfante, P. (2015). The phosphate transporters *Lj*PT4 and *Mt*PT4 mediate early root responses to phosphate status in non mycorrhizal roots. *Plant, Cell & Environment, 39*, 660–671.

Wang, E., Schornack, S., Marsh, J. F., Gobbato, E., Schwessinger, B., Eastmond, P., et al. (2012). A common signaling process that promotes mycorrhizal and oomycete colonization of plants. *Current Biology, 22*, 2242–2246.

Wewer, V., Brands, M., & Dörmann, P. (2014). Fatty acid synthesis and lipid metabolism in the obligate biotrophic fungus *Rhizophagus irregularis* during mycorrhization of *Lotus japonicus*. *The Plant Journal, 79*, 398–412.

Willmann, A., Thomfohrde, S., Haensch, R., & Nehls, U. (2014). The poplar NRT2 gene family of high affinity nitrate importers: Impact of nitrogen nutrition and ectomycorrhiza formation. *Environmental and Experimental Botany, 108*, 79–88.

Xu, H., Cooke, J. E. K., Kemppainen, M., Pardo, A. G., & Zwiazek, J. J. (2016). Hydraulic conductivity and aquaporin transcription in roots of trembling aspen (Populus tremuloides) seedlings colonized by Laccaria bicolor. *Mycorrhiza, 26*, 441–451.

Xu, H., Kemppainen, M., El Kayal, W., Lee, S. H., Pardo, A. G., Cooke, J. E. K., et al. (2015). Overexpression of *Laccaria bicolor* aquaporin *JQ585595* alters root water transport properties in ectomycorrhizal white spruce (*Picea glauca*) seedlings. *The New Phytologist, 205*, 757–770.

Yang, S. Y., Grønlund, M., Jakobsen, I., Grotemeyer, M. S., Rentsch, D., Miyao, A., et al. (2012). Nonredundant regulation of rice arbuscular mycorrhizal symbiosis by two members of the phosphate transporter 1 gene family. *The Plant Cell, 24*, 4236–4251.

Yang, W., Pollard, M., Li-Beisson, Y., Beisson, F., Feig, M., & Ohlrogge, J. (2010). A distinct type of glycerol-3-phosphate acyltransferase with *sn*-2 preference and phosphatase activity producing 2-monoacylglycerol. *Proceedings of the National Academy of Sciences of the United States of America, 107*, 12040–12045.

Zhang, Q., Blaylock, L. A., & Harrison, M. J. (2010). Two *Medicago truncatula* half-ABC transporters are essential for arbuscule development in arbuscular mycorrhizal symbiosis. *The Plant Cell, 22*, 1483–1497.

Zhang, L., Feng, G., & Declerck, S. (2018). Signal beyond nutrient, fructose, exuded by an arbuscular mycorrhizal fungus triggers phytate mineralization by a phosphate solubilizing bacterium. *The ISME Journal, 12*, 2339–2351.

Zhang, H., Wei, S., Hu, W., Xiao, L., & Tang, M. (2017). Arbuscular mycorrhizal fungus *Rhizophagus irregularis* increased potassium content and expression of genes encoding potassium channels in *Lycium barbarum*. *Frontiers in Plant Science, 8*, 440.

Zhang, T., Wen, X., & Ding, G. (2017). Ectomycorrhizal symbiosis enhances tolerance to low phosphorous through expression of phosphate transporter genes in masson pine (*Pinus massoniana*). *Acta Physiologiae Plantarum*, *39*, 101.

Zhang, H., Yin, W., & Xia, X. (2010). Shaker-like potassium channels in *Populus*, regulated by the CBL–CIPK signal transduction pathway, increase tolerance to low-K^+ stress. *Plant Cell Reports*, *29*, 1007–1012.

Ziegler, J., Schmidt, S., Strehmel, N., Scheel, D., & Abel, S. (2017). *Arabidopsis* transporter ABCG37/PDR9 contributes primarily highly oxygenated coumarins to root exudation. *Scientific Reports*, *7*, 3704.

Kolibri 360 GmbH
Ferdinand-Jühlke-Straße 7
99095 Erfurt, Deutschland
produktsicherheit@kolibri360.de